Conversion of Système International to Imperial Units

Variable	SI Unit	Imperial Equivalent
Length	1 km	
	1 m	1.094 yd
	1 m	39.37 in.
	1 m	3.281 ft
Area	1 ha	2.471 A
	1 km^2	0.39 mi^2
	1 m^2	1.196 yd^2
	1 cm^2	0.15 in.^{2}
Volume	1 m^3	1.308 yd^3
	1 m^3	35.3 ft^3
	1 cm^3	0.06 in.^{3}
	1 km^3	0.24 mi^3
	1 L	1.06 U.S. qt
	1 L	0.88 imperial qt
	1 L	0.264 U.S. gal
	1 L	0.22 imperial gal
Velocity	1 kph	0.621 mph
	1 m/s	3.281 ft/sec[1]
	1 m/s	2.24 mph
Pressure	1 standard atmosphere	14.7 psi
	1 bar	14.5 psi
Mass	1 kg	2.205 lb
	1 g	0.035 oz
Temperature	°Celsius	$°C = 5/9(°F - 32)$

Common Abbreviations Used in the Imperial System

mi	mile	mph	miles per hour	gal	gallon
yd	yard	oz	ounce	qt	quart
in.	inch	lb	pound	A	acre
ft	feet	psi	pounds/in.^{2}		

CLIMATOLOGY
AN ATMOSPHERIC SCIENCE

THIRD EDITION

CLIMATOLOGY
AN ATMOSPHERIC SCIENCE

John J. Hidore
University of North Carolina at Greensboro

John E. Oliver
Indiana State University

Mary Snow
Embry-Riddle Aeronautical University

Rich Snow
Embry-Riddle Aeronautical University

Prentice Hall

New York Boston San Francisco London Toronto Sydney
Tokyo Singapore Madrid Mexico City Munich Paris
Cape Town Hong Kong Montreal

Library of Congress Cataloging-in-Publication Data
Hidore, John J.
    Climatology: an atmospheric science/John J. Hidore, Mary Snow, Rich Snow. —3rd ed.
        p.   cm.
    Includes bibliographical references and index.
    Previous edition cataloged under Oliver, John E.
    ISBN-13: 978-0-321-60205-3 (alk. paper)
    ISBN-10: 0-321-60205-6 (alk. paper)
    1. Climatology.   2. Atmospheric physics.   I. Snow, Mary B.   II. Snow, Richard E.
III. Oliver, John E. Climatology.   IV. Title.

    QC981.O393 2010
    551.5—dc22   2009023188

Acquisitions Editor: Christian Botting
Editor in Chief, Geosciences and Chemistry: Nicole Folchetti
Project Manager, Editorial: Tim Flem
Editorial Assistant: Christina Ferraro
Marketing Assistant: Keri Parcells
Managing Editor, Geosciences and Chemistry: Gina M. Cheselka
Project Manager, Science: Beth Sweeten
Art Editor: Connie Long
Art Studio: Spatial Graphics
Art Director: Jayne Conte
Cover Design: Margaret Kenselaar
Senior Operations Supervisor: Nick Sklitsis
Operations Specialist: Amanda Smith
Photo Researcher: Elaine Soares
Production Supervision/Composition: GGS Higher Education Resources, A Division of Premedia Global, Inc.
Cover Credit: NASA/GSFC Scientific Visualization Studio

Prentice Hall
is an imprint of

www.mygeoscienceplace.com
www.pearsonhighered.com

IN MEMORIAM

John E. Oliver, 1933–2008

Dr. John E. Oliver was born October 21, 1933, in Dover, Kent, England, the son of Albert and Florence Oliver. He received a bachelor's degree from London University and his master's and doctoral degrees from Columbia University in New York. After teaching at Columbia University, he joined the faculty at Indiana State University in Terre Haute, Indiana, in 1973. During his tenure at Indiana State University, John served terms as Chair of the Geography-Geology Department and as the Associate Dean, College of Arts and Sciences. He was emeritus professor at the time of his death.

John distinguished himself as the author of numerous books and articles in peer-reviewed scientific journals. He wrote extensively in the areas of climatology and physical geography. Among his major works is the *Encyclopedia of World Climatology* (2005), which he edited. John was co-founder of the journal *Physical Geography,* and served as climatology editor of the journal from 1979–2000. In honor of his teaching, research, and university administration, he received the Distinguished Professor award from the College of Arts and Sciences at Indiana State University. The Climate Specialty Group in the Association of American Geographers presented him the first Lifetime Achievement Award (1998) for his many contributions to the study of climatology and to the AAG.

Perhaps John's most lasting significance resides in those who knew him. We remember him as a true scholar, a genuine and caring person, a skillful professor, and to many, a great friend.

BRIEF CONTENTS

CONTENTS

PREFACE

The origin of *Climatology* goes back many years, when John Oliver was on the faculty of Indiana State University and I was on the faculty of Indiana University. We joined forces and began teaching joint graduate seminars in climatology and physical geography. In passing, we talked of the need for a different book on climatology. The plan came to fruition with the publication of the first edition.

This book is written for students seeking an introduction to atmospheric process through time and space. We have made a conscious effort over the years to retain both the literary and scientific components of the study of climate. In some chapters, there is considerable descriptive information of climatic events without numerical data or quantitative analysis. In other select chapters, we present quantitative material for more mathematically inclined students. *Climatology* is an appropriate book for college students taking their first course in atmospheric science. It contains four major sections covering some of the main areas of climatology. They are:

1. Physical and dynamic climatology
2. Climate change
3. Regional climatology
4. Applied climatology

Much has changed since the inception of this book. The volume of climatic data has exploded. Automated weather stations are now present in many isolated areas of the landmasses. Data from the atmosphere over the ocean has also grown dramatically. Not only is more data available from ships, but floating buoys record and transmit data to collection centers. The number of satellites that monitor atmospheric data has grown rapidly, and the variety of data has expanded. Both polar orbiting satellites and geostationary satellites are in use. New environmental oriented satellites are coming online and many are in planning. Many countries around the world operate weather satellites and share the data collected. Most chapters in *Climatology* have new sections to reflect new developments in data collection, data analysis, and in the data itself.

Perhaps the biggest change in recent decades has been the international concern with evidence that Earth's climate is warming and the impact of that warming. Climate change and global warming are common items in the news media. El Niño and La Niña are also frequently in the news. Studies going back more than a century suggest that carbon dioxide was increasing in the atmosphere and that this would affect global temperatures. In recent decades, monitoring stations have come online. What became apparent almost immediately was the rapid change in carbon dioxide that has taken place.

Global conferences, such as the Kyoto Conference, and the work of the Intergovernmental Panel on Climate Change (IPCC) present a consensus that the problem of climate change is a very serious one. Closely paralleling the viewpoint of many world scientists was the recent action taken by the *Bulletin of the Atomic Scientists*. The group of scientists developed the Doomsday Clock in 1947 based on the threat of nuclear war. Midnight on the clock is the projected time of global catastrophe. In January of 2007, the clock was moved forward from seven minutes to midnight to five minutes before the hour. The acceleration towards midnight was made when global warming was added to the nuclear threat.

Over the years *Climatology* has evolved to reflect the changes that have taken place both in the global climate and in the field of climatology. The third edition carries on this effort. There is now a major section of the book devoted to climate change. This section includes chapters on the following topics:

- The natural causes of climate change
- The reconstruction of past climates
- Greenhouse gases and global warming
- The effects of climate change on the physical environment
- The effects of climate change on the living world

Additional features new to the third edition include:

- New material on the measurement of total solar irradiance in Chapter 2.
- Information on new satellites and the variables they are monitoring in Chapters 1 and 7.
- A new combined chapter on wind and circulation patterns replaces Chapters 5 and 6 from the previous edition.
- Ocean-atmosphere interactions are now explored in Chapter 6, which contains new material on the thermohaline circulation and the Pacific Decadal Oscillation (PDO).
- The chapter on atmospheric storms has been updated, and now includes a number of new tables on a variety of characteristics of atmospheric storms.
- The Köppen system of climate classification is now emphasized in the section on Regional Climates.

We are saddened that John Oliver died unexpectedly in 2008. John has had major input on each edition—including this edition— and he is sorely missed. On a positive note, two new authors have made major contributions to this edition: Dr. Mary Snow and Dr. Richard Snow of Embry-Riddle Aeronautical University, each of whom received their Ph.D in Physical Geography from Indiana State University with the support and guidance of John Oliver. Finally, we extend our thanks to the following reviewers for their comments as we prepared this revision: Kenneth Parsons, Embry-Riddle Aeronautical University; Stephen Stadler, Oklahoma State University; Alfred Stamm, State University of New York, Oswego.

John J. Hidore

CLIMATOLOGY
AN ATMOSPHERIC SCIENCE

About Our Sustainability Initiatives

This book is carefully crafted to minimize environmental impact. The materials used to manufacture this book originated from sources committed to responsible forestry practices. The paper is Forest Stewardship Council (FSC) certified. The binding, cover, and paper come from facilities that minimize waste, energy consumption, and the use of harmful chemicals.

Pearson closes the loop by recycling every out-of-date text returned to our warehouse. We pulp the books, and the pulp is used to produce items such as paper coffee cups and shopping bags. In addition, Pearson aims to become the first climate neutral educational publishing company.

The future holds great promise for reducing our impact on Earth's environment, and Pearson is proud to be leading the way. We strive to publish the best books with the most up-to-date and accurate content, and to do so in ways that minimize our impact on Earth.

Mixed Sources

Product group from well-managed forests, controlled sources and recycled wood or fiber

www.fsc.org Cert no. SCS-COC-00648
© 1996 Forest Stewardship Council

FSC

Prentice Hall
is an imprint of

PART I

Physical and Dynamic Climatology

Climatology in the World Today

Until quite recently, the study of climate was not a highly visible part of the everyday scene. Usually, climate was investigated when someone was going on vacation and wanted to know weather conditions at the location they were to visit. Much has changed over the past 20 years or so. Today the topic of climate is perhaps the most widely publicized modern science. Discussion and reports of global warming, climate change, and the influences of El Nino appear in newspapers, television programs, and award–winning films. The significance of climate was truly recognized when the Intergovernmental Panel on Climate Change (IPCC) and Al Gore were awarded the Nobel Peace Prize. Information derived from the IPCC reports is extensively used in this book's chapters on climate change.

The climatology-related topics the media cover are but a small part of climatology. Without a good grounding in the many aspects of the long-term behavior of the atmosphere, it is often difficult to completely understand the full meaning of the current concerns.

Any study of climatology requires a basic understanding of the physical controls of climate. Essentially, two sets of factors determine the nature of climate over the earth's surface and through time:

1. The *astronomical factors* include the shape of the earth and its motion relative to the sun and solar radiation.
2. The *terrestrial factors* include atmospheric composition, the nature of the earth's surface, its oceans and continents, and the physiographic/geographic features of the surface.

The combined influence of these two sets of factors results in variations in the amount of solar energy available at a location and major climatic differences over the globe. The global inequalities in energy distribution lead to global exchanges of energy and mass. These are accomplished through the movement of winds over the surface, oceanic circulation, and the global exchanges of water through processes involved in the hydrologic cycle. These mass and energy exchanges then affect the environment and living things within the environment, including humans. The study of each of these sets of processes and interactions provides the content of climatology.

To provide a backdrop for the material covered in the following chapters, Table 1.1 lists some of the major changes that have taken place in the study of climate in recent years.

TABLE 1.1 Some Developing Topics of Climate Studies

Area of Development	Examples
Computer technology	The continued development of computer technology has enabled construction of meaningful global climate models, while desktop computing permits climatic data from a wide variety of sources to be instantly available and analyzed.
Data improvement	Availability of, for example, the U.S. Historical Climatology Network and the technique of reanalysis greatly improves data use and utility.
Dating techniques	The development of sophisticated methods to accurately date past events has greatly enhanced the understanding of climates in the past.
Economic studies	The great improvement of climate forecasts enables future benefits and drawbacks to be estimated.
Environmental impact studies	The relationship of climate to many aspects of the environment is demonstrated by the impacts on, for example, future water supplies.
International cooperation	It has long been appreciated that understanding the atmosphere requires global cooperation. The IPCC provides evidence of the success achieved when global cooperation occurs.
Media	The general population is becoming more aware of climate events. Many aspects of climate, ranging from the greenhouse effect to the character of monsoons, are widely reported.
People studies	The influence of climate on human activities and events is seen in development of new areas of study such as cultural climatology and well-known relationships such as sports and climatology.
Satellites	There is now a long enough history of satellite coverage to enable long-term climate studies to be completed.
Severe weather	Long-term studies by using such tools as Doppler radar and advanced statistical analysis have led to important findings concerning the nature and frequency of severe storms.
Teleconnections	The influence of climate conditions in an area remote from another became widely publicized with El Nino. Further studies have identified many other teleconnections.

THE ATMOSPHERE AND ATMOSPHERIC SCIENCES

The atmosphere serves several major functions on the planet. Beyond providing the reservoir of gases required for life, it also plays a critical role in the distribution or redistribution of energy over the planet. The earth's atmosphere is a fundamental and critical part of the environment of planet Earth. Together with the lithosphere (the solid earth), the hydrosphere (water in all its forms), and the biosphere (living organisms), it makes the earth the habitable, hospitable place that it is. The atmosphere provides an insulating layer around the earth that raises the mean temperature of the surface from –23°C (–9°F) to nearly 15°C (59°F). It shields the surface from the large doses of ultraviolet radiation that would destroy most life forms. It also serves as a major transporter of heat horizontally across the earth's surface and vertically away from the surface. Tropical regions receive far more energy than polar regions. Atmospheric circulation helps equalize this imbalance by moving heat from the warmer to the colder areas. In addition to transferring energy, the atmosphere transfers water from low to high latitudes and from ocean to land.

Because of the complexity of the gaseous envelope that surrounds the planet, atmospheric scientists often divide its study into specific areas of interest. One such division identifies the three fields aerology, meteorology, and climatology. **Aerology** (or aeronomy) is essentially the study of the free atmosphere through its vertical extent. Initially the major goal of aerology was the identification of atmospheric structure and the amount and distribution of its parts. Today aerology deals mostly with the chemistry and physical reactions that occur within the various atmospheric layers. The word *aerology* is less widely used now than it once was, and its content is frequently considered part of meteorology. We deal with this aspect of the atmosphere only briefly in later chapters.

Meteorology is the science that deals with motion and phenomena of the atmosphere, with the view to both forecasting weather and explaining the processes involved. It deals largely with the status of the atmosphere over a short period and uses principles of physics to reach its goals. **Climatology** is the study of atmospheric conditions over periods of time measured in years or longer. It includes the study of the kinds of weather that occur at a place. It is concerned not only with the most frequently occurring types, the average weather, but the infrequent and unusual types as well. Dynamic change in the atmosphere brings about variation and occasional extremes that have long- as well as short-term impacts. As a result, climatology is the study of all the weather at a given place over a given period.

CLIMATOLOGY: A BRIEF HISTORY

The study of climate began at least as early as the ancient Greek culture. In fact, the word *climate* comes from a Greek word meaning "slope." In this context, it refers to the slope, or inclination, of the earth's axis. It specifies an earth region at a particular place on that slope—that is, the location of a place in relation to parallels of latitude. This mathematical derivation represents one of many contributions to mathematical geography by philosophers such as Eratosthenes and Aristarchus. The Greek search for knowledge about the world resulted in written works on the atmosphere. The first book on climate was *Airs, Waters, and Places* written by Hippocrates in 400 B.C. In 340 B.C., Aristotle wrote *Meteorologica*, the first treatise on meteorology.

The academic world did not resume the Greek interest in the atmosphere for many hundreds of years. Although Arab scholars of the 9th and 10th centuries expanded on the Greek writings, the renewed interest came in the middle of the 15th century, with the Age of Discovery. Extended sea voyages and development of new trading areas led to descriptive reports of climates outside Europe. Many of these descriptions were quite fanciful; they provided the basis for misconceptions about parts of the world that prevailed for centuries.

Scientific analysis of the atmosphere began in the 17th century with the design of instruments to measure atmospheric conditions. These provided data from which

TABLE 1.2 Significant Events in the Development of Climatology

Date	Event
ca. 400 B.C.	The influence of climate on health is discussed by Hippocrates in *Airs, Waters, and Places*.
ca. 350 B.C.	Weather science is discussed in Aristotle's *Meteorologica*.
ca. 300 B.C.	The text *De Ventis* by Theophrastus describes winds and offers a critique of Aristotle's ideas.
ca. 1593	Galileo Galilie describes the thermoscope. (The first thermometer is generally attributed to Santorre, 1612.)
1662	Francis Bacon writes a significant treatise on wind.
1643	Evangelista Torricelli invents the barometer.
1661	Robert Boyle's law on gases is propounded.
1664	Weather observations begin in Paris; although often described as the longest continuous sequence of weather data available, the records are not homogeneous or complete.
1668	Edmund Halley constructs a map of the trade winds.
1714	The Fahrenheit scale is introduced.
1735	George Hadley writes his treatise on trade winds and effects of the earth's rotation.
1736	The Centigrade scale is introduced. (It was first formally proposed by du Crest in 1641.)
1779	Weather observation begins at New Haven, CT, the longest continuous sequence of records in the United States.
1783	The hair hygrometer is invented.
1802	Jean Lemark and Luke Howard propose the first cloud classification system.
1817	Alexander von Humboldt constructs the first map showing mean annual temperature over the globe.
1825	August devises the psychrometer.
1827	Heinrich W. Dove begins to develop the laws of storms.
1831	William Redfield produces the first weather map of the United States.
1837	Pyrheliometer for measuring insolation is constructed.
1841	James Espy describes movement and development of storms.
1844	Gaspard de Coriolis formulates the idea of the "Coriolis force."
1845	H. Berhaus constructs first world map of precipitation.
1848	Heinrich Dove publishes the first maps of mean monthly temperatures.
1862	Louis Renou drafts the first map (showing western Europe) of mean pressure.
1879	Alexander Supan publishes a map showing world temperature regions.
1892	Beginning of the systematic use of balloons to monitor free air.
1900	The term *classification of climate* is first used by Wladimir Köppen.
1902	Existence of the stratosphere is discovered.
1913	The ozone layer is discovered.
1918	Beginning of the development of the polar front theory by Vilhem Bjerknes.
1925	Beginning of systematic data collection using aircraft.
1928	Radiosondes are first used.
1940	Nature of jet streams is first investigated.
1960	United States launches the first meteorological satellite, Tiros I.
1978	U.S. Congress passes the National Climate Program Act.
1978	The United States bans the use of chlorofluorocarbons (CFCs) as aerosol propellants.
1987	The Montreal Protocol limiting the production of CFCs is signed by more than 30 nations.
1989	The state of Vermont bans the use of CFCs in automobile air conditioners.
1990	Doppler radar network introduced in the United States.
1992	Rio de Janeiro Earth Summit leads to "Framework" Convention on Climate Change.
1998	Kyoto Protocol on climatic change is held.
2000	Climate change concerns lead to disruption of UN Climate Conference in The Hague.

TABLE 1.3 Elements of the National Climate Program Act of 1978

The programs shall include, but not be limited to, the following elements:

1. assessment of the effect of climate on the natural environment, agricultural production, energy supply and demand, land and water resources, transportation, human health, and national security

2. basic and applied research to improve the understanding of climatic processes, natural and human-induced, and the social, economic, and political implications of climatic change

3. methods for improving climate forecasts

4. global data collection on a continuing basis

5. systems for the dissemination of climatological data and information

6. measures for increasing international cooperation in climatology

7. mechanisms for climate-related studies

8. experimental climate forecast centers

9. submission of five-year plans

laws applying to the atmosphere were derived. Galileo invented the thermometer in 1593, and Torricelli invented the barometer in 1643. In 1662, Boyle discovered the basic relationship between pressure and volume in a gas.

Instruments were improved and standardized during the 18th century, and extensive data collection and description of regional climates began. Explanation of phenomena through the study of the physical processes began in the 19th century. Climatic data are the measurements of the earth's climate system. The most widely recorded data are surface temperature and precipitation. The longest complete climatic records exist for temperature and precipitation. These range in length from some 325 years for measurements in central England to about 200 years for stations in Europe and the United States. Most of the stations that make up today's observational network have records of less than 100 years. In 1817, von Humboldt constructed the first map that showed temperatures using isotherms. Soon after, in 1827, Dove explained local climates using polar and equatorial air currents.

In the 20th century, contributions became more frequent, and ideas essential to understanding the atmosphere evolved. The significant contributions of individuals since the beginning of the 19th century are too many to present here, although Table 1.2 lists some key developments.

An important event in the development of climatology was the National Climate Program Act. Signed into law in September 1978, this act has the stated purpose "to establish a national climate program that will assist the nation and the world to understand and respond to natural and human-induced climate processes and their implications." This program, as shown in Table 1.3, is a statement about the importance of climate and our understanding of it.

THE CONTENT OF CLIMATOLOGY

Table 1.4 presents some subdivisions of climatology. Climatography consists of the basic presentation of data in written or map form. As the names imply, physical and dynamic climatology relate to the physics and dynamics of the atmosphere. Physical climatology deals largely with energy exchanges and physical processes. Dynamic climatology is more concerned with atmospheric motion and exchanges that lead to and result from that motion. Some scientists consider the study of past and future climates to be a subgroup of climatic studies. However, methods used in studying past and future climates are the same as those in other aspects of climatology.

TABLE 1.4 Subdivisions of Climatology

Applied climatology
Bioclimatology
Climatography
Climate Change
Dynamic Climatology
Paleoclimatology
Physical climatology
Urban climatology

Within each area of climatology specialized analytic methods are used. For example, the exchanges of energy are examined using an energy budget approach. The various pathways and exchanges of energy in various forms are accounted for. Similarly, a water budget analysis examines the phases of water and the concomitant energy exchanges. Of particular note is the expansion of studies such as bioclimatology (the relationship between climate and the living world) and urban climatology (the nature of climate and climate change within cities) to become recognized specialties in climatology. Currently one of the fastest-growing areas within climatology is climate change on a global scale.

ATMOSPHERIC VARIABLES AND DATA ACQUISITION

Table 1.5 lists the atmospheric variables measured most frequently today. Almost 10,000 stations throughout the world record at least one of these variables. They are part of a primary land-based system of observational stations. Unfortunately, the distribution is not even over the earth. Large parts of the world have but a sparse network of recording stations.

Because the oceans occupy almost three-quarters of the earth's surface, climatic data over the oceans is of prime importance. Archives of surface data from the earth's oceans began in 1854. Through international agreement, the major maritime nations began a regular program of recording atmospheric and oceanic data from merchant and military ships. Nonetheless, long-term data from the oceans exist only for the popular sea lanes, and for large oceanic areas only limited data are available.

In the early 1800s, the only data representing conditions at high altitudes came from mountain observatories. In 1885, balloons became more widely used to monitor air currents and temperatures. During World War I (1914–1918), the use of balloons, kites, and aircraft multiplied rapidly. A similar impetus occurred in World War II (1939–1945). The need for upper air observations prompted a group of countries to establish a worldwide network of upper air observation stations. Balloon-carrying instruments, called **radiosondes**, are released into the atmosphere at specified times to simultaneously sample atmospheric conditions. Today, a network of about 1000 stations using radiosondes routinely measures air temperature, dew-point temperature, and wind direction and velocity.

Besides the parameters listed in Table 1.5, observations of other variables take place at selected weather stations. These include solar radiation and sunshine, soil temperature and evaporation, air pollution data, and water quantity and quality.

A new dimension in the study of the atmosphere was added when satellites were placed in orbit with the express purpose of sensing the atmosphere. The first was TIROS I in 1960. This was followed in 1966 by the launching of the geostationary ATS-1 (Applications Technology Satellite), the first in a series of weather satellites. A geostationary (or Earth-synchronous) satellite orbits the earth at a height of about 35,900 km (22,300 mi). At this altitude, the velocity necessary to maintain orbit is equal to the rotational velocity of the earth. Even though the earth and satellite are both in motion, the effect is that the satellite appears to remain in place over a given point on Earth. The working satellites are the Geostationary Operational Environmental

TABLE 1.5 Commonly Observed Atmospheric Variables

Air temperature
Barometric pressure
Cloud type, height, and amount
Current or prevailing weather
Dew-point temperature
Precipitation
Sunshine
Wind velocity and direction

Satellite (GOES)/Synchronous Meteorological Satellite (SMS) series. The significance of satellite imagery in climatology is the subject of Chapter 7.

While weather data are gathered at many places, the data are of little value unless they are readily available to potential users. Fortunately, there are agencies and organizations that control the collection and processing of original data. The international agency responsible for worldwide climatic data is the **World Meteorological Organization (WMO)**, located in Geneva, Switzerland.

The IPCC Data Distribution Center (DDC) acts as a gateway to many data sets used to assess changes in the global climate. The DDC acts in concert with various research centers to archive and distribute much of the data that the IPCC provides. The DDC also maintains another data set that is arranged by country and contains both averages and month-by-month variations in the data.

The World Climatic Data Centers is a network of data centers that was originally designed to archive the data derived during the International Geophysical Year (IGY). It has now expanded to include climatological, meteorological, astronomical, oceanographic, and geophysical data sets. Table 1.6 lists centers that have data that are directly related to research in climatology.

The main library for records generated by government weather services in the United States is the National Climatic Data Center in Asheville, North Carolina. Reporting to the National Climatic Data Center are the Regional Climate Centers. Of singular importance in relation to using surface data is the U.S. Historical Climatology Network (HCN). This center maintains a set of high-quality data for some 1000 stations in the contiguous United States.

The availability of data for climatic analysis has been greatly enhanced by the development of reanalysis. An uneven distribution had developed in the previous 150 years. Almost all data came from 26% of the land surface with a few data from Antarctica and Greenland, which was deemed unsuitable for in-depth analysis. In addition, because of changes in location and the local environmental station, data were not always comparable. Computer development enabled statistical manipulation of data so that local inconsistencies were removed and the data were homogenized. Data sets covering the entire earth's surface were generated. The data were again manipulated to account for the uneven distribution. The end result is reanalysis that has been of enormous value in climate research. Of course, it must be remembered that the actual accuracy depends on the quality of the original measurements.

Data Representation

As part of the geographic aspect of climatology, maps play a significant role in the depiction of climates and climatic data. Many of these maps use isolines—lines joining locations of equal value to show distribution of the elements. The name given to an isoline depends on the climatic element shown on the map (Table 1.7).

The presentation of numeric values is often most clearly seen in graphs and diagrams. Throughout this text, as in most other climatology works, graphs are used to show trends and patterns because they are easier to comprehend than a long list of numbers. But a graph is a statistical tool based on a mathematical grid that has one or

TABLE 1.6 Selected Members of the World Data Center Network that are Directly Related to Research in Climatology

World Data Center	Affiliation and Location	Data Emphasis
Atmospheric Trace Gases	Carbon Dioxide Information Center, Oak Ridge National Laboratory, Oak Ridge, Tennessee	Data related to atmospheric trace gases that affect and contribute to the earth's energy budget
Glaciology–USA	National Snow and Ice Data Center, Boulder, Colorado	Snow cover, snow pack, sea-ice extent, sea-ice thickness, images and pictures of historical glacial extent
Glaciology–China	Lanzhou Institute of Glaciology and Geocryology, Chinese Academy of Sciences, Lanzhou, China	Glacial Atlas of China, snow cover, glacial extent and variation, periglacial data, and hydrologic data
Glaciology–UK	The Royal Society and the Scott Polar Research Institute, University of Cambridge	Data related to glaciers, periglacial processes, satellite imagery, snow and ice chemistry
Airglow–Japan	National Astronomical Observatory, Tokyo, Japan	Airglow data, solar radiation data
Meteorology–China	Climate Data and Applications Office, Beijing, China	Real-time synoptic data, historical surface climate data, dendrochronology data, glacial data, and atmospheric chemistry data
Meteorology–Russia	Federal Service of Russia for Hydrometeorology and Monitoring of the Environment, Obninsk, Russia	Both surface observations and gridded surface and upper air meteorology data, marine ship observation data, and aerology data
Meteorology–USA	National Climatic Data Center, Asheville, North Carolina	Archives of data from many national and international research projects and experiments, including data from the IGY 1957–1958 and International Quiet Sun Year, 1964–1965, among many others. Synoptic surface and upper air climate data, Global Historic Climate Data Network, ozone data for the world since 1965
Paleo-climatology, USA	National Geophysical Data Center (NGDC), Boulder, Colorado	Dendrochronology data, ice-core data, seafloor sediment cores, coral data, proxy data on climatic forcing, including volcanic aerosol data, ice volume, atmospheric composition, etc.; numeric model simulation experiments data, climate reconstructions and maps

more numeric data sets, and care must be used in its preparation. Most graphs are made up of two axes that need to be scaled with calibrated graduations and clearly numbered. By convention, a graph consists of a horizontal axis (x-axis, or abscissa) and a vertical axis (y-axis, or ordinate). The x-axis is associated with the independent variable, that which influences the other variable, the dependent variable. The y-axis represents the variable that is dependent on that of the x-axis. The x-axis on many climate graphs shows time.

Some graphs may actually be classed as diagrams, with histograms and climagrams providing apt examples. A histogram is used to illustrate the relative frequency

TABLE 1.7 Isolines Used in Climatology	
Isoline	**Lines of**
Isallobar	Equal pressure tendency showing similar changes over a given time
Isamplitude	Equal amplitude of variation
Isanomaly	Equal anomalies or departures from normal
Isobar	Equal barometric pressure
Isocryme	Equal lowest mean temperature for specified period (e.g., coldest month)
Isohel	Equal sunshine
Isohyets	Equal amounts of rainfall
Isokeraun	Equal thunderstorm incidence
Isomer	Equal average monthly rainfall expressed as percentages of the annual average
Isoneph	Equal degree of cloudiness
Isonif	Equal snowfall
Isophene	Equal seasonal phenomena (e.g., flowering of plants)
Isoryme	Equal frost incidence
Isoterp	Equal physiologic comfort
Isotherm	Equal temperature

of particular values. A climagram, sometimes referred to as a hythergram, relates mean values of two climatic variables as a loop. The examples shown in Figure 1.1 use the average monthly temperature and precipitation values at Terre Haute, Indiana. Figure 1.1a is a continuous graph with the months of the year on the x-axis and temperature on the y-axis. The bar graph in Figure 1.1b is average monthly precipitation. The intersection of monthly average temperature and monthly precipitation is plotted in Figure 1.1c. This climagram is typical of a midlatitude continental station; other climate regimes show different shapes and patterns.

Units

The historical development of science has provided a legacy of units in which variables may be measured. In some cases this has led to a confusing array, which is well illustrated by the units used for temperature. The Kelvin, Celsius, and Fahrenheit scales used to indicate temperature provide an appropriate example. (Note: Conversion factors for this and other units are given in the Appendix while each unit is defined in the Glossary.)

The influence of historical development on the use of units is well illustrated using length. Among the earliest relationships established in the English system was that three barleycorns laid end to end or eight laid side by side equal 1 inch. Twelve of these measurements were then equal to 1 foot. The latter, however, was open to question because it is said that during the reign of James I, the length of his foot became the official length. The use of barleycorns to measure an inch is still reflected in present-day measure. A number 7 shoe is 1/3 inch (1 barleycorn) longer than a number 6.

In such a way, the English system developed over many years and was most prevalent until about 1790, when the metric system was introduced in France. In much of Europe, this became the most widely used system. Based on decimal relationships, basic units (meter, gram, and liter) were given appropriate prefixes based on increments of 10.

Clearly, for any system of units to be valuable, there must be a consistent set of standards. On a worldwide basis, the International Bureau of Weights and Measures,

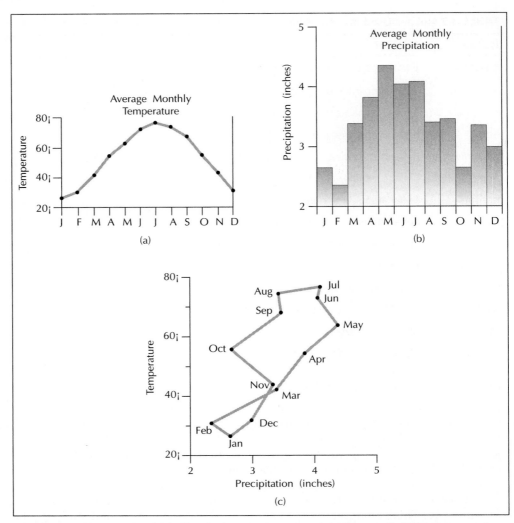

FIGURE 1.1 Some ways in which climatic data are represented: (a) line graph, (b) histogram or bar graph, and (c) climagram or hythergram.

located in Paris and established by the Treaty of the Meter in 1875, sets the standard. This treaty called for International Conferences to meet periodically. In 1960, the conference adopted the International System of Units (or SI, for Systéme International). SI units comprise the metric system based on fundamental physical quantities. Initially, six quantities were designated—mass (kilogram), length (meter), time (second), temperature (Kelvin), current (ampere), and luminous intensity (candela), with a seventh, the mole (the number of grams to which the molecular weight is numerically equal)— and these were adopted in 1971. Units frequently used in atmospheric studies—such as temperature, pressure, and energy—are derived from these basic quantities. These are dealt with as each is encountered in later chapters.

THE STANDARD ATMOSPHERE

Earlier in this chapter, it was stated that climatology is partially meteorological. Weather is the basic ingredient in climate. Therefore, it is essential to begin the study of climate with an introduction to the atmosphere and its workings.

The atmosphere is continually changing and is never exactly the same at any two different points in time or space. Although it is constantly changing, there is a set of conditions that describes the atmosphere most of the time. It is helpful to know the most frequent conditions in a vertical column through the atmosphere to use these conditions as a basis for examining the extent of change. Therefore, scientists constructed a model atmosphere. In the United States, this model is the U.S. **standard atmosphere**. This standard atmosphere represents an idealized, steady-state representation of Earth's atmosphere from the surface to 1000 km (600 mi) because it is assumed to exist in a period of moderate solar activity. The standard atmosphere does not represent an average condition but a steady-state condition. This is a condition where the atmosphere is in balance with the processes that govern it. The standard atmosphere defines profiles of chemistry, temperature, pressure, density, and several other variables.

Atmospheric Chemistry

The atmosphere consists of a mixture of gases that in its normal state is colorless, odorless, and flavorless. The gases contained are governed by gas laws (considered later). Some of the particles in the atmosphere are single atoms, such as argon and helium. Others are molecules consisting of atoms of two or more elements, such as water vapor and carbon dioxide.

Knowledge that the atmosphere includes many gases began with the work of chemists in the 18th century. The first gas studied in detail was carbon dioxide (CO_2). Its discovery in 1752 is somewhat surprising because, compared with nitrogen and oxygen, there is only a small amount of carbon dioxide in the atmosphere. Gaseous nitrogen, discovered by Rutherford in 1772, at first was called *Mephitic air*. Shortly after this, Joseph Priestly isolated oxygen, which he called *dephlogisticated air*.

Over time, other gases were discovered, the last of which was argon, isolated in 1894. Eventually the gases were given their modern names, and their relative proportion by volume in the atmosphere was determined. As shown in Table 1.8, nitrogen and oxygen make up the bulk of the atmosphere. There are other gases present in very small, but important, amounts. Of singular importance in understanding the ways in which atmospheric gases act is an understanding of the gas laws. These laws, which show the relationships among temperature, pressure, and volume, are outlined in Box 1.1.

TABLE 1.8 Chemical Composition of the Dry Atmosphere Below 80 km

Gas	Parts per Million	Percentage of Total
Nitrogen	780,840.0	78.1
Oxygen	209,460.0	20.9
Argon	9,340.0	0.9
Carbon dioxide	390.0	
Neon	18.0	
Helium	5.2	
Methane	1.4	
Krypton	1.0	
Nitrous oxide	0.5	
Hydrogen	0.5	
Xenon	0.09	
Ozone	0.07	

Note: Data from U.S. Department of Commerce, NOAA, *United States Standard Atmosphere* (Washington, DC: Government Printing Office).

Quantitative Expression: Box 1.1

The Gas Laws

All gases respond to temperature and pressure changes. A series of *gas laws*, some of which are explained here, describe the relationships between these variables.

1. *Boyle's laws* describe relationships among pressure, volume, and density of gases. In each case, the equations assume temperature to be held constant. The first law states

$$P_0 V_0 = P_1 V_1 = K,$$

where P_0 and P_1 and V_0 and V_1 are the pressure and volume at two times, 0 and 1, respectively. K is a constant. The relationship indicates that an increase in pressure results in a decrease in volume and vice versa.
    The second law relates pressure *(P)* to density *(D)*:

$$P/D = K \text{ (at a constant temperature)}.$$

As pressure increases, so does density.

2. If pressure is considered constant, the volume of a gas can be related to temperature by *Gay–Lussac's law*:

$$V_t = V_0(1 + t/273),$$

where V_t is the volume at a temperature t (in °C) and V_0 is the volume at 0°C.

3. *Charles' law* provides the relationship between pressure and temperature when volume is held constant:

$$P_t = P_0(1 + t/273).$$

Here P_t is pressure at a given temperature, t (in °C); and P_0 is pressure at 0°C.

4. The *combined gas law* uses a gas constant, r, to combine the relationships of gas, pressure, volume, and temperature:

$$PV = rT.$$

This *equation of state* can be written in a form that draws on the molecular weight of a mass to derive a universal gas constant, R, and the density of the gas

$$P = \rho RT,$$

in which P is pressure, ρ is density, R is the universal gas constant (2.87×10^6 erg/g°K), and T is temperature. This equation allows derivation of any one variable if the other two are known.

The gases making up the atmosphere can be divided into two groups: constant gases (those relatively constant by volume) and variable gases.

Constant Gases

Constant gases remain in the same proportion in the atmosphere upward to an altitude of about 80 km (49 mi). The three most important constant gases are nitrogen (78% by volume), oxygen (21%), and argon (0.93%). Nitrogen is by far the most abundant of the gases, but it is relatively inactive in the atmosphere. Argon is also inactive, but it is present in rather small amounts. Oxygen is present in large quantities and is very active in the chemical processes of the physical and biological environments. There are many other constant gases in lesser amounts.

Variable Gases

Variable gases, as the term implies, vary in proportion of total atmospheric gases from time to time and place to place. The most important variable gases are water vapor

TABLE 1.9 Major Ingredients in a Steady-State Tropospheric Aerosol

Source	Total (tons)	Percentage of Total
Vegetation	1.7×10^7	25.8
Dust rise by wind	1.6×10^7	24.1
Sea spray	7.6×10^6	11.9
Forest fires	6.2×10^6	9.9
Sulfur cycle	5.5×10^6	8.6
NO_x NO_3	5.5×10^6	7.7
Nitrogen cycle (ammonia)	3.9×10^6	6.0
Combustion and industrial	1.7×10^6	2.6
Anthropogenic sulfates	1.7×10^6	2.6

and carbon dioxide. Water vapor content varies from nearly zero to a maximum of about 4% by volume, but this small relative volume is extremely important. Carbon dioxide exists in amounts that average nearly 0.04%. Of course, water vapor is the source of all precipitation that falls on the earth. Water vapor, carbon dioxide, and dust all absorb solar radiation. Ozone, another variable gas, is in the lower atmosphere in small amounts, the average being about one part per million. Another variable element of the atmosphere, which in many ways acts like a gas, is the particulate matter suspended in the air (**aerosol**). This includes soil particles, smoke residue, ocean salt, bacteria, seeds, spores, volcanic ash, and meteoric particles (see Table 1.9). The primary source of the solid particles is the earth's surface. Particulate matter decreases rapidly with altitude. High-altitude particles occur from meteoric dust, volcanic eruptions, and nuclear explosions in the atmosphere. The overall amount of particulate matter in the atmosphere varies from as little as one hundred parts per cubic centimeter to several million parts per cubic centimeter.

Both particulate matter and water vapor are quite important in the atmosphere. They are largely responsible for the day-to-day variation in solar energy reaching the surface of the earth. Solid particles, moreover, serve as nuclei for condensation of water vapor and thus are necessary for precipitation.

VERTICAL STRUCTURE OF THE EARTH'S ATMOSPHERE

Some changes in the atmosphere occur with height. The atmosphere can be divided vertically into several different zones, based on a variety of changes that occur with altitude (Figure 1.2).

Troposphere

The lowest zone is the **troposphere**. The troposphere is a turbulent zone that has a rather uniform decrease in temperature with height. Worldwide, the average rate of change in temperature with altitude (the lapse rate) is 6.5°C/1000 m (3.5°F/1000 ft). Similarly, as shown in Figure 1.3, the vertical distribution of other variables shows a varied decrease with height.

The troposphere has two subzones—the lower and upper troposphere. The lower troposphere extends upward to about 3 km above the surface; it is the zone in which there is maximum friction between the earth and the atmosphere. It is the zone most affected by the daily changes in surface conditions. Another characteristic of this zone is the frequent existence of temperature inversions. A temperature inversion exists when the temperature increases rather than decreases with height.

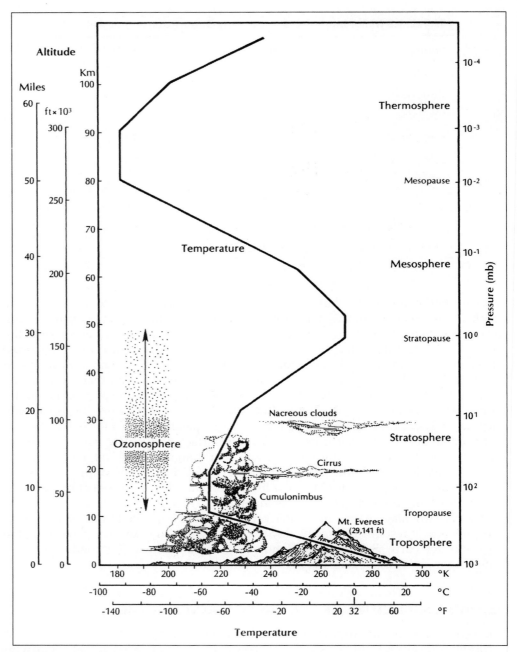

FIGURE 1.2   Vertical structure of the atmosphere.

The upper troposphere extends to a mean height of 11 km (7.2 mi). It is less affected by the daily changes that take place near the surface and from the effects of friction at the surface. The primary changes that take place in this zone result from the secondary circulation (atmospheric storms) and the seasonal change in energy.

The water vapor content of the atmosphere at any point in time and space depends on the temperature of the air and to a lesser extent on atmospheric pressure, the proximity to a moisture source, and the history of the air mass. Water vapor content is normally highest close to the surface for two reasons. One is that most of the water in the atmosphere gets there as a result of evaporation from the ocean and from evapotranspiration. The second reason is that normally air temperature is highest near the surface. The most water vapor recorded in the atmosphere is on the

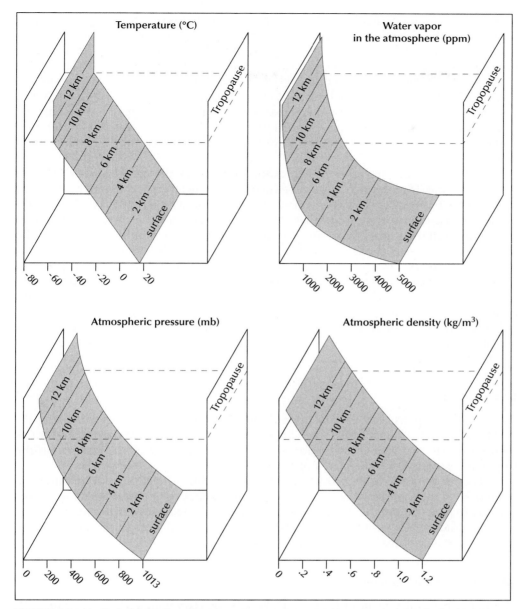

FIGURE 1.3 Vertical distribution of temperature, water vapor, pressure, and density in the lowest 14 km of the atmosphere.

shores of the Red Sea, where temperatures average 34°C (93°F). Water temperature reaches 35°C (95°F) here in summer. At air temperatures from 35°C down to –5°C (95°F–23°F), the amount of water vapor the air can hold decreases by a half for every 10°C (18°F) drop in temperature. The recorded range of water vapor in the atmosphere near the surface varies from a low of 0.1 ppm in the Antarctic and Siberia to a high of 35,000 ppm along the Persian Gulf. Due to the combination of atmospheric temperatures and altitude above the moisture source, the mean water vapor content decreases rapidly with altitude to a height of 12 to 15 km (7.4 to 9.3 mi). Here it reaches a level of 2 to 3 ppm.

The atmosphere is an unconfined gaseous fluid resting on the surface of the earth. As a result of the air having mass and being compressible, the mass and pressure of the air decrease with height. Atmospheric pressure at any point on the surface or in the atmosphere is the force per unit area exerted by the mass of the atmosphere above that point. Air close to the surface is subject to the mass of gases above. Hence, the greatest portion of the mass of the atmosphere lies near the surface. Atmospheric pressure decreases with height in a geometric fashion. That is, it decreases at a decreasing rate. The pressure data in Figure 1.3 show that surface pressure is on the order of 1000 mb. In the lower stratosphere, only 15 km (9 mi) above sea level, the pressure is only 100 mb. Of the total mass, about 50% lies below 5,500 m (18,000 ft), 84% lies below 13 km (8 mi), and 99% lies below 30 km (18 mi). The density of the atmosphere decreases with height at a geometric rate, but the gases stay in roughly the same proportion up to heights of at least 80 km (48 mi). The gases decrease in density away from the earth and extend for a distance of at least 1000 km (600 mi).

A distinct boundary layer known as the **tropopause** marks the upper reaches of the troposphere. There are actually a series of overlapping layers at different heights that make up the tropopause. This boundary zone is significant in several ways. Basically, it marks the upper limit of most turbulent mixing started from the surface. It represents a cold point in the vertical temperature structure of the atmosphere. It also marks the upper limit of most of the water in the atmosphere. Very little moisture penetrates the tropopause except through severe thunderstorms, which will go as high as 20 km (12 mi). The amount of water vapor in the atmosphere above the tropopause is very small. At the equator, the tropopause is at a height of 16 to 17 km (9.9–10.5 mi). The temperature averages –70 to –85°C (–94 to –121°F), and the pressure averages only 100 mb, only one-tenth of sea-level pressure. Above the north and south poles, the tropopause is at a height of 9 to 12 km (5.5–7.4 mi), with a temperature averaging –50 to –60°C (–58 to –76°F) and at a pressure of some 250 mb.

Stratosphere

Above the tropopause is the stratosphere, which is named for the layered nature of the air at these levels. The **stratosphere** is relatively stable and relatively dry, and it has relatively little vertical motion. Some high-velocity winds occur just above the tropopause; otherwise, winds are noticeably absent.

The temperature at the tropopause is about –58°C (–72°F). The temperature remains nearly the same up to about 20 km (12 mi). The upper region of the stratosphere extends to some 50 km (31 mi) and has temperature increases with height that range up to as much as 4°C/km (2.2°F/1000 ft). However, the temperature at the top of the stratosphere is near that at the bottom.

Above the Stratosphere

The upper boundary layer of the stratosphere is the **stratopause**. As is the case with the tropopause, there may not be a single boundary layer but instead a series of overlapping layers making up a transition zone.

Above the stratopause is a layer identified by a temperature decrease with altitude (**mesosphere**). Beginning at an elevation of about 48 km (29 mi), the decline in temperature continues outward to the **mesopause** near 80 km (49 mi). Up to the mesopause, the mixture of gases is about the same as at the surface. For this reason, the lower 80 km of the atmosphere is termed the **homosphere**. Beyond the 80-km altitude, the composition of the air begins to change. Above the homosphere atmosphere, gases tend to stratify on the basis of molecular weight. This layer is termed the **heterosphere**. To a height of 220 km (130 mi), molecules of nitrogen largely make up the atmosphere. Layers of helium and oxygen exit outward from there (Figure 1.4).

The mesosphere coincides with a zone called the *ionosphere*. As the name implies, the ionosphere contains ionized gases and free electrons resulting from absorption of solar radiation. Short-wave radiation from the sun is absorbed, and the energy causes the electrons to split from the atoms of nitrogen and oxygen. The proportion of ionized

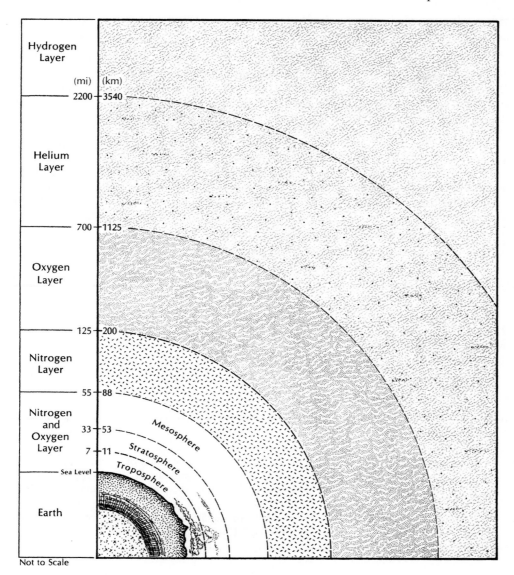

FIGURE 1.4 Chemical zones in the atmosphere.

particles is large because the density of particles is low and solar energy high. As height in the ionosphere increases, the fewer particles there are, the more solar energy is available, and the more ionization that takes place.

Incoming solar radiation begins to interact with the atmosphere at altitudes beyond 500 km (310 mi). This outer region is also where gaseous particles escape the earth's gravity. Particles are far enough apart that some of them have a high enough velocity in a direction away from the earth to escape to space. The atmosphere is very thin at this altitude, and it is here that incoming space vehicles and meteorites begin to heat due to friction.

Above this, to perhaps 1,125 km (675 mi), atomic oxygen is prevalent. Beyond this layer of atomic oxygen, helium is most common out to 3,540 km (2,124 mi). Still farther out, hydrogen atoms predominate. The boundaries among the zones are not clearly defined. The heights represent the altitudes above which different chemistries predominate.

Summary

Climate is the total of weather at a place for a particular period. Study of climate is often divided into climatography, synoptic climatology, dynamic climatology, applied climatology, and climate forecasting.

The study of climatology began in ancient times but was stimulated considerably by information gained in the great voyages during the Age of Discovery. During this period, instruments were developed and used to make observations that were recorded. The development of weather satellites now aids climatic interpretation and understanding. Similarly, the creation of the National Climate Program has promoted study.

Climatologists gather, store, and analyze climatic data. Most climatic data come from weather monitoring stations and are stored by both government and private agencies. Computers and computer techniques aid in data analysis, including summarizing data and analyzing data through time. The modeling of past and future climates is growing rapidly in importance.

The atmosphere changes with time and from place to place. Important changes occur with altitude; to provide a baseline for atmospheric studies, the standard atmosphere was developed as a model. This model depicts atmospheric variables, such as pressure and temperature, as they are most frequently found. It is thus a model of mean conditions with height.

Key Terms

Aerology, *5*

Aerosol, *15*

Climatology, *5*

Heterosphere, *18*

Homosphere, *18*

Mesopause, *18*

Mesosphere, *18*

Meteorology, *5*

Radiosondes, *8*

Standard atmosphere, *13*

Stratopause, *18*

Stratosphere, *18*

Tropopause, *18*

Troposphere, *15*

World Meteorological Organization (WMO), *9*

Review Questions

1. Explain the difference between weather and climate.
2. Why was the Age of Discovery important for climatology?
3. What variables are measured in atmospheric studies?
4. What types of data are available today?
5. What types of international cooperation are found in the atmospheric sciences?
6. What are the main areas of study of climatology?
7. What is the standard atmosphere? Why is it of value?
8. Differentiate between constant and variable atmospheric gases.
9. What are the major characteristics of the troposphere? The homosphere? The ionosphere?

2

Energy and the Climate System

The earth–atmosphere system is sustained by the supply of energy from the sun. Solar radiation is the ultimate source of energy that results in all the varied atmospheric conditions experienced on Earth. It is fitting that a study of climatology is initiated by examining energy transferred and exchanged in the earth–atmosphere system.

THE SOLAR SOURCE

The sun is a gaseous mass with a diameter 109 times that of the earth. The source of the sun's energy is nuclear fusion produced in the core of the sun, where a nuclear reaction takes place which changes hydrogen into helium. Because the sun is a gaseous mix, no sharp boundaries exist within it. Since different sections of the sun have different characteristics, it can be described using arbitrary divisions. Major parts are the core, the **photosphere** (or visible surface), the **chromosphere**, and the **corona** (Figure 2.1).

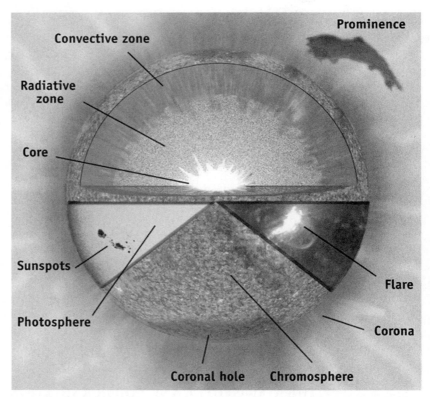

FIGURE 2.1   Internal structure and surface features of the sun (from NASA, at http://learn.arc .nasa.gov/planets/0/sunpartsfull.jpg).

Photosphere

The photosphere is the bright outer layer of the sun, which emits most of the radiation, particularly visible light. It consists of a zone of burning gases 300 km (200 mi) thick. The photosphere is an extremely uneven surface. There are many small bright areas called *granules*. Their brightness is a result of bursts of extremely hot gases formed in the photosphere that well up to the surface. Darker areas around each granule are cooler gases. The granules average 1000 km (620 mi) in diameter. The hot gases spread out, cool on reaching the surface, and settle back below the surface. The net effect of all these granules is convection of hot gases outward and cooler gases inward. It is much like the boiling of a large kettle. The elements in the photosphere are hydrogen (90%) and helium (10%). At the radiative surface, the effective temperature is on the order of 6000°K (11,000°F). The surface temperature determines both the amount of energy emitted and the wavelengths at which it occurs.

Chromosphere

Just above the photosphere is the chromosphere. It is a relatively thin layer of burning gases. This region of gases is under little pressure and extends away from the sun as much as 1 million km (600,000 mi). It is here that the particles reach velocities high enough to leave the sun. The electrons and protons escape the sun in a stream called the *solar wind*. When they reach the earth environment, they meet the magnetic field that surrounds Earth, and the magnetic field diverts most of these particles toward the two poles.

   A feature of solar activity that occasionally affects the earth is the solar flare. Solar flares are sudden and explosive bursts of energy near sunspot clusters. They emit huge quantities of radiant energy and atomic particles. The atomic particles add to the solar wind. When these particles reach Earth, they disturb the ionosphere. This in turn interrupts radio and satellite communications and triggers the auroras. A solar flare has a life span of only a matter of hours.

Sunspots

Prominent visible features of the surface of the sun are dark areas known as **sunspots**. Sunspots appear as dark areas because they are about 1500°C (2700°F) cooler than the surrounding chromosphere. Science credits Galileo with reasoning that they were a feature of the sun and not clouds between the earth and the sun. They start as small areas about 1600 km (1000 mi) in diameter. Each spot has a black center, or umbra, and a lighter region, or penumbra surrounding it. A sunspot has a life span ranging from a few days to a few months.

Because records of sunspots have existed for many years, and sunspots are relatively easy to see, they are one indicator of solar activity. The number of sunspots appears to follow an 11-year cycle. First, the sunspots increase to a maximum, with 100 or so visible at a given time. Over a period of years, the number diminishes until only a few or none occur. Change from the maximum to the minimum number takes about 5.5 years, and the complete cycle is thus 11 years. This is not an absolute periodicity, as the interval between maxima cannot be forecast with certainty; however, 11 years represent a mean interval.

Another measure of solar activity is the number of faculae (Latin for "small torch"), which is closely associated with sunspots: They brighten as the number of dark sunspots increases. Their increased brightness is the dominant factor in changing solar output rather than sunspot darkening.

A series of satellites made measurements that provide unequivocal observations of how the sun's output varies with the 11-year cycle in solar activity. These results show that the level of total solar irradiance (TSI) is greatest around the solar maxima, and the amplitude of the solar cycle variation is less than 1%.

Models of the TSI that combine changes in sunspot number and faculae brightness produce a good fit with satellite observations of the TSI. These models explain not only the observed changes in solar output since 1980 but also the longstanding hypothesis that the cold period known as the Little Ice Age might be related to changes in solar activity. In particular, the colder weather of the 17th century appears to have been the result of an almost complete absence of sunspots during this period, known as the "Maunder Minimum." The small change in the TSI during the last two solar cycles is an order of magnitude too small to explain observed correlation between solar activity and global temperature trends since the late 19th century.

Just how the sunspot cycle influences weather and climate is a matter of controversy. Some researchers claim to have found that weather patterns in a particular part of the earth follow the sunspot cycle closely. In other cases, no relationships exist. Even the physical processes by which sunspots affect weather are not clear. The wavelengths of these emissions correspond to the wavelength of x-rays, and they have different effects on different parts of the atmosphere. A further complication is that sunspot activity coincides with solar wind intensity, and it is difficult to separate the effects of each. What is discernable from observations is that the amount of solar energy received at the outer atmosphere, the solar constant, has steadily declined since the mid-1970s; however, average global temperatures have increased.

ENERGY TRANSFER

Energy is the capacity for doing work. It can exist in a variety of forms and change from one form to another (Table 2.1). The transfer of energy from place to place is of major importance in climatology. There are basically three ways in which this transfer can take place: conduction, convection, and radiation.

Conduction consists of energy transfer directly from molecule to molecule, where the molecules are densely packed and contact one another. Energy always moves from an area of more energy to an area of lesser energy. In a sense, it is much as water flows from high places to lower places.

Convection involves the transfer of energy by the movement of an energy-laden substance from one location to another. Both convection and conduction depend on the existence of a physical substance in which to operate. This substance might be a solid, liquid, or gas.

TABLE 2.1 Energy Forms and Transformations

Forms

Radiation	The emission and propagation of energy in the form of waves
Kinetic energy	The energy due to motion: one-half the product of the mass of a body and the square of its velocity
Potential energy	Energy that a body possesses by virtue of its position and that is potentially converted to another form, usually kinetic, energy
Chemical energy	Energy used or released in chemical reactions
Atomic energy	Energy released from an atomic nucleus at the expense of its mass
Electrical energy	Energy resulting from the force between two objects having the physical property of charge
Heat energy	A form of energy representing aggregate internal energy of motions of atoms and molecules in a body

Examples of Transformations

Atomic Energy (Sun)	$\longrightarrow$	Radiation (Sunlight)	$\longrightarrow$	Heat (Earth surface)	$\longrightarrow$ Radiation (Terrestrial)
		Radiation (Sunlight)	$\longrightarrow$	Chemical energy (Photosynthesis)	$\longrightarrow$ Food chain
		Potential energy (Water vapor)	$\longrightarrow$	Kinetic energy (Raindrop)	$\longrightarrow$ Heat (Friction)

Radiation is the only means of energy transfer through space without the aid of a material medium. The major source of energy on our planet is the sun. Between the sun and the earth, where a minimum of matter exists, radiation is the only important means of energy transfer. So important is this solar input that most climatological phenomena are related to it.

Fully comprehending the role of solar energy in the functioning of the atmosphere requires three stages. First is understanding the nature of the energy emitted by the sun. Second is understanding the effect of solar radiation on the earth–atmosphere system. Third is understanding changes it undergoes in the system.

THE NATURE OF RADIATION

Every object at a temperature above absolute zero, $-273°C$ ($-459°F$), radiates energy to its environment in the form of electromagnetic waves that travel at the speed of light. Energy transferred in the form of waves has characteristics that depend on **wavelength**, amplitude, and frequency (Figure 2.2). Radiant energy emits along a spectrum from very short to very long wavelengths (Figure 2.3). The characteristics of the radiation emitted by an object depend mainly on its temperature.

The amount of radiation varies as the fourth power of absolute temperature (°K). The Kelvin scale is based on the concept of absolute zero. Absolute zero is the theoretical temperature at which all molecular motion would cease. It is equivalent to $-273°C$ ($-459°F$). The hotter an object is, the greater the flow of energy from it. The Stefan-Boltzmann Law expresses this relationship as follows:

$$F = \sigma T^4$$

where

F = flux of radiation emitted per square meter
σ = constant (5.67×10^{-8} W/m^2 K^4 in SI units)
T = object's surface temperature in degrees Kelvin

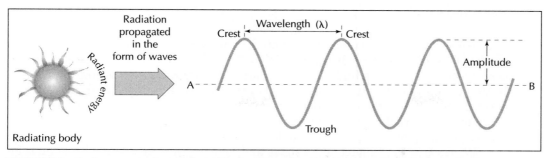

FIGURE 2.2 Radiant energy is transferred in the form of waves. The waves are defined in terms of wavelength, amplitude, frequency, and speed. Wavelength is the distance between wave crests. Amplitude is half the height difference between crest and trough. Frequency is the number of waves past a point in space per unit time. Speed is the distance a wave travels per unit time.

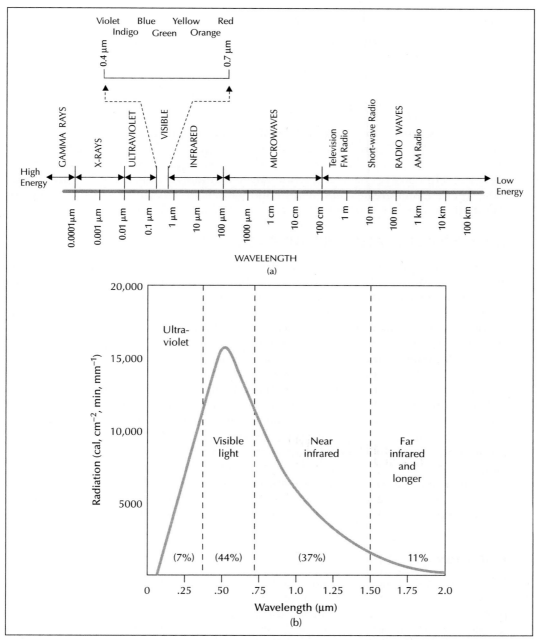

FIGURE 2.3 (a) The electromagnetic spectrum of solar radiation passes from short-wave high-energy to long-wave low-energy waves. In a range between 0.1 and 1 μm (10^{-4} and 10^{-5} cm), the energy is visible to the human eye. (b) The portion of the electromagnetic spectrum containing most of the energy emitted by the sun.

For example, the average temperature at the surface of the sun is 6000 K. The average temperature of Earth is 288 K (59°F). The temperature at the surface of the sun is more than 20 times as high as that of Earth. Twenty raised to the fourth power is 160,000. Therefore, the sun emits 160,000 times as much radiation per unit area as Earth. The sun emits radiation in a continuous range of electromagnetic waves. They range from long radiowaves with wavelengths of 10^5 m (62.5 mi) down to very short waves such as gamma rays, which are less than 10^{-4} μm (4^{-10} in.) in length.

Wien's Law of radiant energy states the wavelength of maximum radiation is inversely proportional to the absolute temperature. Thus the higher the temperature, the shorter the wavelength at which maximum radiation occurs:

$$\lambda_{max} = \frac{2897}{T}$$

where

T = temperature in degrees Kelvin
For the sun:
$\lambda_{max} = 2897/6000 = 0.48$ μm

For Earth:
$\lambda_{max} = 2897/288 = 10$ μm

At the temperature of the surface of the sun, the maximum radiation is in the range from 0.4 to 0.7 μm. This is precisely the range of radiation that the human eye perceives, and we call it the *visible range*. This is a very good example of evolutionary processes. Human eyes have evolved to take maximum advantage of the part of solar radiation that is in greatest abundance. Individuals do not all see the same range of radiation, as some see shorter and longer wavelengths than others.

THE ATMOSPHERE AND SOLAR RADIATION

The energy emitted by the sun passes through space until it strikes some object. The intensity of radiation reaching a planet is in proportion to a basic physical law known as the inverse square law. This law states that the area illuminated, and hence the intensity, varies with the squared distance from the light or energy source. Thus if the intensity of radiation at a given distance X is one unit, at a distance of $2X$, the intensity of the light will be one-fourth that received at distance X. This law applies to the intensity of solar energy intercepted by planets in the solar system. Earth receives only about 1/2,000,000,000 of the sun's energy output.

We know the amount of energy radiated by the sun, and we know that the mean earth–sun distance is 149.5 million km (93 million mi). We also know the amount of radiation intercepted by a surface at a right angle to the solar beam at the outer limits of the atmosphere. This quantity of radiation is approximately 1367 W/m^2 (1.97 cal/cm^2/min) and is known as the **solar constant**. This value is often cited in another measure of solar energy, the Langley, which is one calorie per square centimeter per minute.

The solar constant is the basic amount of energy available at the outer limits of Earth's atmosphere. Several processes deplete the solar radiation as it passes into the atmosphere. These processes include reflection, scattering, absorption, and transmission.

Reflection

The earth and its atmosphere reflect part of the solar radiation back to space. There is considerable variation in reflection of natural surfaces. Reflectivity, or **albedo**, is expressed as a percentage of the incident radiation reaching the surface. Clouds are by far the most important reflectors in the earth environment. Cloud reflectivity ranges from 40 to 90%, depending on the type and thickness.

Water covers the largest area of Earth's surface. The reflectivity of water depends on the angle of the solar beam and the roughness of the water surface. Reflectivity of

TABLE 2.2 Typical Albedos of Various Surfaces to Solar Radiation

Type of Surface	Albedo (%)
Fresh snow	75–95
Clouds	
Cumuliform	70–90
Stratus	60–84
Cirrostratus	44–50
Planet Venus	78
Old snow and sea ice	30–40
Dry sand	35–45
Planet Earth	30
Desert	25–30
Concrete	17–27
Savanna	
Dry	25–30
Wet	15–20
Grass-covered meadow	10–20
Tundra	15–20
Dry, plowed field	5–25
Asphalt road or parking lot	5–17
Green field crops	3–15
Deciduous forest	10–20
Coniferous forest	5–15
Moon	7
Water	
Angle of inclination of the sun:	
0°	99+
10°	35
30°	6
50°	2.5
90°	2

water decreases as the sun gets higher and higher in the sky. When the surface is smooth and the sun is near the horizon, reflectivity is high. People out in boats before 10 A.M. and after 2 P.M. can sunburn even when wearing broad-brimmed hats because hats do not protect from radiation reflected from the water. Water reflects as little as 2% of the radiation when the solar angle is 90° and the water is choppy.

The land surface of the earth reflects only 40 to 50% of solar radiation. Reflectivity of land surfaces varies with the type of surface cover. Fresh snow reflects more than 75%, and dry sand reflects over 35% of the incident radiation. Table 2.2 provides examples of the reflectivity of various natural surfaces. Planet Earth has an albedo of 30% that represents the mean reflectivity from the ocean, land, and atmosphere.

Scattering

Scattering is the process by which small particles and molecules of gases diffuse part of the radiation in different directions. The process changes the direction of the radiation in a relatively random fashion. The English scientist Lord Rayleigh (1842–1919) developed the explanation for scattering, and the effect is therefore called *Rayleigh scattering*.

The amount and direction of scatter depends on the ratio of the radius of the scattering particle to the wavelength of the energy. Furthermore, the amount of scatter is inversely proportional to the fourth power of the wavelength. This means that in a given set of conditions, the shorter wavelengths scatter more readily than do long wavelengths of radiation. For example, radiation of wavelengths twice as long as that of another wavelength scatters only one-sixteenth as much.

The most obvious effect of scattering in the atmosphere is sky color. The only reason our sky appears blue is because of scattering of radiation in the shorter wavelengths of visible light. As radiation in the visible range enters the outer regions of Earth's atmosphere, small gas molecules scatter the shortest wavelengths, so the violet scatters first. Normally, the atmosphere scatters and absorbs the violet so the sky doesn't have a violet color. Blue scatters next. This randomly diffused radiation in the blue range scattered through the lower atmosphere gives it its color. Most of the radiation in the visible range has wavelengths greater than the diameter of particles in the dry atmosphere. This energy passes through without being changed in any way.

Scattering also explains the orange and red colors seen at dawn and sunset (Figure 2.4). At these times, the radiation passes through the atmosphere at a very low angle. Thus it passes a long distance through air close to the ground. The lower atmosphere contains not only the dry gases but water vapor, solid particles, organic material, and salt. These larger particles scatter the longer wavelength radiation. In fact, the

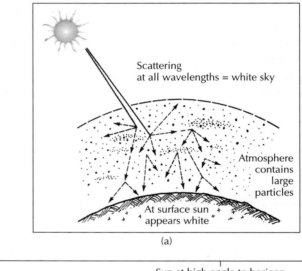

(a)

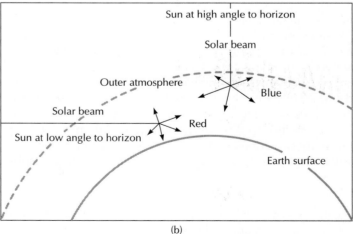

(b)

FIGURE 2.4   (a) Scattering of the full visible spectrum on a hazy day. The radiation is all scattered such that white light seems to come from all directions. (b) Sky color depends on the lengths of the paths solar radiation takes through the atmosphere. The shortest wavelengths are scattered first. Thus, when the radiation is perpendicular to the earth, the sky appears blue. When the sun is low in the sky, the blue is all filtered out, and yellow, orange, and red predominate.

atmosphere absorbs most of the radiation in the shorter wavelengths (violet to green). Only the longest wavelengths, orange and red, pass through.

In early morning and late evening, the bottoms of clouds often have a red tint. This phenomenon results from red light reflected from the clouds. The colors associated with shorter wavelengths are scattered before the light reflects from the clouds. Spectacular sunsets occur when there is a lot of dust in the atmosphere. Dust scatters the radiation, as do gas particles. Dust from storms on the surface or from volcanic eruptions causes unusually colorful sunsets. There were colorful sunsets following the eruption of El Chichon, Mexico, on April 4, 1985. Following the eruption of Krakatoa in 1883, there were brilliant sunsets for several years until the dust settled.

Rayleigh scattering is not the only scattering theory. Gustave Mie developed another theory for scattering in 1908. This theory states that molecules that have a larger ratio of diameter to wavelength than those that give rise to Rayleigh scattering scatter light at all wavelengths. More light scatters in a forward or continuing direction than in a backward direction. The sky is often the darkest blue shortly after a rain because the rain washes out the larger particles of debris and removes much of the moisture. There is then less scattering and a darker sky. Similarly, on a clear day, the sky is a darker blue directly overhead than near the horizon. This is because there is less overall scattering in the direct path through the atmosphere overhead. Large particles scatter the light coming from near the horizon. The bright white color of the sky on a hazy day is due to the scattering of all of the visible light as it passes through the haze. When the sky is cloud covered, only scattered radiation reaches the ground.

In polar regions, solar radiation scattered downward towards the earth (sky radiation) is a significant part of the total radiation. During the season when the sun does not come above the horizon, sky radiation is the main source of radiant energy.

Absorption

Absorption retains incident radiation and converts it to some other form of energy. Most often, it changes to sensible heat which raises the temperature of the absorbing object. For example, sunlight striking the side of a house is absorbed and heats the wall.

Gas molecules, cloud particles, haze, smoke, and dust absorb part of the incoming solar radiation. Such absorption is selective, for gases absorb only in certain wavelengths. Each gas has a characteristic absorption spectrum and so different gases absorb different portions of the electromagnetic spectrum. The two most common gases in the atmosphere, nitrogen and oxygen, absorb ultraviolet radiation. Triatomic oxygen (ozone) absorbs shorter wavelengths than nitrogen and oxygen.

Nitrogen does not absorb much incoming solar radiation. The maximum solar radiation is at 0.5 μm, and nitrogen does not absorb well in this frequency. Oxygen (O_2) and ozone (O_3) absorb well at wavelengths below 0.3 μm, with most of the absorption occurring in the ionosphere. Water vapor absorbs fairly well in the infrared range but not in the range of maximum solar radiation. These three gases make up over 99% of atmospheric gases. Because none are good absorbers in the visible range where most solar radiation occurs, there is a major window that lets in solar radiation.

Transmissivity

Reflection, scattering, and absorption impede the solar beam as it passes through the atmosphere. **Transmissivity** is the proportion of the solar radiation ultimately passing through the atmosphere. Transmissivity depends on both the state of the atmosphere and the distance the solar beam must travel through it. The relative distance the solar beam travels through the atmosphere is the path length or optical air mass. The path length has a value of 1 when the sun is directly overhead, or 90° above the horizon. The path length increases as the angle of the sun above the horizon decreases, and the length can be calculated for any sun angle. For example, if the sun is 30° above the horizon, the path length has a value of 2 because the solar beam has twice the distance to travel through the atmosphere.

THE PLANETARY ENERGY BUDGET

The amount of solar radiation reaching a unit area of the surface, the insolation, is made up of energy transmitted directly through the atmosphere and scattered energy. These two radiation sources make up the global solar radiation. On days of thick cloud cover, no direct radiation reaches the surface. Solar energy reaches the surface only as scattered or diffuse radiation.

Figure 2.5 is a simple model of the planetary energy balance. This model is similar to that of the standard atmosphere. It represents a composite of the planet over a period of years. The model shows the percentages of the total annual energy influx that is reflected, scattered, absorbed, and transmitted. Reflected energy plays no part in planetary heating. For the planet taken as a whole, a composite or average reflectivity of the atmosphere, ocean, and land masses is about 30%. This is termed the *planetary albedo*.

Clouds are the most important element in reflecting solar radiation back to space. The brilliant white of cloud tops you see when flying above them is reflected radiation of all wavelengths. Both Earth and Venus are very bright reflectors due to cloud cover. It is the cloud cover of Venus which makes it the brightest of the planets as seen from Earth. It is the second brightest object in the sky after the moon. While the moon appears bright on a clear night, it has an albedo of only 7%. The moon lacks a cloud cover, and the rocks of the surface are a relatively dark color. The earth appears much brighter from space than does the moon as it reflects more than four times as much sunlight.

Twenty-two percent of the incoming solar radiation is scattered, with most of it reaching the ground. However, of the 22% scattered, over one-fourth goes back to space. Like reflected energy, this scattered radiation that goes back to space plays no part in heating the planet. In the model, the atmosphere absorbs 17% of the incoming radiation. The stratosphere absorbs much of the shorter wavelengths, such as ultraviolet. Water vapor and carbon dioxide in the troposphere absorb the longer wavelengths. The remainder of the energy budget flow considers the long-wave energy emitted by the earth and the interchanges of that energy in the atmosphere.

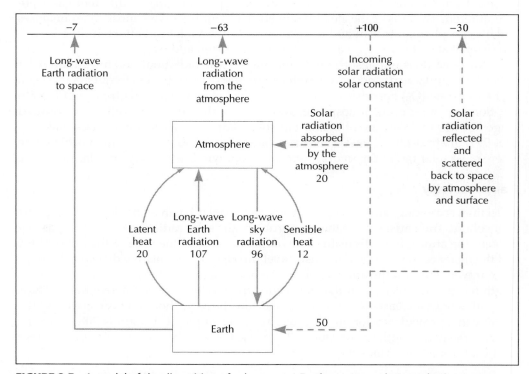

FIGURE 2.5 A model of the disposition of solar energy, Earth energy, and atmospheric energy.

The Greenhouse Effect

Life is possible on planet Earth solely because of a process known as the **greenhouse effect**. It is more properly called the *atmosphere effect* because the analogy to a greenhouse is imperfect. However, the name is firmly entrenched, and its use is continued here. The process assumes that the solar energy passing through the greenhouse glass (analogous to the atmosphere) is absorbed by materials (analogous to Earth's surface) inside. The radiated heat from inside the greenhouse cannot pass freely through the glass (analogous to greenhouse gases), and the air of the greenhouse (analogous to the lower atmosphere) is warmed.

If not for a complex interchange between the earth's surface and the atmosphere, as described by the greenhouse effect, the mean temperature of the atmosphere near the surface would be −20°C (−4°F). There would be no water in the liquid or gaseous forms. A selective absorption process raises the mean temperature of the atmosphere near the surface to 15°C (59°F). Solar radiation reaches Earth in fairly short wavelengths, and the atmosphere is relatively transparent to this radiation. The solar energy received at the surface of the earth is processed and eventually radiated or otherwise released to the atmosphere.

Nitrogen and oxygen are poor absorbers of both short-wave solar radiation and long-wave Earth radiation. Several of the variable gases, including water vapor and carbon dioxide, are relatively transparent to solar radiation in the visible range but good absorbers of Earth radiation. Clouds are even better absorbers of Earth radiation. Figure 2.6 shows the radiation curve for the earth and the bands in which the atmosphere absorbs this radiation. The earth radiates energy at wavelengths in the infrared band of 3 to about 30 μm. However, from 5 to 8 and beyond 13 μm in length, Earth radiation is transmitted into space virtually unabated. Carbon dioxide efficiently absorbs radiation that is 4 μm and that is from 13 to 17 μm in wavelength.

The result of these selective absorption processes is that most Earth radiation that escapes to space does so in a narrow band from 8 to 13 μm. This is the **atmospheric window** for Earth radiation. Figure 2.6 also shows that there is even some absorption of outgoing Earth radiation by upper atmospheric ozone at about 10 μm. Thick clouds are effective absorbers of Earth radiation over the entire range, even in the bands from 8 to 13 μm. So when layers of cloud over 100 m (330 ft) thick cover the sky, the atmosphere is able to absorb most of the earth's radiation.

The gases and liquid water that absorb Earth radiation also are good radiators of energy. The atmosphere radiates part of the energy absorbed to space and part back to

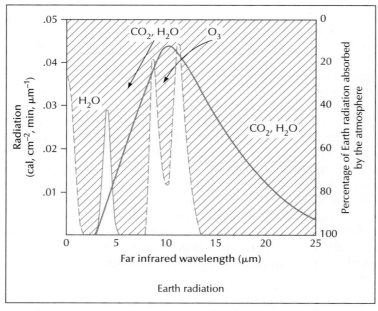

FIGURE 2.6 Earth radiation and atmospheric absorption.

the earth's surface. Nearly two-thirds of atmospheric radiant energy is directed back to the surface. This provides an additional energy source to direct solar radiation. It is the largest source of radiant energy absorbed at the earth's surface. It is nearly double the amount of energy received directly from the sun. It is this additional atmospheric radiation that raises the mean temperature at the surface to 15°C (59°F).

THE EARTH'S SURFACE AND SOLAR ENERGY

To maintain a steady-state temperature over a period of years, the earth disposes of the energy it receives. It does this through one of three processes: radiation, evaporation (with transpiration), or conduction–convection.

The earth, like all other objects with a temperature above absolute zero, radiates energy. The planet has a mean temperature of 15°C (59°F). This is a much lower temperature than the sun, and so the wavelengths of radiation are much longer. Earth radiation is at its maximum at 10 μm. Radiation from the surface of the earth is in wavelengths some 20 times as long as the incoming solar radiation. Energy gained by the earth that is not returned to the atmosphere by radiation is transferred by **latent heat** or by **sensible heat** transfer.

Changing liquid water to water vapor (evaporation) requires a considerable amount of energy. This absorbed energy is stored as latent heat in the water vapor until it changes back to liquid water. Condensation in the atmosphere then adds this energy to the air.

Conduction moves some energy from the earth to the air above. Because air is a very poor conductor of heat, the process heats only a few centimeters of air in this fashion. However, this warmed layer of air moves upward by convection, and the conduction process continues at the surface. The conduction–convection process is the main heating process in the lower atmosphere and provides the environmental temperatures that people experience, the sensible temperature. This process is discussed in more detail in a later chapter.

THE STEADY-STATE SYSTEM

The energy flow is a continuous process. Energy from the sun enters the earth–atmosphere system and ultimately returns to space. To remain at a steady-state temperature, incoming energy and outgoing energy must be in balance. If more energy came in than went out, the earth would get progressively hotter. If more energy went out than came in, it would get cooler.

The earth's surface receives energy from two main sources. One source is direct and scattered solar radiation. Fifty percent of the solar radiation reaching the planet passes through the atmosphere to the surface. The atmosphere transmits 33% of incoming radiation, and an additional 17% reaches the surface as scattered energy. The second source of energy is atmospheric radiation. This is part of the greenhouse effect. The total amount of energy the earth receives from these two sources is 146 units. Each unit is equivalent to 1/100 of the total solar radiation reaching the outer atmosphere.

Radiation, evaporation, and conduction balance the incoming 146 units of energy. Radiation to the atmosphere removes most of the energy received at the surface. A small amount (7 units) is radiated directly back to space. Evapotranspiration removes another 20 units of energy. Conduction–convection removes the remaining 12 units. If we consider the net loss of heat beyond the exchange of the greenhouse effect, the earth loses nearly equal amounts of heat by radiation and evaporation of water. Conduction removes a smaller amount. Convection moves heat rapidly away from the surface. Convection and evaporation–condensation move heat upward from the surface. In greenhouses, air warmed by conduction cannot escape, and so it continues to warm.

TABLE 2.3 Energy Balances of the Earth's Surface, Atmosphere, and Planet*

Energy Balance of Earth's Surface

Inflow		Outflow	
Solar radiation	50	Earth radiation	114
Sky radiation	96	Latent heat	20
Total	146	Conduction	12
		Total	146

Energy Balance of the Atmosphere

Inflow		Outflow	
Solar radiation	20	Radiation to space	63
Condensation	20	Radiation to surface	96
Earth radiation	107	Total	159
Conduction	12		
Total	159		

Energy Balance of Earth

Inflow		Outflow	
Solar radiation	100	Reflected and scattered	30
Total	100	Sky radiation to space	63
		Earth radiation to space	7
		Total	100

*Units are in hundredths of the incoming solar radiation over a year.

The atmosphere is also in balance. The atmosphere receives energy by absorbing direct solar radiation, by absorbing Earth radiation, by conduction of heat at the surface, and by the heat of condensation. Seventy-seven percent of the energy received by the atmosphere comes from the earth. The largest single source of energy is Earth radiation. Only 13% of the total energy comes from the direct absorption of solar radiation. Therefore, we must understand that as a result of the greenhouse effect, the prime heat source for the atmosphere is the earth's surface. Radiation to space and radiation back to the surface balances the energy received by the atmosphere.

Energy received by Earth is counterbalanced by an equal amount of radiation to space. The largest amount of energy released is by radiation from the atmosphere. This is primarily from cloud tops. Reflection and scattering of solar radiation and direct radiation from the surface remove the remaining energy. Table 2.3 shows the inflow and outflow of energy for the surface, the atmosphere, and for the planet.

Earth is in a steady-state balance of incoming and outgoing energy. The temperature of the earth undergoes small changes, but the mean temperature stays nearly the same. It varies only slightly around 15°C (59°F) at the surface, although there are daily, seasonal, and year-to-year changes. There are mechanisms at work that prevent major change and keep the system in a steady state. For instance, if the earth receives more energy than usual, or if more outgoing energy is absorbed, the temperature must go up. If the temperature goes up, the earth radiates away energy at a higher rate (Wien's Law). By the same token, if the earth starts to cool, it will lose less heat by evaporation, conduction, and radiation. These processes would offset any reduction in incoming energy.

Because the energy flow is in a steady state, the planetary climate is steady over a period of years. One of two events must occur in order for the steady-state temperature of the lower atmosphere to change and hence for the climate to change: There must be a change in either the flow of energy to and from the planet or a change in the internal greenhouse effect. We will see later in this book that such changes do occur.

Summary

The flow of energy into, through, and out of the atmosphere is dynamic and complex. The source for the energy that drives the atmosphere is solar radiation. The supply of solar radiation is steady but not constant. Changes occur in the amount of radiant energy ejected from the sun. However, this amount of energy is steady enough to refer to it as the solar constant. Energy reaching the atmosphere is disposed of in several ways. Some is reflected and scattered back to space, some is absorbed, and some is transmitted through to the surface. Earth's surface absorbs and reradiates this energy, which becomes the major heat source for the atmosphere. There is a built-in storage mechanism known as the greenhouse effect, which raises the temperature of the atmosphere such that life can exist. The energy balance of the planet is a steady-state system that maintains a mean temperature at the surface of about 15°C (59°F).

Key Terms

Albedo, *26*	Convection, *23*	Photosphere, *21*	Solar constant, *26*
Atmospheric window, *31*	Corona, *21*	Radiation, *24*	Sunspots, *23*
Chromosphere, *21*	Greenhouse effect, *31*	Scattering, *27*	Transmissivity, *29*
Conduction, *23*	Latent heat, *32*	Sensible heat, *32*	Wavelength, *24*

Review Questions

1. How is energy transferred? Give examples of processes and exchanges.
2. Why does the sun emit energy mostly in the form of light, while Earth radiates at longer wavelengths?
3. How much energy does the sun emit? How do we know this?
4. What are the main features of the sun's structure?
5. What process results in colorful sunsets? Explain how it works.
6. Differentiate between reflection and scattering in the atmosphere.
7. Outline the absorption characteristics of common atmospheric gases.
8. Describe the main features of the greenhouse effect on Earth.
9. By what processes is energy transferred from the surface to the atmosphere?
10. Why is the earth–atmosphere system considered a steady-state system?

Atmospheric Temperatures

Temperature is the most widely used atmospheric measurement. It is a measure of the quantity of heat energy present in a substance. More basically, it is a measure of the kinetic energy of the motion of the molecules in the body. The temperature of a substance measures the amount of heat per unit volume. The total heat in a substance depends not only on the temperature, but on the mass. Raising the temperature of 25 g (0.9 oz) of water from 20°C to 25°C (68°F–77°F) requires five times more energy than raising 5 g (0.15 oz) of water the same amount. The temperature in the container with 5 g (0.15 oz) of water is the same as that in the container with 25 g (0.9 oz), but the total heat is substantially different.

The temperature of substances, or within substances, determines the direction of the flow of energy. The flow of heat is always from the area of most heat to that of the least heat. Or the flow of heat is from the substance with the highest temperature to that of the lowest. The study of atmospheric temperatures is the study of the ebb and flow of energy at a place and the variation in energy from place to place.

THE SEASONS

Temperature varies with time over the earth's surface due to changes in radiation received at the surface. There are two periodic patterns of radiation influx and temperature that are due to the motions of the earth. One produces the **seasons**, and the other produces the daily changes in radiation and temperature. There are few places, if any, on the earth's surface that are truly without seasons. Seasons are most often thought of as summer, fall, winter, and spring. This is because most people live in the midlatitudes where temperature change from summer to winter is very large. However, over large areas of the earth, the change from hot to cold is not as important as the change from a rainy season to a dry season. The most important differences in energy received at various locations on Earth result from basic motions of the earth in space. These are rotation on its axis and revolution around the sun. Figure 3.1 is a schematic diagram of these Earth–sun relationships over a period of a year.

In its revolution around the sun, the earth follows an elliptical orbit so that the distance from Earth to the sun varies. Earth is usually closest to the sun (**perihelion**) on January 4 and most distant (**aphelion**) on July 4. The date varies as a result of leap year. The amount of solar radiation intercepted by Earth at perihelion is about 7% higher than at aphelion. This difference, however, is not the major process in producing the seasons.

The most important element in producing the seasons is the amount of radiation received at a place through the year. The amount of radiation varies as the angle of the sun above the horizon (intensity), and the number of hours of daylight (duration) change through the year. The intensity of solar radiation is largely a function of **angle of incidence**—the angle at which the solar energy strikes the earth. The angle of incidence directly affects both the energy received per unit area of surface and the amount of energy absorbed. Since the earth is nearly spherical, a curved surface is exposed to the radiation. Intensity of radiation is maximum in latitudes where solar radiation is perpendicular to the surface; this is the **solar equator** (Figure 3.2). The intensity of radiation decreases north and south of the solar equator as the angle of incidence decreases.

The seasonal changes in energy result from the inclination of the earth on its axis. The axis of the earth is tilted 23°30' from being perpendicular to the **plane of the ecliptic** (Figure 3.3). As the earth revolves about the sun, the solar equator moves north and south through a range of 47°. The geographic equator (0° latitude) is the mean location of the solar equator. The solar equator moves north and south through the year between 23°30' north (Tropic of Cancer) and 23°30' south (Tropic of Capricorn). The two tropics represent the latitudes farthest from the equator where solar radiation is perpendicular to the surface sometime during the year. As the earth travels around the sun during a year, the intensity of solar radiation varies at all latitudes.

The geographic equator has the least variation in the angle of incidence. Here the sun is never more than 23°30' from the zenith (the point directly overhead). All latitudes between the two tropics experience a variation in the angle of incidence. The variation reaches 43° at the Tropics of Cancer and Capricorn. All places between 23°30' and 66°30' of latitude experience a change in the angle of solar radiation through a year's time.

The primary factor responsible for the hot and cold seasons as well as the wet and dry seasons is the revolution of the earth about the sun and the inclination of the earth's axis relative to its orbital plane. This results in an imbalance of energy over the earth's surface through the year.

The change in the length of the daylight period strengthens the seasonal variation in temperature. Only at the time of the spring and fall equinoxes are the length of daylight and darkness equal everywhere over the earth. There is an imbalance between daylight and darkness the rest of the year. This imbalance is such that the daylight period is longer in the hemisphere where there is maximum intensity of solar radiation. Thus, when the sun's vertical rays are north of the equator, the daylight period in the northern hemisphere is longer than 12 hours. On the summer solstice, the length of daylight increases from 12 hours at the equator to 24 hours at the Arctic Circle (Tables 3.1, 3.2, and 3.3 provide data related to latitude and various Earth–sun

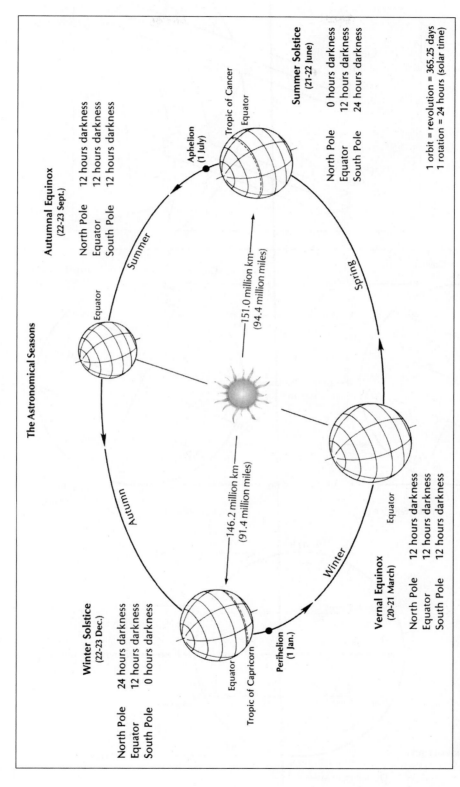

FIGURE 3.1 Orbit of the earth around the sun, the seasons, and change in length of day.

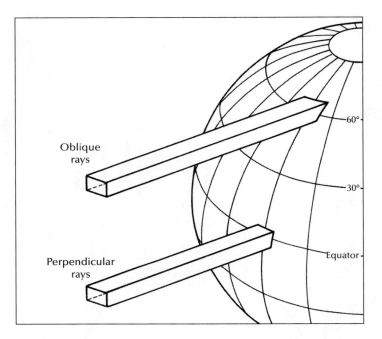

FIGURE 3.2 Angle of the rays of the sun and intensity of radiation. The same amount of energy is contained in both beams of radiation. The lower the angle of the beam, the larger the area over which the beam is spread.

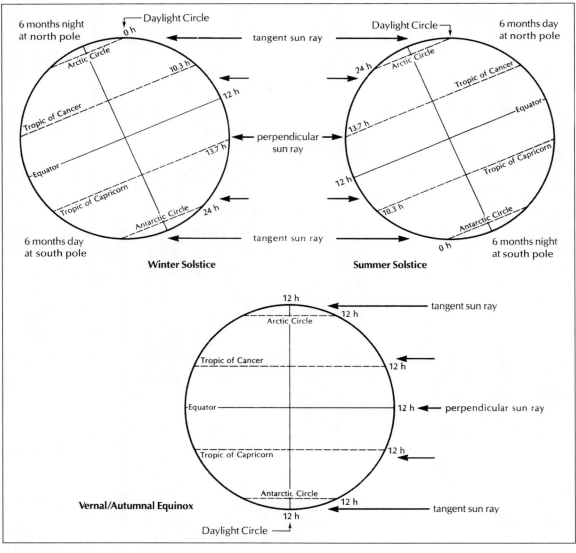

FIGURE 3.3 Earth's tilted axis results in variations of the angle of overhead sun and length of daylight. Diagram shows conditions at the equinoxes and solstices.

TABLE 3.1 Ephemeris of the Sun (Declination of the Sun on Selected Days of the Year)

Day	Jan.	Feb.	Mar.	April	May	June
1	−23°04′	−17°19′	−7°53′	+4°14′	+14°50′	+21°57′
5	22°42′	16°10′	6°21′	5°46′	16°02′	22°38′
9	22°13′	14°55′	4°48′	7°17′	17°09′	22°52′
13	21°37′	13°37′	3°14′	8°46′	18°11′	23°10′
17	20°54′	12°15′	1°39′	10°12′	19°09′	23°22′
21	20°05′	10°50′	−0°05′	11°35′	20°02′	23°27′
25	19°09′	9°23′	+1°30′	12°56′	20°49′	23°25′
29	18°08′	–	3°04′	14°13′	21°30′	23°17′

Day	Jul.	Aug.	Sept.	Oct.	Nov.	Dec.
1	+23°10′	+18°14′	+8°35′	−2°53′	−14°11′	−21°40′
5	22°52′	17°12′	7°07′	4°26′	15°27′	22°16′
9	22°28′	16°06′	5°37′	5°58′	16°38′	22°45′
13	21°57′	14°55′	4°06′	7°29′	17°45′	23°06′
17	21°21′	13°41′	2°34′	8°58′	18°48′	23°20′
21	20°38′	12°23′	+ 1°01′	10°25′	19°45′	23°26′
25	19°50′	11°02′	− 0°32′	11°50′	20°36′	23°25′
29	18°57′	9°39′	2°06′	13°12′	21°21′	23°17′

Source: U.S. Naval Observatory, *The American Ephemeris and Nautical Almanac for the Year 1950.* Table by R. J. List (Washington, DC, 1950).

TABLE 3.2 Length of Daylight for Intervals of 1° of Latitude*

	0°	1°	2°	3°	4°	5°	6°	7°	8°	9°
0°	12:07	12:11	12:15	12:18	12:22	12:25	12:29	12:32	12:36	12:39
	12:07	12:04	12:00	11:57	11:53	11:50	11:46	11:43	11:39	11:36
10°	12:43	12:47	12:50	12:54	12:58	13:02	13:05	13:09	13:13	13:17
	11:33	11:29	11:25	11:21	11:18	11:14	11:10	11:07	11:03	10:59
20°	13:21	13:25	13:29	13:33	13:37	13:42	13:47	13:51	13:56	14:00
	10:55	10:51	10:47	10:43	10:39	10:35	10:30	10:26	10:22	10:17
30°	14:05	14:10	14:15	14:20	14:26	14:31	14:37	14:43	14:49	14:55
	10:12	10:08	10:03	9:58	9:53	9:48	9:42	9:37	9:31	9:25
40°	15:02	15:08	15:15	15:22	15:30	15:38	15:46	15:54	16:03	16:13
	9:19	9:13	9:07	9:00	8:53	8:46	8:38	8:30	8:22	8:14
50°	16:23	16:33	16:45	16:57	17:09	17:23	17:38	17:54	18:11	18:31
	8:04	7:54	7:44	7:33	7:22	7:10	6:57	6:42	6:27	6:10
60°	18:53	19:17	19:45	20:19	21:02	22:03	–	–	–	–
	5:52	5:32	5:09	4:42	4:18	3:34	2:46	1:30	–	–

*The upper figure in each pair of figures represents the longest day, and the lower figure represents the shortest day. To find the length of daylight for 37 degrees, read down to 30° in the left column and across the row to 7°. The longest day at 37° is 14 hr, 43 min. Note that the two periods do not add up to 24 hr because daylight is measured when the rim of the sun, rather than the center of the sun, is visible.

TABLE 3.3 Total Daily Solar Radiation Reaching the Ground (cal/cm², Transmission Coefficient = 0.6)

	Approximate Date							
Latitude	Mar. 21	May 6	June 22	Aug. 8	Sept. 23	Nov. 8	Dec. 22	Feb. 4
90° N		127	299	125				
80°	6	158	309	156	5			
70°	47	234	349	232	46			
60°	120	312	406	308	118	10		10
50°	202	376	450	372	199	58	19	58
40°	282	426	477	421	278	130	75	131
30°	350	453	481	449	345	213	152	215
20°	404	459	465	454	398	293	237	296
10°	436	444	428	439	430	366	323	370
0°	447	407	372	404	440	422	397	427
10° S	436	353	303	349	430	461	457	465
20°	404	282	222	279	398	475	497	480
30°	350	206	143	204	345	470	514	475
40°	282	125	70	124	278	441	509	445
50°	202	56	18	55	199	391	481	395
60°	120	10		10	118	323	434	327
70°	47				46	242	373	245
80°	6				5	164	330	166
90°						131	319	133

Source: Smithsonian Institution, Smithsonian Meteorological Tables, 1966. Values apply to a horizontal surface.

relations). The imbalance in daylight and darkness results in a greater accentuation of the seasons than the position of the solar equator alone would produce.

There are two major aspects of the energy balance that distinguish tropical environments from those of midlatitude and polar areas. The first aspect is that the influx of solar energy in tropical areas is high throughout the year even though it does vary some. From this point of view, these are indeed environments without a winter. The intensity of solar radiation is high all year because the rays of the sun are always at a high angle. Also there is very little change in the length of daylight, ranging from 11 to 13 hours throughout the year.

Polar areas differ from tropical areas and midlatitudes in the annual distribution of solar energy. During the winter months in polar regions, direct solar radiation is absent. But some radiation, including some in the visible range, continues to find its way to the surface. In summer, solar intensity is low, but the duration is long and so the total energy is quite large. Much of the energy goes to evaporate water or melt snow and ice.

The intensity of radiation in the polar regions is never very high for any length of time when compared with midlatitude or tropical systems. At the Arctic and Antarctic

Circles, the sun never appears more than 47° above the horizon, and at the poles it is never more than 23.5° above the horizon. Both polar areas receive more hours of radiation during the year than other parts of the earth due to refraction of the sunlight. When the sun is below the horizon, reflection and refraction of the sunlight rays produce twilight. These processes bring some sunlight to the surface until the sun is 18° below the horizon (astronomical twilight). Thus, at high latitudes, there may be several months with no direct sunshine but almost continuous twilight. The stars, moon, and auroras are also sources of light for these areas so darkness is not so intense as it might be and total darkness seldom exists.

The northern hemisphere has the most land area; it is 40% land and 60% water. The southern hemisphere is only 20% land and 80% water. Temperatures outside the Antarctic continent vary much less on a seasonal basis in the southern hemisphere than in the northern hemisphere.

DAILY TEMPERATURE CHANGES

The rotation of Earth on its axis produces alternating periods of day and night, and daily variations in temperature result.

Daytime Heating

After the sun comes above the horizon in the morning, the surface begins to heat. If the air is relatively calm, conduction rapidly moves heat from the surface to the boundary layer of air. This heat transfer takes place in a very limited laminar layer of air, often only a few millimeters deep. Gradually, the heat is distributed upward by diffusion of heated molecules of air. On a hot, clear, still day in summer, there may be a thermal gradient of 20°C (36°F) in the lower 2 m (6.6 ft) of air. In tropical deserts, it may become impossible to see through a surveying instrument by 10 A.M. The upward movement of heat so disturbs the air as to make sighting impossible. If wind is blowing, the movement of heat upward is much more rapid, and the gradient in the lower 2 m (6.6 ft) is much less. Even with turbulence carrying heat away from the surface, daily changes in temperature normally do not extend above 1 km (0.6 mi).

As expected, temperatures are highest during the day, when large amounts of energy are flowing to Earth, and lowest at night. However, there is not a one-to-one relationship in the time of highest solar input and highest temperature because air temperature largely results from absorbing earth radiation.

Highest daytime temperatures usually occur several hours after the time of maximum solar input, which is solar noon. Despite the common notion that the hour of highest temperature occurs at the time when solar energy input equals the flow of outgoing energy, this is not true. Measurements show that equilibrium between incoming and outgoing radiation often occurs about an hour and a half before sunset and not at the time of maximum temperature. During the early afternoon, the earth receives a steady flood of incoming long-wave radiation from the lower atmosphere (the greenhouse effect). It is when this flow of long-wave radiation from the atmosphere reaches a maximum that highest temperatures occur.

The extent of the daily lag in maximum temperature varies. Under normal conditions, the lag will be greatest when the air is still and dry and the sky is free of clouds. In these instances, the maximum temperature of the day may not occur until an hour or so before sunset. When humidity is high or the atmospheric aerosol is unusually thick, the lag will not be as great. Minimum lags occur when there is a cloud cover such that the incoming and outgoing radiation are reduced.

Actual daily high and low temperatures can occur at any time of day if there is a shift in wind direction and cold or warm air is fed into the area. This is typical of temperature changes brought about by circulation around midlatitude lows and also by the development of land and sea breezes.

Moist soil or a cover of vegetation also reduces the maximum temperatures. The process is that of evaporation or transpiration. Evaporation is a major cooling mechanism. Five hundred and ninety calories of heat are taken from the environment for each gram of water that evaporates. This is why it never gets as warm in summer over water as it does over land. Cities east of the Mississippi River generally have mean daily high temperatures and extreme highs that are 10°C (18°F) lower than cities in the western desert. The higher atmospheric humidity and a surface covered by vegetation reduce the temperature of the lower atmosphere.

Nighttime Cooling

Once the sun passes the zenith in its western arc, radiation intensity starts to decrease. Sometime near sunset on a clear day, the ground surface and the boundary layer of air begin to receive less energy than they emit and begin to cool. The ground surface cools most rapidly since it is a better radiator than air. Soon the coolest air is close to the ground and temperature increases with height. This is counter to the standard atmospheric state.

Earth radiation decreases through the hours of darkness and the surface cools. How much cooling occurs depends on how long the night is, how high the humidity is, and on wind velocities. Cooling is greatest on a night when humidity is low, the sky is clear, and the wind is calm. A wet surface retards cooling just as it retards warming. Dew, frost, and fog are major feedback mechanisms that prevent the atmosphere from getting still colder. When dew forms, 590 calories of heat are added to the atmosphere for each gram of water. This heat added by condensation offsets the heat lost by radiation. Frost is an even more effective means to control cooling. For each gram of water that sublimates as frost, 680 calories of heat are added to the boundary layer of air. Fog likewise is a major block to cooling. Not only does the condensation add heat to the air, but the fog acts as a blanket, reducing radiative heat loss. In desert regions, dew and frost are very important in reducing nighttime cooling. Minimum daily temperatures occur in the early morning when incoming solar energy and outgoing earth radiation balance. Shortly after dawn, incoming solar rays provide enough energy to balance outgoing terrestrial rays.

Daily Temperature Range

The daily range in temperature is a function of both daytime heating and nighttime cooling. When conditions are good for rapid inflow of radiant energy in the daytime and outgoing radiation at night, the range will be large. The conditions that favor high daily ranges in temperature are clear skies, low relative humidity, and calm winds.

The magnitude of the daily change in temperature varies a great deal. Near the equator, the daily range exceeds the annual range. Near the two poles, the daily range is reduced to almost zero since there is generally only one daylight and one nighttime period each year.

Over the ocean, the daily range is also small. There are several reasons for the low range. First is the high specific heat of water. It heats slowly and cools slowly. Second, there is mixing of the surface water with the water below that modifies heating and cooling. Third, solar radiation penetrates deeper into the ocean than into the land. This distributes the heat more evenly with depth.

The daily range is a result of solar heating during the day and Earth radiation at night. The daily variation decreases rapidly with height above the ground. The greatest changes are in the boundary layer and it decreases upward to about 1 km (0.6 mi). Above 1 to 2 km (0.6–1.2 mi), the daily variation is minimal.

In North America, daily ranges are a function of cloud cover and humidity. In the drier regions, daily ranges may exceed 22°C (40°F), whereas in more humid regions, the normal range is nearer 17°C (30°F). The greatest known recorded daily range in temperature occurred in North Africa, where the temperature dropped from a high of 56°C (132°F) in the afternoon to 0°C (32°F) the following morning. The difference is an amazing 56°C (100°F).

SEASONAL LAG AND EXTREME TEMPERATURES

There is a seasonal lag of maximum and minimum temperatures. The revolution of the earth causes maximum and minimum solar energy (outside the tropics) to occur at the time of the solstices in each hemisphere. Thus, June and December represent the time of maximum and minimum solar energy receipts in the northern hemisphere. The reverse holds true in the southern hemisphere. The months of maximum and minimum solar energy are not the warmest or coldest. Table 3.4 provides examples showing that there is a month or more lag between the time of maximum and minimum solar radiation and the warmest and coldest months.

Temperature extremes are those temperatures farther from normal and that occur least frequently. For extreme temperatures of either heat or cold to be reached, conditions for incoming and outgoing radiation must be optimum. For high temperatures, there must be high-intensity solar radiation, long hours of sunlight, clear dry air, little surface vegetation, and relatively low wind velocities. These conditions are optimal near the Tropics of Cancer and Capricorn near the time of the solstice. Here the atmosphere is dominated by high pressure and clear skies. The highest recorded surface temperature was measured in El Aziz, Libya, in September 1922, where the temperature reached 58°C (136°F). In North America, the highest temperature recorded is 57°C (134°F) in Death Valley, California, in July 1913. At any location, the record high temperatures occur when the air is clear and dry.

Record lows occur under conditions that favor the loss of energy from the surface and lower atmosphere by direct radiation to space. These conditions occur with high-pressure systems near the poles. The coldest temperature ever recorded is –89°C (–128°F) on the Antarctic continent at Vostock in 1983. The cold temperature is in part due to the high altitude of the ice mass. In the northern hemisphere, the lowest temperature recorded is –68°C(–90°F) at Verkoyansk, Russia. In North America, the coldest temperature on record is –62°C (–80°F) at Prospect Creek, Alaska. The records of both cold and heat occur away from the oceans. The ocean greatly modifies temperature extremes.

TABLE 3.4 Sample Data Illustrating Seasonal Temperature Lag

	Average Temperature at Solstice		Average Monthly Temperature	
	June	Dec.	Warmest Month	Coldest Month
Charleston, SC (33°N)	26°C (79°F)	11°C (51°F)	July 28°C (82°F)	Jan. 10°C (50°F)
Urbana, IL (40°N)	22°C (72°F)	0°C (32°F)	July 25°C (77°F)	Jan. –1°C (30°F)
Naples, Italy (41°N)	22°C (72°F)	11°C (51°F)	Aug. 25°C (77°F)	Jan. 9°C (48°F)
Moscow, Russia (43°N)	19°C (66°F)	–6°C (22°F)	July 21°C (70°F)	Jan. –8°C (17°F)
Edmonton, Canada (53°N)	14°C (57°F)	–8°C (18°F)	July 17°C (63°F)	Jan. –14°C (7°F)

FACTORS INFLUENCING THE VERTICAL DISTRIBUTION OF TEMPERATURE

In the standard atmosphere, the decrease of temperature with height in the troposphere (the **lapse rate** or **environmental lapse rate**) is given as an average of 6.5°C per 1 km (3.5°F per 1000 ft). This value reflects the difference in the average temperature of the surface and the temperature at 11 km (15°C and –59°C, respectively). However, the measured lapse rate varies appreciably from the mean, depending on the nature of the air mass and the surface over which the lapse rate is measured.

The nature of the underlying surface influences the vertical distribution of temperature. For example, temperature decreases most rapidly with altitude over continental areas in summer. Figure 3.4 shows a cross-section of the atmosphere up to 23 km (14 mi). The profile is along the 80th meridian (which passes through eastern Canada, the United States, Cuba, and Panama) from the north pole to the equator in January and July. The heavy lines on the graphs show the tropopause. Note the following features:

1. The tropopause is lower over high latitudes than over low latitudes. A well-marked break occurs in the middle latitudes.
2. The north–south temperature gradients are much steeper in winter than in summer.
3. The strongest horizontal gradients are in the midlatitudes in both summer and winter. This is the region of maximum storm activity.
4. The coldest part of the troposphere occurs over the equator in the region of the tropopause.

The **diurnal range** of temperature at higher elevations is more than at sea level. Figure 3.5 provides a graphic model of this effect. The reason for the greater range is that the atmosphere is less dense at higher altitudes. Maximum daily temperatures are

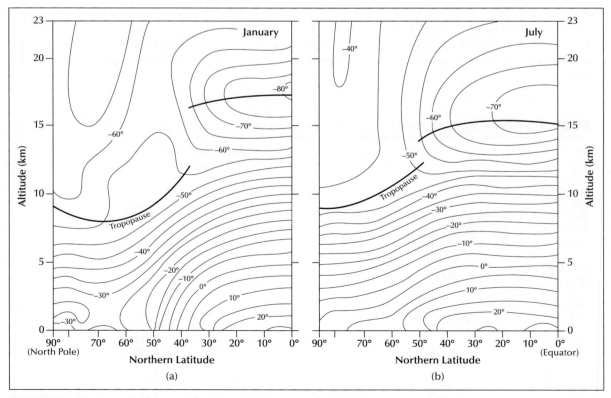

FIGURE 3.4   Mean vertical cross sections through the atmosphere in (a) January and (b) July. Both sections are along the meridian 80° W in the northern hemisphere.

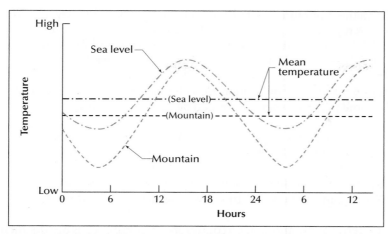

FIGURE 3.5   Daily range of temperature in a mountain and a lowland station at places of similar latitude.

about the same or slightly less than at low altitudes. The main difference occurs at night, when heat escapes much more readily at high elevations because of the lower density of gases.

Decrease of pressure with altitude changes our everyday interpretation of some values on temperature scales. The gas laws state the relationships between temperature and pressure. At higher elevations, pressure changes become important. Under reduced pressure, molecules of water vapor escape more easily from a water surface. Thus, at sea level, water boils at a temperature of 100°C (212°F). At Quito, Ecuador, at an elevation of 2824 m (9350 ft), water will boil at 90.9°C (196°F). At the top of Mount Everest, the boiling point of water is only 71°C (160°F).

FACTORS INFLUENCING THE HORIZONTAL DISTRIBUTION OF TEMPERATURE

The temperature that occurs at any location is essentially a result of the net radiation available and the way that radiation is budgeted. It has already been pointed out that net radiation depends on gains of solar and terrestrial energy; the available energy is then used for sensible heat transfer and evaporation The amount of net energy, and the disposition of that energy at the surface vary depending on a number of factors. These factors include latitude, surface properties, position with respect to warm or cool ocean currents, and elevation.

Latitude

The highest temperatures on Earth do not occur at the equator but near the Tropics of Cancer and Capricorn. A partial explanation of this occurrence is the migration of the vertical rays of the sun between 23.5° N and 23.5° S. The vertical rays of the sun move quickly over the equator, but slowly as they progress north and south. Thus, between 6° N and S, the sun's rays are near vertical for 30 days around the equinoxes. Between 17.5° and 23.5° N and S, near vertical rays occur for 86 days around the solstice. The longer period of high sun and the concurrent longer days allow time for surface heat to accumulate. Thus, the zone of maximum heating and highest temperatures is near the Tropics of Cancer and Capricorn. The predominance of clear skies near the two tropics enhances heating even more compared with the very cloudy equatorial belt.

If the earth were a homogenous body without the land–ocean distribution, its temperature would change evenly from the equator to the poles. It is possible to compare actual temperatures with hypothetical data for a uniform surface and identify areas where temperature anomalies occur. Anomalies represent deviations from temperatures that would only result from solar energy. Isolines called **isanomals** are drawn through points of equal temperatures to produce isanomalous temperature maps of the world.

Figures 3.6 and 3.7 show isanomalous temperature maps for January and July, respectively. Some general observations can be made from studying these maps:

1. In winter in each hemisphere, the land masses have large negative anomalies of as much as –16°C (3°F) below the hypothetical mean. In contrast, ocean areas show no anomalies or show slightly positive ones.
2. In summer in each hemisphere, the largest anomalies are also over the continents, but they are positive. In some areas, a small negative anomaly occurs over the oceans.
3. The patterns of the anomalous conditions over the oceans are distinctive in shape and are associated with ocean currents.
4. Small anomalies occur in the equatorial realms, and the highest anomalies occur in upper middle latitudes.

The major factors that alter the zonal solar climate and hence the patterns of temperature over the globe are identified from these observations. The controls are of two types. First are factors that are due primarily to geographic location on the earth's surface. This basically determines the amount of energy received from the sun. In contrast to these location factors are temperature characteristics that result from the transport of energy by the mobile atmosphere and ocean. This dynamic effect can substantially change the temperatures of a place.

Surface Properties

The disposition of solar energy striking a surface largely depends on the type of surface. Of particular note is **albedo**. Surfaces with high albedo absorb less incident radiation, so there is less total energy available. Polar icecaps are maintained because they reflect as much as 80% of the solar radiation falling on them.

Even if two surfaces have similar albedos, the incident energy does not always result in similar temperatures because the heat capacities of the surfaces may differ. The heat capacity of a substance is the amount of heat required to raise its temperature. *Specific heat* is the amount of heat (number of calories) required to raise the temperature of 1 g (0.035 oz) of a substance 1°C (1.8°F). As Table 3.5 shows, the specific heat of substances can vary appreciably.

Of particular significance is that the **specific heat** of water is some five times greater than that of rock material and the land surface in general. This means that the amount of heat required to raise the temperature of water 1°C (1.8°F) is five times greater than required for the same temperature increase on land. The same amount of energy applied to a land surface and a water surface would result in the land becoming much hotter than the water. The difference increases due to the different heat conductivity of the earth materials and water (Table 3.6). Loose, dry soil is a very poor conductor of heat. Only a superficial layer will experience a rise in temperature from solar radiation. Water has only a fair conductivity, but its general mobility and transparency permit heat to circulate well below the surface. A natural undisturbed soil with a vegetation cover may have daily temperature changes to a depth of 1 m (3.3 ft). A quiet pool of water has daily temperature variations that can be measured to a depth of perhaps 6 m (20 ft).

Land masses heat much more rapidly than oceans in summer. However, because this heat concentrates near the surface, it rapidly radiates to the atmosphere as winter approaches. Hence, land masses tend to experience extreme temperatures, whereas water bodies are more equable and show less change (Figure 3.8).

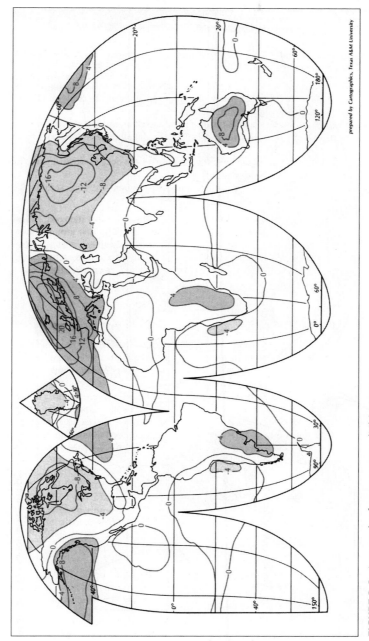

FIGURE 3.6 Isanomals of temperatures (°C) in January.

FIGURE 3.7 Isanomals of temperatures (°C) in July.

TABLE 3.5 Specific Heats of Selected Substances

Substance	Specific Heat (cal/g/°C)
Water	1.0
Ice (at freezing)	0.5
Air	0.24
Aluminum	0.21
Granite	0.19
Sand	0.19
Iron	0.11

TABLE 3.6 Thermal Conductivity of Selected Substances

Substance	Heat Conductivity*
Air	0.000054
Snow	0.0011
Water	0.0015
Dry soil	0.0037
Earth's crust	0.004
Ice	0.005
Aluminum	0.49

* The number of calories passing through an area 1 cm^2 in a second when temperature gradient is 1°C/cm.

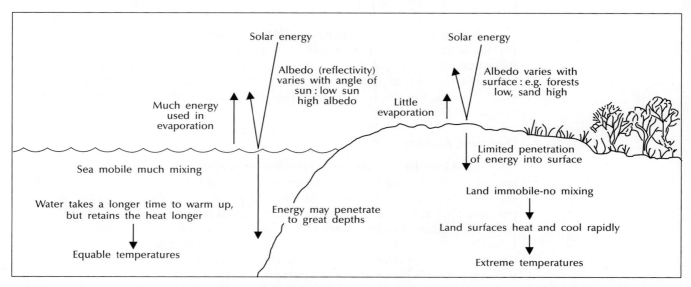

FIGURE 3.8 Solar energy striking land and sea surfaces divides in different ways. The water surfaces are conservative, warming and cooling slowly; the land masses experience large temperature changes.

As noted in chapter 2, absorbed energy raises the temperature of the absorbing surface, which then radiates according to its temperature. Net energy—that received by the surface but not radiated back to the atmosphere—is then either passed to the air in the form of sensible heat or used to evaporate moisture from that surface and transferred as **latent heat**. The amount of sensible heat available to pass to the atmosphere depends in part on the fraction of net energy that goes to evaporate water. Over the oceans, evaporation is continuous. On land, it depends on the amount of water available at the surface. The significance of this process is determined by comparing the ratio of sensible heat to latent heat for different parts of Earth. The Bowen Ratio (chapter 2), which expressed this relationship, is H/LE. The larger the ratio, the more sensible heat, rather than latent heat passing to the atmosphere.

The pattern of temperature follows sensible heat flow. Thus temperatures of places located in moist environs will never experience the very high temperatures found in arid locations. Cities along a coast have smaller ranges in temperature than do inland cities. Cities that have the wind blowing onto the land have a lower range than coastal cities, where the dominant wind direction is offshore.

Inland cities, particularly in the northern hemisphere, have the highest annual ranges in temperature. In North America, ranges are greatest in the Great Plains of northern United States and southern Canada. The most extreme annual range is found in Russia. At Yakutsk, the annual range is 62°C (112°F). This represents a change in mean temperature of more than 13°C (24°F) per month from January to July. So distinctive is this effect that meteorologists use the term **continentality**, a measure of the continental influence on weather and climate. Various indices to measure continentality have been devised, some of which are considered in the regional setting of later chapters.

Aspect and Topography

The combined influences of the steepness and direction a slope faces determine its aspect. Differences that occur on north-facing and south-facing slopes in the northern hemisphere illustrate the importance of aspect. A north-facing slope may have snow on it while a south-facing slope is bare. A north-facing slope gets less intense radiation and, as the sun gets lower in the sky, will be in shadow long before the south-facing slope.

Slope aspect influences many natural phenomena. For example, the height of permanent snow and ice on mountains varies from one slope to another, as does the tree line. Similarly, the depths of snow and frost differ on north- and south-facing slopes.

Topography also plays an important role in climates of some lowlands. On a continental scale, mountain ranges that run north–south have a very different effect from those that run east–west. The lack of any extensive east–west mountain barrier in the United States permits polar and tropical air to penetrate long distances into the continent. One result of this unobstructed flow of air is the high incidence of tornadoes in the United States.

Dynamic Factors

The largest imbalance of energy is between tropical regions and the poles. This imbalance is partly alleviated by the transfer of latent heat (LE), sensible heat (H), and heat stored in the water of the oceans (S). Table 3.7 provides the theoretical planetary temperature for sea level and the actual mean annual temperature for every 10° of latitude. The biggest differences are at the equator and for latitudes above 60°. Tropical latitudes are cooler than the theoretical value, whereas high latitudes are warmer. The differences between the actual and theoretical values result from the transport of energy over the earth by air and ocean currents. Every storm system, circulation pattern, and evaporation/precipitation event aids the moving of heat from tropical regions toward the poles.

TABLE 3.7 Theoretical Temperatures in a Stationary Atmosphere and Actual Temperatures by Latitude

	Equator			Latitude					
		10°	20°	30°	40°	50°	60°	70°	80°
				Temperature in °C (°F)					
				Northern Hemisphere					
Planetary temp.	33(91)	32(89)	28(83)	22(72)	14(57)	3(37)	−11(12)	−24(−11)	−32(−26)
Actual temp.	26(79)	26(80)	25(78)	20(69)	14(57)	5(42)	−1(30)	−10(13)	−18(−1)
Difference	−7(−12)	−6(−9)	−3(−5)	−2(−3)	0	+2(+5)	+10(+18)	+14(+24)	+14(+25)
				Southern Hemisphere					
Planetary temp.	33(91)	32(89)	28(83)	22(72)	14(57)	3(37)	−11(12)	−24(−11)	−32(−26)
Actual temp.	26(79)	25(78)	22(73)	17(62)	11(53)	5(42)	−3(26)	−13(8)	−27(−17)
Difference	−7(−12)	−7(−11)	−6(−10)	−5(−10)	−3(−4)	+2(+5)	+8(+14)	+11(+19)	+5(+9)

TEMPERATURES OVER THE EARTH'S SURFACE

As discussed in this chapter, a number of factors play a role in determining the distribution of average temperatures over the earth's surface. Figures 3.9 and 3.10 illustrate the distribution of mean July and January temperatures, respectively. Figure 3.11 shows the annual average range of temperature.

In each case, the general decline in temperatures from equator to poles is clear, illustrating the basic influence of latitude. However, significant variations from a simple zonal pattern exist, and these result from a combination of other temperature controls. Some of these variations are:

1. The maps show the temperature extremes that occur over continental land masses. The largest land mass, Asia, experiences average temperatures that range from −50°C (−58°F) in January to more than 20°C (68°F) in summer. Figure 3.11 clearly shows the effect of the size of the continent on the average annual range of temperature.
2. The oceans exhibit the results of heat transport by ocean currents. The displacement of isotherms toward the poles shows the warming effects of the North Atlantic and North Pacific Drifts. Cold ocean currents, such as the California and Canary currents, cause temperatures to be lower.
3. The maps show actual temperatures that include the effect of altitude. Some world temperature maps have isotherms reduced to sea level, which eliminates the effect of elevation. Notice, for example, the isotherms over South America. The Andes show as a tongue of colder temperatures extending toward the equator.
4. The closest approximation to zonal temperatures occurs over the southern ocean and Antarctica. This area is the most extensive location where homogenous surfaces encircle the globe without interruption.
5. Compare Figures 3.9 and 3.10 and note that the hemispheric temperature gradient is steepest in winter. For example, in January, the northern hemisphere gradient is more than 60°C (110°F). In July, it is about 10°C (18°F). This pattern has highly significant results in atmospheric circulation.

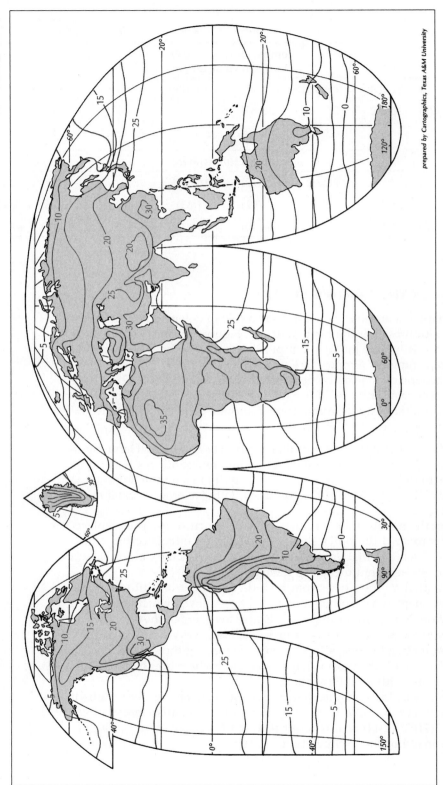

FIGURE 3.9 Distribution of average July temperatures over the earth (°C).

prepared by Cartographics, Texas A&M University

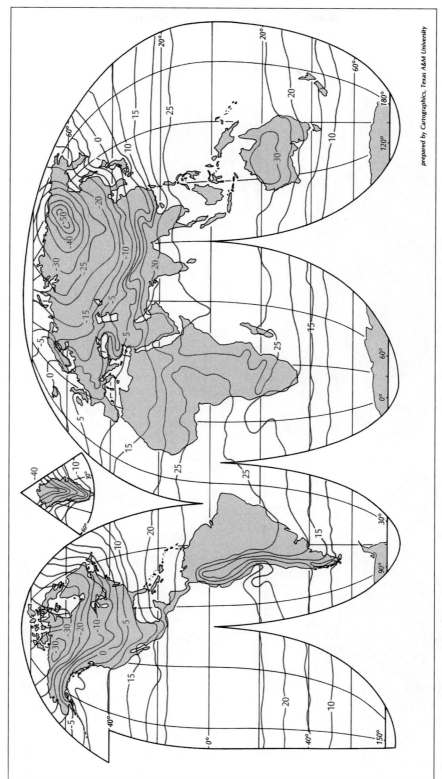

FIGURE 3.10 Distribution of average January temperatures over the earth (°C).

prepared by Cartographics, Texas A&M University

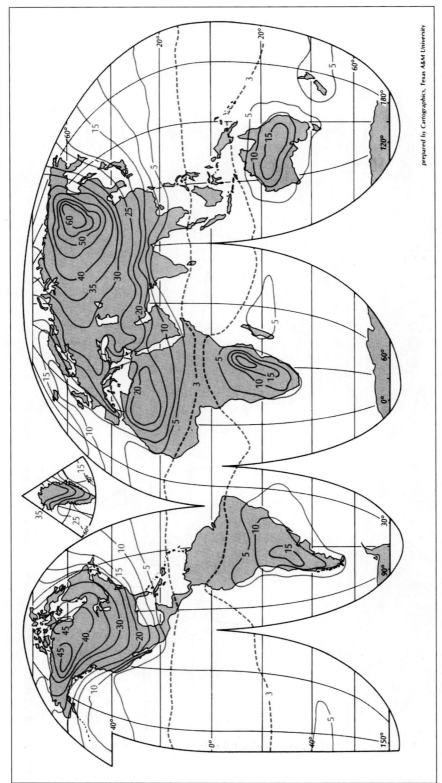

FIGURE 3.11 Distribution of mean annual range in temperature over the earth (°C).

prepared by Cartographics, Texas A&M University

Summary

Temperature measures the amount of heat energy available over the earth's surface and is a function of the net radiation at a location. There is no direct relationship between maximum and minimum solar radiation and highest and lowest temperatures. A lag exists in both daily and seasonal temperatures. Care must be taken when using mean temperature data because means do not provide a full assessment of the temperature regime. Some measure of the variation of data about the mean is necessary for complete description. Similarly, the use of the mean for a given period obscures temperature changes within the time period.

The lapse rate represents the vertical distribution of temperature in the troposphere. In the standard atmosphere, the lapse rate is 6.5°C for each 1 km (3.5°F per 1000 ft). The horizontal distribution of temperature over the earth's surface is a function of both location and the dynamic movement of the atmosphere. The former consists of such factors as latitude, surface properties, and aspect. The latter consists of energy transferred by the atmosphere and the oceans.

Key Terms

Albedo, *46*

Angle of incidence, *36*

Aphelion, *36*

Continentality, *50*

Diurnal range, *44*

Environmental
  lapse rate, *44*

Isanomals, *46*

Lapse rate, *44*

Latent heat, *50*

Perihelion, *36*

Plane of the ecliptic, *36*

Seasons, *36*

Solar equator, *36*

Specific heat, *46*

Review Questions

1. Why does the earth have seasons?
2. What determines the intensity of solar radiation at a location at a given time?
3. What is the solar equator, and how does it differ from the geographic equator?
4. Assuming the same air mass is in place and calm conditions exist, outline the processes involved in temperature change over a 24-hour period.
5. Explain (a) the daily lag and (b) the seasonal lag of temperature.
6. Explain isanomals and their significance.
7. What is specific heat, and what role does it play in the climate of a location?
8. Explain how temperatures in the center of a continent might differ from those at a coastal location at the same latitude.
9. Explain the Bowen Ratio and why it varies from location to location.
10. What are some of the temperature differences caused by aspect?

CHAPTER

4

Moisture in the Atmosphere

The significance of water as an atmospheric variable is a result of its unique physical properties. Water is the only substance that exists as a gas, liquid, and solid at temperatures found at the earth's surface. This special property enables water to cycle over the earth's surface. While changing from one form to another, it acts as an important vehicle for the transfer of energy in the atmosphere.

CHANGES OF STATE

The chemical symbol of water, H_2O, is probably the best known of all chemical symbols. It tells us that the water molecule is made up of two atoms of hydrogen for one of oxygen. Water in all of its states has the same atomic content. The only difference is the

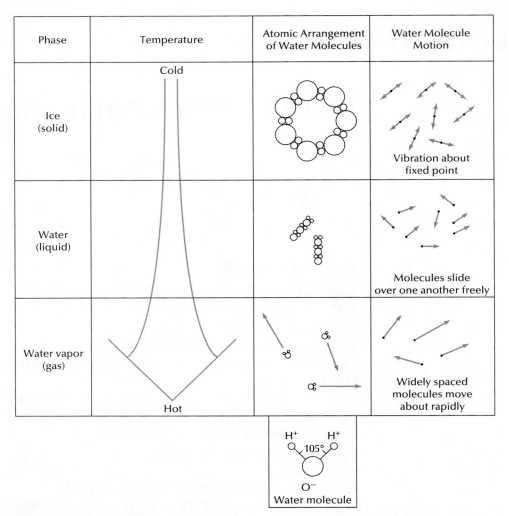

FIGURE 4.1   Schematic diagram of the arrangement and motion of water molecules in different phases.

arrangement of the molecules (Figure 4.1). Little energy is available at low temperatures, and the bonds binding the water molecules are firm. The water molecules pack tightly in a fixed geometric pattern in the solid phase. As temperature increases, the available energy causes the bonds of the ice phase to weaken. Because they are not firmly set, bonds form, break, and form again. This permits flow to occur and represents the liquid phase of water. In the liquid stage, there is still bonding, but it is much less compact than in the ice phase. At higher temperatures and with more energy, the bonding of the water molecules breaks down, and the molecules move in a disorganized manner, which is the gas phase. If the temperature decreases, the molecules will revert to a less energetic phase and reverse the processes. Gas will change to liquid and liquid to solid.

Figure 4.2 illustrates these changes of phase. Note that the processes of melting, evaporation, and sublimation from the solid to liquid phase absorb energy. This added energy causes the molecules to change their bonding pattern. The amount of energy incorporated is large for the changes to the water vapor stage and much lower for the change from ice to water.

The energy absorbed is latent energy and goes back to the environment when the phase changes reverse. When water vapor changes to liquid, it releases the energy

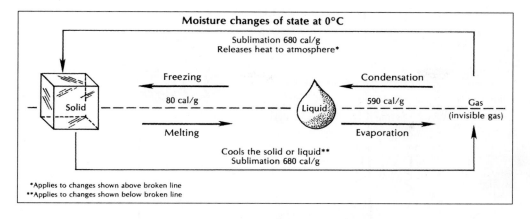

Heat transfers associated with phase changes of water

Phase Change	Heat Transfer	Type of Heat
Liquid water to water vapor	540–590 cal absorbed	Latent heat of vaporization
Ice to liquid water	80 cal absorbed	Latent heat of fusion
Ice to water vapor	680 cal absorbed	Latent heat of sublimation
Water vapor to liquid water	540–590 cal released	Latent heat of condensation
Liquid water to ice	80 cal released	Latent heat of fusion
Water vapor to ice	680 cal released	Latent heat of sublimation

FIGURE 4.2   Changes of state of water.

originally absorbed and retained as latent heat. The same is true when water freezes and water vapor sublimates to ice.

The significance of the release of latent heat shows in many ways. It provides energy to form thunderstorms, tornadoes, and hurricanes. It also plays a critical role in the redistribution of heat energy over the earth's surface. Because of the high evaporation in low latitudes, air transported to higher latitudes carries latent heat with it. This condenses and releases energy to warm the atmosphere in higher latitudes.

THE HYDROLOGIC CYCLE

The **hydrologic cycle** is a conceptual model of the exchange of water over the earth's surface. Figure 4.3 shows one model of the hydrologic cycle. It shows large-scale changes of state, with evaporation providing the moisture that condenses and becomes precipitation. In addition, the diagram shows that a process of transpiration occurs. Transpiration is, in some ways, a special form of evaporation in that moisture returns to the air through evaporation but it is via vegetative processes. In dealing with evaporation and transpiration amounts, climatologists often combine them into a single parameter called **evapotranspiration**.

Some of the precipitation that falls to the surface passes to the soil to become soil moisture, which growing plants use; some passes deeper into the ground to become groundwater. Other precipitation runs off the surface and is collected in ponds, lakes, and reservoirs or flows as surface water in streams and rivers. Eventually water finds its way to the oceans and starts the cycle over again.

How is water distributed over the earth? The divided circle in Figure 4.4 shows that, of all the water available on Earth, 97% occurs in the oceans. The remaining 3% is mostly ice found in the large ice caps of the world. Almost all the rest is in groundwater.

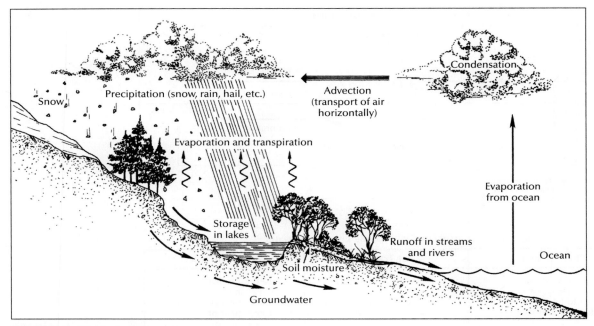

FIGURE 4.3   The hydrologic cycle.

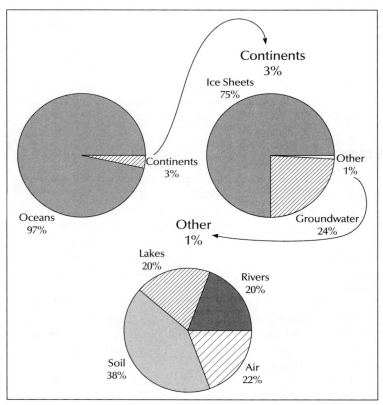

FIGURE 4.4   The proportional distribution of water over the earth's surface.

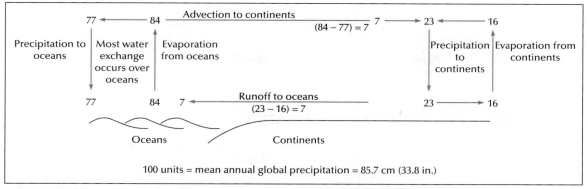

FIGURE 4.5 Estimates of water transfer within the hydrologic cycle. A value of 100 is used to denote average global precipitation.

Rivers, lakes, and soil moisture account for less than one percent of the total water. The atmosphere contains only 0.35% of all the water available. Yet this small amount is the reservoir that provides the moisture for clouds and precipitation that occurs over the earth's surface.

The actual exchanges of water provide some surprising data. Assume that the average precipitation of the world (the amount that would fall if every place on Earth got the same amount) is 85.7 cm (33.8 in.) and let this amount equal 100%. Most water evaporates from the ocean, and the greater part of all precipitation falls over the ocean. As illustrated in Figure 4.5, the amount of water that falls on the land is appreciably less (23%). This is expected, however, for the oceans provide much of the available moisture and occupy a much greater area than the continents.

RELATIVE HUMIDITY

One widely used measure of water vapor in the atmosphere is relative humidity. The calculation of relative humidity depends on the maximum amount of moisture that air can hold—the saturation level:

$$\text{Relative humidity} = \frac{\text{Vapor pressure in the air}}{\text{Saturation vapor pressure}} \times 100.$$

The relative humidity is stated as a percentage, so the fraction is multiplied by 100.

The maximum amount of water vapor that may be in the air is mainly a function of the temperature of the air. Warm air may contain more moisture than cold air. In fact, the maximum amount of moisture that may be in the air increases rapidly with increasing temperature. This is because evaporation rates increase rapidly with temperature.

Of special significance is the **dew point**, or dew-point temperature. This is the temperature to which a parcel of air must cool before condensation exceeds evaporation. The name provides a good guide to the meaning of the concept because dew provides visible evidence that the air has reached a critical temperature in terms of the physical state of water in the atmosphere.

On a cool, calm night, the air near the ground loses heat by radiation to the air above. With little mixing of the air, a thin layer of air next to the ground becomes cold. If the temperature drops far enough, condensation will exceed evaporation, and moisture condenses on the ground in the form of dew. The temperature at which this transition occurs is the dew-point temperature. At temperatures below freezing, it is called the frost point rather than the dew point. This is the temperature to which air must cool for the formation of ice crystals from water vapor. The formation of layers of ice on a car window after a cold night is a clear sign of cooling to the frost point. Often because of the way in which heat escapes from an object such as a car, ice appears on the windows and nowhere else. This is particularly the case with the sloping windshield and rear window.

EVAPORATION AND TRANSPIRATION

Evaporation is the process by which water changes from a liquid to a gaseous state. Water is sufficiently volatile in solid and liquid states to pass directly into the gaseous state at most environmental temperatures. Sublimation is the process of change from ice to water vapor. Because the source of water vapor is at the earth's surface, the amount of water vapor present in the atmosphere decreases with height. Most atmospheric moisture is found below 10,000 m (33,000 ft).

The amount of water that actually evaporates from a given water surface in a given period depends on the following factors:

1. Vapor pressure of the water surface. This factor depends on water temperature. The higher the water temperature, the greater the surface vapor pressure. When water temperature is higher than air temperature, evaporation takes place.
2. Vapor pressure of the air. The greater the vapor pressure of the air, the less evaporation there will be. The rate of evaporation varies directly with the difference between vapor pressure of the water surface and vapor pressure of the air.
3. Wind. Air movement is usually turbulent, with moist air removed from near the water surface and replaced by dry air from above. Evaporation thus varies directly with the velocity of the wind. The higher the wind velocity, the more evaporation there is.

The analysis of terrestrial water balances is not quite as simple as that of water bodies. For land areas, the net atmospheric transfer of water vapor depends on the amount of water available and the processes of evaporation and transpiration. Transpiration refers to the loss of water from living plants. Transpiration cools the plant and keeps its temperature within the tolerable limits. Transpiration is difficult to measure in the field so climatologists use evapotranspiration. It is the combined loss from the surface through evaporation and transpiration. A wide variety of factors affect evapotranspiration. Among them are the following:

1. Radiation intensity
2. Atmospheric temperature
3. Atmospheric dew point
4. Length of day (photoperiod)
5. Wind velocity
6. Type of vegetation
7. Soil moisture conditions
8. Type of precipitation

Evapotranspiration from a mixed land-and-water surface is nearly always less than evaporation would be from a similar-sized water surface. On a global scale, evapotranspiration for the continents is some 470 mm per year. From the ocean, it is 1300 mm per year. Average evapotranspiration from the continents varies through time and space. The major variables are the amounts of water and energy available. In tropical areas, where there is ample water, evapotranspiration rates are very high. In the lower Amazon Valley and the central Congo River basins, rates of 1200 mm a year nearly approach the evaporation over the open ocean. In parts of the Atlantic and Gulf Plain of the United States, the amount is almost as high.

CONDENSATION NEAR THE GROUND: DEW, MIST, AND FOG

Water vapor in the atmosphere eventually condenses to form water droplets. It is the condensation process that leads to deposition of water at the earth's surface. The greater part of the condensation process occurs in the formation of clouds, but some does occur near the surface in the form of dew, mist, and fog.

In order for water vapor to condense, some solid or other liquid surface is required in order to change its state from gas to liquid. In the case of dew, vegetation or another surface provides the needed base. Dew is often seen first on car tops or windows as the metal cools faster than vegetation or soil material.

In the atmosphere, there are large numbers of cloud condensation nuclei (CCNs). There are many different types of particles that can serve as nuclei. They include dust, soot from fires, and exhaust from industrial processes and engines. Salt, sulfates, and organic matter also serve as nuclei.

Typically CCN particles are very small. Most are less that 0.0002 mm in size. Without nuclei present, water can be cooled far below the freezing point before condensation will occur. Recent studies indicate that bacteria are the most common nuclei for ice to form in clouds. Bacteria also serve as nuclei for raindrops. Dead bacteria are currently used as nuclei in snow-making machines on ski slopes.

Cooling to the condensation level is the main means by which saturation occurs. If condensation of water droplets occurs in the layers of air immediately above the

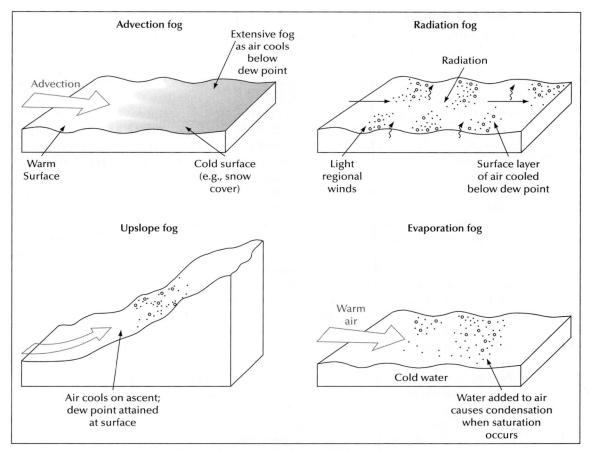

FIGURE 4.6 Schematic diagrams illustrating types of fog formation.

ground, it forms mist or fog. Mist is the suspension of microscopic water droplets that reduces visibility at the earth's surface. It forms a fairly thin grayish veil that covers the landscape. In contrast, fog is the suspension of very small water droplets in the air that reduces visibility to less than 1 km (0.625 mi). In fog, the air feels raw and clammy; given the correct illumination, the fog droplets are often visible to the naked eye. Mist does not provide the same damp, raw feeling, and the individual droplets are too small to see.

While all fog looks the same, its causes are quite variable. The most common types other than those associated with frontal systems (see Chapter 7) are air mass fogs.

Advection is the horizontal transport of air. Advection fogs can form either by the transport of warm air over a cold surface or the transport of cold air over a warm, wet surface. In the former, the warm air in contact with the cold surface cools. If the dew point is reached, condensation will occur in the form of a fog. This can occur in several ways, as illustrated in Figure 4.6. Evaporation fogs are often referred to as steam fogs. Steam fogs, or arctic "sea smoke," form when cold air blows over a warm sea surface, which results in rapid evaporation from the water. This saturates the cold air, resulting in the steam fogs commonly seen in arctic regions.

Radiation fog is an extension of the way in which dew forms. It occurs under clear skies in relatively still air and forms when nocturnal cooling of the ground results in a chilling of the layers of air near the surface. If the temperature drops to the dew point, then a layer of fog forms. Compared with advection fog, radiation fog lasts but a short time and "burns off" in the early morning when the sunshine heats the layer of chilled air.

Another type of fog is that which forms when air rises up the side of a mountain. This **upslope** fog is due to adiabatic expansion.

FOGGY PLACES

The conditions favorable for the formation of fog occur in some areas more frequently than in others. The result is that the occurrence of fog is quite variable over both space and time.

Advection fogs are often widespread and persistent. The foggiest place in the United States is in the Libby Islands off the coast of Maine. This is part of the extensive fog belt associated with the Newfoundland coast and the offshore Grand Banks. Here, air flowing off the relatively warm waters of the North Atlantic Drift crosses the cold water of the Labrador Current. The warmer air cools below its dew point and fog results. A similar situation occurs off the Aleutian Islands in the North Pacific. Warm air from the North Pacific Drift comes in contact with the cooler waters of the Bering Current to give widespread advection fog. Such fogs are most frequent from March to September.

Fogs are common along the coastal areas of California. Cold **upwelling** water occurs close to the coast and air moving in from the warmer Pacific waters comes in contact with it. Often the fog starts here as low stratus clouds because wind action causes the air to mix. Air that has come in contact with the cold water cools further. When it rises, it results in fog forming above the sea surface. When this low layer of moisture reaches the hilly coast, people see it as fog at ground level. When fog actually forms at the sea level, it is usually very dense and results in minimal visibility.

Like the Newfoundland fogs, the fogs of California also occur during the summer months. During the dry summer, the moisture that condenses provides some moisture for the coastal redwood trees and, significantly, provides a humid environment that reduces loss of moisture through evaporation. In coastal Chile, South America, a similarly formed fog produces enough moisture to maintain green plants—the Garoua—in a desert region.

Upslope fogs are a result of adiabatic expansion. On the eastern side of the Rockies, the land slopes gradually upward from the Great Plains. When an east wind blows in the area, warm moist air slowly ascends the slope, cools, and condenses.

Sometimes known as "Cheyenne fog," the right conditions can lead to a thick blanket of fog extending from Amarillo, Texas, to Cheyenne, Wyoming.

CONDENSATION ABOVE THE SURFACE: CLOUDS

Clouds provide the most readily visible weather phenomenon. Their beauty is not overlooked. Poems, songs, and paintings use clouds in highly expressive ways. Although atmospheric scientists also see beauty in clouds, they look to the form and appearance of clouds as important keys to both understanding and predicting atmospheric conditions.

All water droplets begin as microscopic particles. Some of the optical phenomena in the atmosphere are a result of these particles. Spectacularly beautiful rays of the sun occur when the sun is low in the sky and partially hidden by clouds. These rays, known as **crepuscular rays**, appear to fan out as a series of giant searchlight beams. The intensity of the rays is a function of how much light is scattered. This is an indication of the concentration of fine particles suspended in the air. Where air is free from debris, such as in polar areas, crepuscular rays do not occur. We can demonstrate this using a bright flashlight. In most places, it is possible to see the beam of a flashlight quite distinctly at night. In clean air, the light beam is invisible.

CLOUD FORMS AND CLASSIFICATION

Before 1800, clouds had no formal names, and there was little knowledge of cloud mechanics. A young Englishman named Luke Howard (1772–1864) provided a new perspective on clouds in 1803, when he presented a classification of clouds into main and secondary types and gave them Latin names. He distinguished three principle cloud forms:

Stratus (from Latin *stratum* = "layer") cloud—lying in a level sheet

Cumulus (from Latin *cumulus* = "pile") cloud—having flat bases and rounded tops and being lumpy in appearance

Cirrus (from Latin *cirrus* = "hair") cloud—having a fibrous or feathery appearance

The classification is so well designed that it remains the system in use today.

As knowledge of clouds increased, a more comprehensive system was required. To gain worldwide uniformity, in 1895, the International Meteorological Committee published a system for naming and identifying clouds. This system was revised several times. The international standard is now under the auspices of the World Meteorological Organization (WMO), which publishes *The International Cloud Atlas*.

In the international classification system, the main types are genera (singular genus), which contain species and, in turn, several varieties. Despite the infinite variety of clouds that occur, they group into 1 of 10 basic types, or genera. The genera are listed in Table 4.1.

A second part of cloud classification is the altitude at which clouds occur. Clouds of similar shapes occur at different levels in the troposphere. Table 4.2 shows clouds grouped as high, middle, and low clouds. The approximate height at which these occur varies with latitude. As already noted, the structure and thickness of the atmosphere varies from equator to pole. To some extent, the 10 genera of clouds relate to height. Those with the term *cirrus* (or prefix *cirro-*) are high clouds. Those with the prefix *alto* are middle clouds. Names for low clouds lack prefixes. An exception to this is the **nimbostratus** cloud, which is classified as both a middle and low cloud. The word *nimbus* (or prefix *nimbo-*) applies to a cloud from which rain is falling. It derives from the Latin for "violent rain."

Figure 4.7 is a generalized diagram illustrating the classification of clouds according to genera and height. Note that cumulonimbus, the thunder cloud, extends from low, through middle, to the high altitudes of the height classification.

TABLE 4.1 Cloud Genera (After WMO)

Cirrus (Ci). Detached clouds in the form of white, delicate filaments, or white or mostly white patches or narrow bands. These clouds have a fibrous (hairlike) appearance or a silky sheen or both.

Cirrocumulus (Cc). Thin, white patch, sheet, or layer of cloud without shading composed of very small elements in the form of grains, ripples, etc., merged or separate, and more or less regularly arranged; most of the elements have an apparent width of less than 1° (approximately the width of the little finger at arm's length).

Cirrostratus (Cs). Transparent, whitish cloud veil of fibrous or smooth appearance, totally or partially covering the sky, and generally producing halo phenomena.

Altocumulus (Ac). White or gray, or both white and gray, patch, sheet, or layer of cloud, generally with shading, composed of laminae, rounded masses, rolls, etc., sometimes partly fibrous or diffuse, and may or may not be merged; most of the regularly arranged small elements usually have an apparent width of between 1° and 5° (approximately the width of three fingers at arm's length).

Altostratus (As). Grayish or bluish cloud sheet or layer of striated, fibrous, or uniform appearance, totally or partly covering the sky, and having parts thin enough to reveal the sun at least vaguely, as through ground glass. Altostratus does not show halo phenomena.

Nimbostratus (Ns). Gray cloud layer, often dark, the appearance of which is rendered diffuse by more or less continually falling rain or snow, which in most cases reaches the ground. It is thick enough throughout to blot out the sun. Low, ragged clouds frequently occur below the layer with which they may or may not merge.

Stratocumulus (Sc). Gray or whitish, or both gray and whitish, patch, sheet, or layer of cloud that almost always has dark parts, composed of tessellations, rounded masses, rolls, etc., that are nonfibrous (except for virga) and may or may not be merged; most of the regularly arranged small elements have an apparent width of more than 5°.

Stratus (St). Generally gray cloud layer with a fairly uniform base, which may give drizzle, ice prisms, or snow grains. When the sun is visible through the cloud, its outline is clearly discernible. Stratus does not produce halo phenomena (except possibly at very low temperatures). Sometimes stratus appears in the form of ragged patches.

Cumulus (Cu). Detached clouds, generally dense and with sharp outlines, developing vertically in the form of rising mounds, domes, or towers, of which the bulging upper part often resembles a cauliflower. The sunlit parts of these clouds are mostly brilliant white; their bases are relatively dark and nearly horizontal. Sometimes cumulus is ragged.

Cumulonimbus (Cb). Heavy and dense cloud, with a considerable vertical extent, in the form of a mountain or huge towers. At least part of its upper portion is usually smooth, or fibrous or striated, and nearly always flattened; this part often spreads out in the shape of an anvil or a vast plume. Under the base of this cloud, which is often very dark, there are frequently low ragged clouds either merged with it or not, and precipitation sometimes in the form of virga.

TABLE 4.2 Approximate Height Range of Cloud Bases

Level	Ranges in Polar Regions (km)	Ranges in Temperate Regions (km)	Ranges in Tropical Regions (km)
High	3–8	5–13	5–18
Middle	2–4	2–7	2–8
Low		from surface to 2 km	

FIGURE 4.7 Cloud forms.

Differences in the structure and shape of clouds permit identification of cloud species. We add a species name to cloud genera to add further information about the cloud. A cloud is classified within only one genus. If it is a cumulus cloud, it cannot be a stratus or cirrus. However, the species name can apply to any genera. For example, the species *castellanus* are clouds that appear to have turrets like a castle. Such a shape applies to many clouds, including cirrus and stratocumulus. If there is no distinct structure or shape in the cloud forms, it is then unnecessary to provide the species name.

A full list of cloud species and their descriptions appears in Table 4.3. The clouds listed here are secondary in importance to the 10 genera, and few persons commit them all to memory. The important factor is the potential for generating precipitation from the clouds.

The word *precipitation* is used to describe any of the various forms of water particles that fall from the atmosphere to reach the ground. It is a useful word for it describes water forms ranging from snowflakes and drizzle to hail and raindrops (Table 4.4). As discussed later in this chapter, the form that precipitation takes determines the nature of its impact on people and the environment. The precipitation process, the process that occurs in clouds to give rise to precipitation, is the end product of a whole set of events.

TABLE 4.3 Cloud Species

Arcus. Shaped like an arc; refers to the lower (dark and threatening) part of a cumulonimbus cloud, particularly of the line-squall type.

Calvus. Bald; refers to the absence of sprouting (cauliflower) structure as well as cirriform appendages in the upper part of a cumulonimbus cloud (see also *congestus* and *incus*).

Capillatus. Hairy; fibrous or striated structure of upper part of cumulonimbus.

Castellanus. Turreted; heap-shaped towers, resembling miniature cumulus protruding from clouds in the middle and upper troposphere, especially altocumulus.

Congestus. Congested, heaped; sprouting, towering structures in the upper portion of a developing cumulus.

Fractus. Fractured, torn; ragged fragments of clouds, notably stratus, cumulus, and nimbostratus.

Humilis. Humble, small, flat; used to characterize nondeveloping cumulus clouds.

Incus. Anvil-shaped; cirriform mass of cloud in the upper part of a developed cumulonimbus.

Intortus. Twisted, entangled; used to describe a type of cirrus.

Mamma. Shaped like udders; protuberances hanging down from the undersurface of a cloud; most pronounced in connection with thundery cumulonimbus.

Pileus. Cap- or hood-shaped; accessory cloud of small horizontal extent above, or attached to, the upper part of a cumulus cloud, particularly during its developing phase.

Spissatus. Spiss, compact; describes cirrus that is sufficiently dense to appear grayish when viewed in a direction toward the sun.

Tuba. Tube-shaped; typical of clouds associated with tornadoes, water spouts, etc.

Uncinus. Hooked; typical of streaky cirrus drifting in strong winds in the upper troposphere.

Velum. Flap; an accessory cloud of considerable horizontal extent, sometimes connecting the upper parts of several cumulus clouds.

Virga. Twig; trails or streaks of precipitation, hanging from the undersurface of a cloud but not reaching the ground.

Source: World Meteorological Organization, *The International Cloud Atlas.*

TABLE 4.4 Types of Precipitation

Rain is precipitation of liquid water particles with diameters over 0.5 mm, although smaller drops are still called rain if they are widely scattered.

Drizzle is a fairly uniform precipitation composed exclusively of fine drops of water with diameters less than 0.5 mm. Only when droplets of this size are widely spaced are they called rain.

Freezing rain or *freezing drizzle* is rain/drizzle that freezes on impact with the ground, with objects at Earth's surface, or with aircraft in flight.

Snow is precipitation of ice crystals most of which are branched. At temperatures higher than about −5°C (23°F), the crystals are generally agglomerated into snowflakes.

Snow pellets are composed of white and opaque grains of ice. The grains are mostly spherical and have a diameter of 2–5 mm. The grains are brittle and when falling on a hard surface bounce and break up. Snow pellets are also known as soft hail and graupel.

Snow grains are very small (less than 1 mm in diameter) grains of white, opaque ice. Snow grains are also called graupel.

Ice pellets are comprised of transparent or translucent pieces of ice that are spherical or irregular and have a diameter of 5 mm or less. They are composed of frozen raindrops or largely melted and refrozen snowflakes.

Hail is precipitation of small balls or pieces of ice (hailstones), with diameters ranging from 5 to 50 mm, falling either separately or agglomerated into irregular lumps. Hailstones are comprised of a series of alternating layers of transparent and translucent ice.

Ice prisms (diamond dust) are ice crystals often so tiny that they seem suspended in air. Such crystals may fall from a cloudy or cloudless sky. Mostly visible when they glitter in sunshine (hence diamond dust), they occur at very low temperatures.

Fog, *ice fog*, and *mist* are also considered forms of precipitation.

VERTICAL MOTION IN THE ATMOSPHERE

The condensation of water vapor to form mist and fog is largely the result of advection and radiation cooling. Condensation of water vapor to form clouds relies on another process—one that depends on vertical motion in the atmosphere. The significance of this process cannot be overstated for the presence or absence of clouds and the occurrence of precipitation ultimately depend on the upward motion of air. Here two essential questions are considered. First, what causes air to rise? Second, what happens to the air as it moves upward?

To help explain many of the processes that occur within the atmosphere, meteorologists often use the concept of a parcel of air. Such a parcel is considered as a volume of air that is small enough to have uniform properties (such as temperature and water vapor content) yet large enough to respond to the meteorological processes that influence it. While it can be any size, visualize it as being about 1 m³ in volume.

The gas laws, defined in Chapter 1, provide a statement of the relationships among pressure, density, and temperature. One of the important facts resulting from the laws is that warm air is less dense than cold air. This density difference is demonstrated every time a hot air balloon rises. The successful ascent of a balloon depends on inflating the balloon with hot air and, as long as the temperature inside the balloon is greater than that of the surrounding air, the balloon will continue to rise. In many ways, the hot air balloon can be considered to act like a parcel of air.

A parcel of warm air surrounded by cooler air will have a tendency to rise. Because pressure decreases with height above the surface, the rising air experiences a pressure decrease. Following the gas laws, the parcel will expand and cool. The rising of air, its expansion, and cooling form the basis for understanding the results of vertical motion in the atmosphere.

For air to move upward requires a mechanism that causes it to rise. One mechanism for lifting occurs when moving air encounters a physical barrier such as a mountain. Air cannot pass through the barrier and is forced to rise over it (Figure 4.8a). Such lifting is termed **orographic** uplift, the term being derived from the word *orography*, which refers to the uneven shape of the earth's surface. Orographic lifting is a mechanical process. Lifting is also a result of the interaction of the properties of air parcels. Such dynamic lifting occurs as a result of convection and frontal activity.

Convective lifting occurs in warm, moist air and is started by heating from the ground surface. When the surface is very warm, the air in contact with the ground is heated. It expands in response to the heating and, having a lower density than the surrounding air, it begins to rise (Figure 4.8b).

Cyclonic lifting occurs along the boundary of air masses of different properties. The boundary between the air masses is termed a *front*, and uplift of air at these fronts is of major meteorological importance. Although dealt with in more detail in later chapters, the basic process involved at a cold front is illustrated in Figure 4.8c.

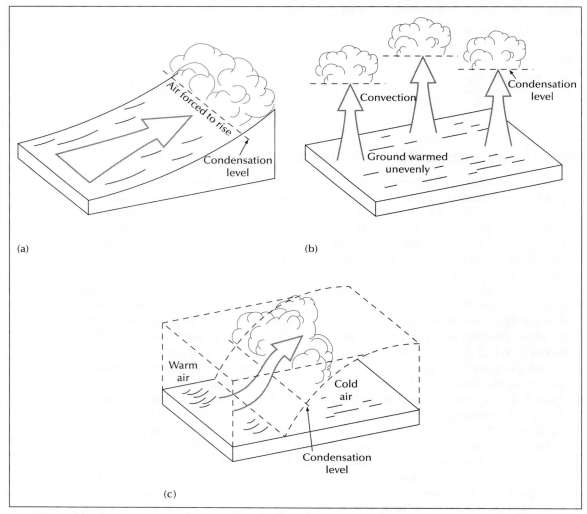

FIGURE 4.8 Processes responsible for the uplift of air.

Adiabatic Heating and Cooling

Once mechanical or dynamic lifting causes air to rise, it undergoes physical changes. Atmospheric pressure decreases with altitude so that a rising parcel of air is subjected to decreasing pressure around it. The parcel of air expands. In the process of expansion, the distances among the individual gas molecules within the parcel are increased. Thus, during expansion of the air, its molecules have to do work (i.e., "push" other molecules out of the way). In doing this work, energy is needed, and this energy comes from the gas molecules. Use of this energy reduces the kinetic energy of the molecules, and the temperature of the parcel decreases. Note that the decrease of temperature is an internal response to expansion of the air parcel; no heat is lost to or gained from the environment surrounding the parcel. Because of this, the process is called *adiabatic*—a process defined as the internal changes within a gas during expansion and contraction when no energy is removed from or added to the gas. Note that the adiabatic process applies to both expansion and contraction. This means that the adiabatic process applies to both rising and sinking air parcels.

The first law of thermodynamics states that the temperature of a gas may be changed by addition (or subtraction) of heat, a change in pressure, or a combination of both. In the adiabatic processes, the addition and subtraction of heat may be disregarded and expressed simply as:

$$\text{Change in temperature} = \text{Constant} \times \text{Change in pressure}$$

This adiabatic form of the first law of thermodynamics is of extreme importance in understanding many atmospheric processes, especially those concerning lapse rates.

Lapse Rates

In the standard atmosphere, temperature decreases with height in the troposphere at an average rate of 6.5°C/km (3.5°F/1000 ft). This standard lapse rate is derived from the difference between the average surface temperature (15°C) and the average temperature at the tropopause (–59°C at 11 km). At any given time, the lapse rate may vary substantially from this mean value. The standard lapse rate is a measured value. On days when a dense cold air mass occurs, the lapse rate measured will be quite different from that on another day when different conditions exist. This lapse rate may be measured when, for example, temperatures are taken at various elevations when ascending a mountain. Because it measures the temperature of the environment, it is called the **environmental lapse rate**.

When a parcel of air is forced to rise over a mountain, it cools at a rate independent of the environment. The cooling of the parcel occurs as a result of the adiabatic cooling process. Because the rate of cooling is based on a physical change within the parcel (a response to decreasing pressure), the rising parcel cools at the adiabatic lapse rate of 10°C/1000 m of ascent (5.5°F/1000 ft). Unless condensation occurs within the rising air, this rate of change is a constant value and is called the **dry adiabatic lapse rate**.

REGIONAL VARIATION IN PRECIPITATION

The amount of precipitation received at the surface varies due to many factors. The basic element is the amount of water vapor in the air, which varies geographically and seasonally. The mean precipitable water content of the atmosphere at a given moment is 25 mm (1 in.), with a maximum near the equator of 44 mm (1.7 in.) and a minimum in the polar regions of 2 to 8 mm (0.08 to 0.3 in.) depending on the season. In latitudes of 40° to 50°, it will range upward of 20 mm (0.8 in.) in the summer and drop to around 10 mm (0.4 in.) in the winter. The presence of water vapor is a necessary but not sufficient condition for precipitation. There is no direct relationship between the amount of atmospheric water vapor over an area and the resulting precipitation. To illustrate this, a comparison can be made between conditions over El Paso, Texas, and St. Paul, Minnesota. The average moisture content above these cities is about the same, and yet the mean annual precipitation is more than three times greater at St. Paul. Other factors must come into play to induce precipitation.

If the total amount of precipitation received over the surface of the earth were spread evenly, it would average 880 mm (35 in.) per year. It varies from near 0 to almost 1200 mm (47 in.). Figure 4.9 shows the mean precipitation over the globe. The total amount of precipitation received depends on several factors:

1. Whether air converges (to give uplift) or diverges (spreads out) in the area.
2. Air mass origin, an indication of the temperature and moisture conditions of the air.
3. Topographic conditions.
4. Distance from the moisture source: The greater the distance from the source of the moisture, the less water vapor will be present in the air because of prior precipitation loss.

Combining these factors, the areas where precipitation is greatest are mountain areas of the tropics, where there is frequent convergence of air from the ocean. Where such conditions exist, rainfall may reach 1200 mm/year (Table 4.5).

Two general locations where precipitation totals tend to be above average for the earth are found relatively far apart. One is near the equator, where there is a zone experiencing convergence of moist tropical air most of the year. These are areas underlying the low pressure convergence area caused by the **intertropical convergence (ITC)**. The trade winds moving toward the equator pick up moisture over the oceans and, when lifted in the ITC, yield abundant moisture. Precipitation is increased over coastal areas by orographic lifting and increased convection started from surface heating. Average precipitation ranges from 1.5 m to 2.0 m (4.5 to 6.6 ft) annually, but in some cases it goes much higher.

The second situation that gives rise to above-average precipitation is found on the west side of the continents in midlatitudes. Precipitation there is due to convergence of maritime air and orographic intensification. The zone is most well defined in latitudes from 50° to 60°. The totals run above 150 mm (6 in.) along the coasts. The amounts are not as high as in the tropics because the moisture capacity of the air is much lower than that of the maritime tropical (mT) air of the tropics.

The arid regions of the world occur in three great realms:

1. Extending in a discontinuous belt approximately between 20° to 30° north and south of the equator. These areas constitute the great tropical deserts, which owe their aridity to large-scale atmospheric subsidence.
2. In the interiors of continents are found the continental deserts, which are arid as a result of their distance from the sea—the major source of the water vapor.
3. The polar deserts of the Arctic and Antarctic constitute the third great area of low precipitation. The low temperatures of these regions, along with subsidence of air from aloft, are probably the most important contributing factor to the low totals.

Mountain ranges play a significant role in the spatial distribution of precipitation. The windward slopes of mountains receive the greatest amount of precipitation. In the lee of the mountain ranges, the precipitation decreases markedly to give a rain-shadow effect. The predominant flow of air along the west coast of North America is from west to east. The mountains produce alternate zones of high precipitation and low precipitation—high on the mountain slopes and low in the

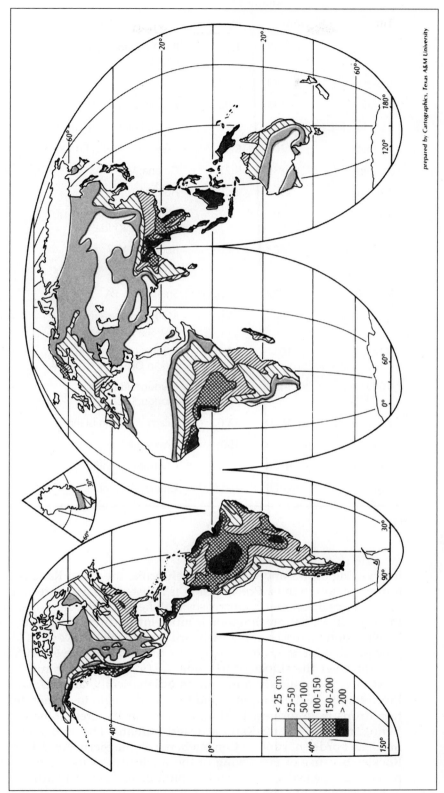

FIGURE 4.9 Mean precipitation over Earth (in cm).

prepared by Cartographics, Texas A&M University

< 25 cm
25-50
50-100
100-150
150-200
> 200

TABLE 4.5 Record Intense Rainstorms

| Time | Amount | | Place | Date |
	mm	in.		
1 min	38	1.50	Barot, Guadeloupe, West Indies	Nov. 26, 1970
5 min	63	2.48	Panama	1911
8 min	126	4.96	Fussen, Bavaria, Germany	May 25, 1920
15 min	198	7.79	Plumb Point, Jamaica	May 12, 1916
20 min	206	8.10	Cuerta de Arges, Romania	Date unknown
30 min	235	9.25	Guiana, VA	Aug. 24, 1906
42 min	305	12.00	Holt, MO	June 22, 1947
1 hr	401	15.78	Muduocaidong, Nei Monggol, China	Aug. 1, 1977
2 hr 10 min	483	19.02	Rockport, WV	July 18, 1889
2 hr 45 min	555	22.00	D'Hanis, TX	May 31, 1935
4 hr 30 min	782	30.80	Smethport, PA	July 18, 1942
6 hr	840	33.07	Muduocaidong, Nei Monggol, China	Aug. 1, 1977
9 hr	1087	42.79	Belouve, La Reunion, Indian Ocean	Feb. 28, 1964
10 hr	1400	55.12	Muduocaidong, Nei Monggol, China	Aug. 1, 1977
24 hr	1870	73.62	Cilaos, Reunion Island	Mar. 15–16, 1952
5 days	3810	150	Cherrapunji, India	Aug. 1841
31 days	9300	366	Cherrapunji, India	July 1861
1 yr	26,460	1041	Cherrapunji, India	Aug. 1860–July 1861

intervening basins and valleys. In California, the Coast Ranges are the first barrier to the onshore winds. Precipitation is substantial, and forests grow on the windward slopes. Precipitation increases with height to the crest at about 760 m (2500 ft). In the Great Valley, precipitation is much lower, and a grassland environment exists. As elevation increases going eastward over the Sierra Nevada Mountains (2600 m), precipitation increases to two or three times that of the Great Valley (Figure 4.10). Forests cover the slopes of the mountains near the summit. The air crossing the crest descends slightly and warms adiabatically, and relative humidity drops. Precipitation declines rapidly, and the clouds begin to evaporate. Annual precipitation drops from around 1300 mm (50 in.) at the crest to about 150 mm (6 in.) at Reno, Nevada, giving Reno a desert climate. The arid zone extends from the Mexican border north into Canada between the Sierra Nevada–Cascade Range and Rockies because of the drying of the westerlies as they cross the mountains. The process is repeated as the currents continue to flow eastward over the Wasatch and Rocky Mountains. The Rockies, with a crest in excess of 3300 m (11,000 ft), cause still further drying of air from the Pacific. Little precipitation falls east of the Rockies

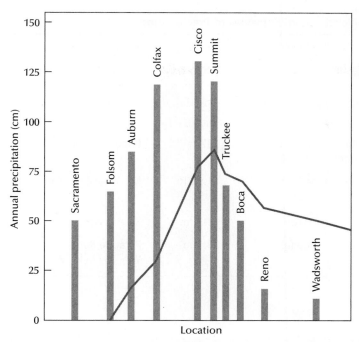

FIGURE 4.10 Precipitation totals across the Sierra Nevada Mountains, illustrating orographic influence on precipitation and the rain shadow in the lee of the mountains. Rainfall totals range from over 125 cm on the windward side of the mountains to less than 10 cm in the valley to the east.

from air originating over the Pacific Ocean. The chief moisture sources for the continent east of the Rocky Mountains are the Gulf of Mexico and the Atlantic Ocean.

VARIATION IN PRECIPITATION THROUGH TIME

Total annual precipitation alone is an insufficient measure of moisture availability because it does not take into account the manner in which the precipitation is distributed throughout the year. It is often the temporal distribution that is the dominant factor in determining use of precipitation. There are major discrepancies between the frequency of precipitation, measured in terms of the number of days per year on which measurable precipitation falls, and mean annual precipitation. In some parts of the world, precipitation falls nearly every day of the year. Bahia Felix, Chile, averages 325 days per year with measurable precipitation. Thus, there is an 89% chance that it will rain or snow on any given day of the year. Buitenzorg, Java, in the tropics, averages 322 days per year with thunderstorms. At the other end of the scale are those desert areas where rain is an oddity. Arica, Chile, averages only about one rain day annually; at Iquique, Chile, not far from Arica, 14 years passed with no measurable rainfall (1899–1913). It may seem strange that one of the areas with the least frequent rainfall is in the same country as one of the rainiest places in the world. The size and shape of the country, the topography of the region, and the general circulation of the atmosphere all play a part in the explanation.

Over most of the earth's surface, precipitation is of a seasonal nature, with a period of the year when precipitation has a high probability of occurring and a dry season when the probability is much lower. Surprisingly, perhaps, those sites that have the highest total annual rainfall are found in areas with a pronounced dry season. Cherrapunji, India, is an example. Although an average of more than 10 m (33 ft) of

TABLE 4.6 World Record Extremes of Precipitation

Longest period without rain	14 years	Iquique, Chile
Lowest average annual precipitation	0.5 mm	Arica, Chile
Greatest number of days per year with precipitation	325	Bahia Felix, Chile
Greatest number of days per year with thunderstorms	322	Buitenzorg, Java
Highest annual average precipitation	11.98 m	Mt. Waialeale, Kauai, HI
Greatest 24-hr snowfall (USA)	1.93 m	Silver Lake, CO, April 14–15, 1921
Highest one-season snowfall (USA)	28.5 m	Paradise Ranger Station, Mt. Rainier, WA, 1971–1972

rain falls each year, there are several months when rain is infrequent. Table 4.6 contains data on some world records of precipitation amounts and frequencies.

WATER BALANCE

To better understand the way in which water is allocated as part of the natural hydrologic cycle of an area, a water balance method proves invaluable. In this, by knowing the amounts that are used in various processes or are stored in the soil, the surplus and deficit can be calculated. The surplus is represented in surface water, and large surpluses lead to flooding. The deficit is a water shortage in relation to plant needs. For water planning and agriculture, it is essential to be aware of both of these values. A number of bookkeeping methods are available to assess the water balance. That introduced by the American climatologist C. W. Thornthwaite is outlined in Box 4.1.

The effect of high summer radiation is apparent in water balances of midlatitude stations (Figure 4.11), where potential evapotranspiration peaks in summer. In some cases, such as along the American west coast, this summer maximum of potential evaporation is out of phase with seasonal rainfall. Here precipitation occurs in the cool season when rainfall is more effective, with less water lost to evaporation.

In the interior areas of the midlatitudes, the summer maximum rainfall coincides with the period of highest rainfall and the time when plants are actively growing. Irrigation is thus an integral part of agricultural practice. In the humid midlatitude areas, precipitation may well exceed evapotranspiration for much of the year except during droughts.

An important aspect of the water balance in midlatitudes is periodic freezing of streams, lakes, ponds, and soil moisture. This takes place where winter temperatures go well below freezing. Most parts of the midlatitudes experience below-freezing temperatures for varying periods during the winter. When such temperatures occur, soil moisture freezes and, depending on the severity of the cold spell, surface water freezes. This freezing stops the flow of runoff. It decreases the water supply available to plants, producing drought. The winter months are a time of year when moisture accumulates in the environment. Even though winter precipitation is generally less than summer precipitation over the northern hemisphere land masses, the

Quantitative Expression: Box 4.1

A Method for Calculating the Water Balance

At any location, a balance between incoming and outgoing moisture is attained, and the balance will reflect the climatic regime that exists. This note outlines Thornthwaite's bookkeeping method using data shown in the following table, which shows data in centimeters:

	Jan	Feb	Mar	Apr	May	Jun	Jul	Aug	Sep	Oct	Nov	Dec	Year
1. PE	1	2	3	5	8	10	12	11	9	5	2	1	69
2. P	14	11	10	6	5	5	1	2	4	9	14	17	99
3. P – PE	13	9	7	1	–3	–5	–11	–9	–5	4	12	16	0
4. ST	0	0	0	0	–3	–5	–2	0	0	4	6	0	0
5. ΔST	10	10	10	10	7	2	0	0	0	4	10	10	0
6. AE	1	2	3	5	8	10	4	2	4	5	2	1	47
7. D	0	0	0	0	0	0	9	9	5	0	0	0	23
8. S	13	9	7	1	0	0	0	0	0	0	6	16	52

Row 1 is the adjusted potential evapotranspiration *(PE)* that is derived by substituting monthly temperature (°C) into an empiric formula derived by Thornthwaite.

Row 2 provides values for monthly precipitation *(P)*.

Row 3 is the difference between precipitation and potential evapotranspiration *(P – PE)*. It provides the amount of moisture available after evapotranspiration requirements have been satisfied. The excess above the difference will either go to soil water or occur as runoff. The amount of water stored by the soil is highly variable and will depend on the nature of the soil. In his initial work, Thornthwaite used a value of 4 in. (10 cm) as a general value to be applied to all water balance studies. Although this does enable comparison of the balance for different stations, it is not realistic in terms of precise water balance studies. Varying soil moisture-holding capacities have been introduced; however, for demonstration purposes, the 4 in. (10 cm) value is retained here.

Rows 4 and 5 give the amount of moisture stored in the soil *(ST)* and the change in storage *(ΔST)* since the previous month, using the 4 in. (10 cm) value as a base. As long as precipitation is greater than potential evapotranspiration, the value remains 4 in. However, as soon as potential evapotranspiration is greater than precipitation *(PE > P)*, plants draw on the available soil moisture and the soil storage falls below capacity.

Row 6 shows the actual evaporation *(AE)*. So long as *P > PE*, then *AE = PE*. The water in storage *(ST)* will then make up the deficit until *(P + ST) < PE*. That is, as soon as the cumulative excess of potential evapotranspiration over precipitation is greater than 4 in. (10 cm), soil moisture is assumed to be totally utilized. A deficit period then exists. At the end of the deficit period, when *P* is greater than potential evapotranspiration *(P > PE)*, the 4 in. (10 cm) must be restored to the soil before any surplus occurs.

Rows 7 and 8 provide the balance of water in terms of deficit *(D)* and surplus *(S)*. It is noted that the deficit does not occur as soon as potential evapotranspiration is greater than precipitation because of the period of soil moisture utilization.

demands on the water supply are also less. Evapotranspiration is reduced in winter because of lower temperatures and reduced plant growth. Over the land masses, some 70% of all precipitation goes directly back to the atmosphere by evaporation. In winter months, this may drop to 40% or less, thus allowing moisture to accumulate on and in the soil. Removing water from circulation by freezing shapes the annual pattern of stream flow. The U.S. Geological Survey uses a water year in all its calculations that is different from the calendar year. For most of North America, streams are at their low point in September or October. During the summer, evapotranspiration uses a big share of the water. As temperatures cool, the vegetation uses less water and moisture begins to collect. During winter, moisture accumulates either as soil moisture or snow. The spring thaw then causes streams to rise. Flooding may occur if a large amount of snow covers the land surface. Spring floods due to snow melt generally characterize poleward locations and mountain regions, but floods also affect other areas.

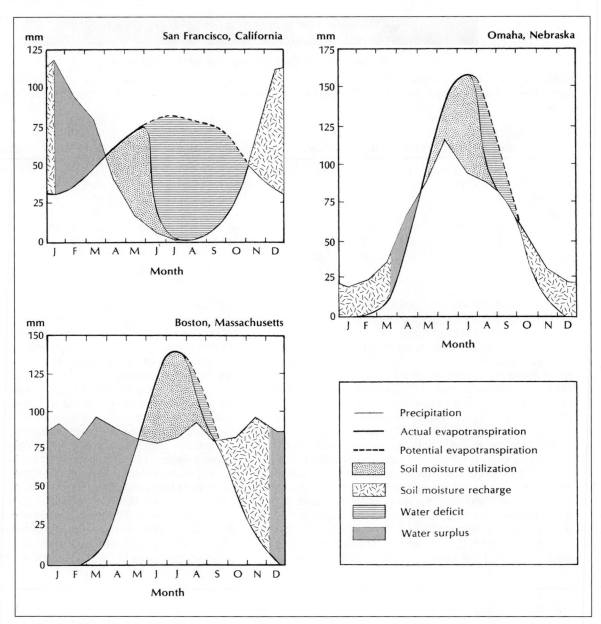

FIGURE 4.11   Selected water balance diagrams for North American stations.

Summary

Condensation occurs near the ground as dew, mist, or fog and at higher levels as clouds. Fog forms as a result of advection, radiation, and adiabatic processes. A high incidence of fog occurs in specific locations. Condensation nuclei play an important role in the condensation process.

The spatial distribution of precipitation shows that two global areas receive rainfall high above the global average while there are three great world realms that are classed as arid deserts. Total annual precipitation is an insufficient measure of moisture availability, and the way in which precipitation is distributed over time is highly significant. Precipitation seasonality forms an important component of the climate of a location.

Key Terms

Crepuscular rays, *65*
Dew point, *61*
Dry adiabatic lapse
    rate, *71*

Environmental lapse
    rate, *71*
Evapotranspiration, *59*
Hydrologic cycle, *59*

Intertropical convergence
    (ITC), *72*
Nimbostratus, *65*
Orographic, *70*

Upslope, *64*
Upwelling, *64*

Review Questions

1. How much energy is absorbed and released when water changes from liquid to gas or gas to liquid at standard atmospheric conditions?
2. What is the relationship among evaporation, transpiration, and evapotranspiration?
3. What percentage of Earth's water is available for human use as groundwater and surface water?
4. What is the relationship between relative humidity and dew point?
5. What are the cloud conditions on most nights when radiation fog develops?

6. Why are salt particles in the air considered hygroscopic?
7. How is lifting condensation level related to dew point?
8. What are the environmental conditions at places on Earth that have a large total annual precipitation?
9. What are the environmental conditions that favor the development of deserts?
10. What latitudinal zones on Earth's surface have above-average precipitation?

5

Wind and Circulation Patterns

Wind is the name given to air moving horizontally over the earth, and *air currents* refer to air moving vertically. Winds transfer heat over the earth's surface and carry water vapor from oceans to continents. Without wind systems, there would be no life as we know it on the land masses. There are two main characteristics of wind: direction and speed. The direction given to a wind is the direction from which it blows. They are named for the main points on the compass rose or assigned a bearing in degrees from north. A wind blowing from east to west is an east wind, or a 90° wind. When the direction of the wind is changing, it is veering or backing. A veering wind is one changing in a clockwise direction, and a backing wind is one changing in a counterclockwise direction.

ATMOSPHERIC PRESSURE

Atmospheric pressure is the force per unit area or the weight exerted by the atmosphere on a surface. There are a number of widely used measures of atmospheric pressure. In the United States, pressure charts are shown using the millibar. This is derived from a bar—a force of 1 million dynes per square centimeter. A dyne is the force needed to accelerate a 1 g mass 1 cm/s^2. A millibar is one-thousandth of a bar. In the standard atmosphere, pressure at sea level at 15°C is 1013.2 mb. Commonly used elsewhere is the standard SI unit the Pascal. Derivation of the Pascal and various pressure equivalents are shown in Table 5.1a.

TABLE 5.1 Pressure Units and Extreme Sea Level Pressure Readings

a. Pressure Units

Pressure = Force per unit area

SI force unit is the Newton (N)

   (the force required to accelerate 1 kg 1 m/s^2)

SI pressure unit = Pascal (Pa)

   (1 Pa is 1 N per square meter)

Bar (1 Bar is 10^6 dynes per square centimeter)

Standard Sea Level Atmosphere

   101,325 Pa which is expressed as 1013.25 hPa

Or   1.01325 Bars which is expressed as 1013.25 mb

Or   Based upon height of column of mercury

   29.92 inches of mercury (76 cm) at sea level

b. Extremes of Sea Level Pressure Readings

Pressure (mb)	Inches of Mercury	Event
1083.8	32.00	Highest recorded. Agata, Siberia. December 31, 1968
1078.2	31.85	Highest recorded in North America. Northway, Alaska. January 31, 1989
1063.9	31.42	Highest in the lower 48 states. Miles City, Montana. December 24, 1983
870	25.7	Lowest recorded. Typhoon Tip, Pacific Ocean near Guam. October 12, 1979

In popular usage in the United States is the equivalent pressure in inches of mercury. A column of mercury about 30 in. high just offsets the weight of a column of atmosphere of the same area. A column of mercury 1 in. on each side and 30 in. high weighs 14.7 lb. This is the weight of a column of the standard atmosphere 1 in.^{2} in area. The measurement of 29.92 linear inches of mercury is equivalent to 14.7 lb/in.^{2} or 1013.2 mb. In the real world, surface pressures vary routinely from about 950 mb (28 in.) to 1050 mb (31 in.). Table 5.1b provides some extremes of pressure.

The variation in atmospheric pressure and density that takes place with height in the standard atmosphere was presented earlier. Like many other atmospheric variables, pressure changes with time and space. For example, at the two poles, pressure is higher than average because of the cold temperatures and subsidence of air. There are other zones where it averages less than the global mean.

There are seasonal changes in pressure in some areas. Central Asia experiences marked differences from summer to winter. These changes are due to large changes in temperature with the seasons. In midlatitudes, there are often day-to-day changes in pressure resulting from traveling storms. There are also regular daily variations. There are maxima around 10 A.M. and 10 P.M. and minima around 4 A.M. and 4 P.M. The exact cause of these daily changes is unknown, but they relate to the daily flood of solar energy that moves around the earth.

Areas below or above normal pressure develop in the atmosphere. Some specific terms refer to these areas. In all cases, the pressures are relative and do not have fixed values. A **low pressure cell** is an area in the atmosphere where pressure is less than the surrounding area, while a **trough** is an elongated area of low pressure. A **high pressure cell** is an area of the atmosphere where barometric pressure is higher than the surrounding area and a **ridge** is an elongated area of high pressure.

Humans are normally not sensitive to small variations in pressure. We are most likely to be aware of them when going up or down in an airplane or driving up or down a long mountain road. In taking off and landing in a modern jet aircraft, the pressure changes are large. Cabins in commercial jets are pressurized so the pressure normally stays near 75% of that at sea level. We are most sensitive to pressure changes when we have a head cold and there is congestion in the eustachian tubes. Air cannot flow into or out of the inner ear to remain balanced with the outside pressure. When there is much difference, it may cause severe earache.

FACTORS INFLUENCING AIR MOTION

Newton's Laws of Motion

Newton's first **law of motion** deals with inertia. It states that a body will change its velocity of motion only if acted on by an unbalanced force. In effect, if something is in motion, it will keep going until a force modifies its motion. On Earth, a parcel of air seldom moves continuously and in a straight line. This is because, as Newton's second law states, the acceleration of any body—in this case, the parcel of air—is directly proportional to the magnitude of the net forces acting on it and inversely proportional to its mass. Note that these laws concern acceleration, which is change of velocity with time.

By identifying the forces that act on a parcel of air, it becomes possible to understand more fully the processes that lead to the acceleration (or deceleration) of air. If we consider a unit parcel of air ($m = 1$), then Newton's second law becomes:

$$\text{Acceleration} = \text{Sum of forces}$$

$$\text{or } F_a = \Sigma F$$

The ΣF is made up of the atmospheric forces so that:

$$\text{Acceleration} = \text{Pressure gradient force } + \text{ Coriolis force}$$
$$+ \text{ Frictional forces } + \text{ Rotational forces.}$$

Pressure Gradient

Differences in heating and internal motion in the atmosphere produce differences in atmospheric pressure. The difference in pressure over space is a **pressure gradient**, and air will move along this gradient from high to low pressure. If no other forces occurred, air on Earth would move in a straight line from high to low pressure. The rate at which air moves depends on the steepness of the gradient. When small differences in pressure occur over large areas, a weak gradient gives rise to weak winds. A steep gradient causes rapid motion or high winds. Figure 5.1 illustrates this effect.

Atmospheric pressure changes slowly in a horizontal direction. The change may be as little as 2 mb in 160 km (100 mi). During the summer over North America, the temperature difference between the Gulf of Mexico and the Canadian border is small. As a result, the density of the air is about the same, and the north–south pressure gradient is weak. Because the pressure gradient is weak, mean wind velocities are low. During the winter, temperatures differ more. It is still warm over the Gulf of Mexico but relatively cold over Canada. Density of the cold air is greater and there is a steeper pressure gradient in the winter. This gradient transports cold air south and wind velocities are higher. The pressure gradient sets air in motion, and once air is underway, other forces take effect.

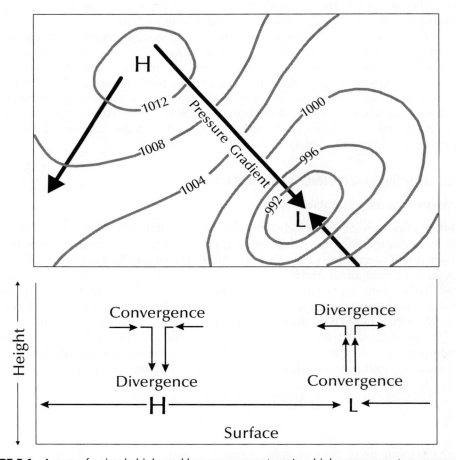

FIGURE 5.1 A map of a simple high- and low-pressure system. In a high-pressure system, pressure decreases outward from the center. In a low-pressure system, pressure decreases toward the center. The lower part of the figure shows convergence and divergence associated with high- and low-pressure systems.

TABLE 5.2 Examples of Drift Due to the Coriolis

1. A person walking at a rate of 10 m/min (4 mph) toward a point in space drifts 40 m/km (250 ft/mi).

2. An auto traveling 100 kph (60 mph) drifts 4.5 m (15 ft) in 1.6 km (1 mi). The tendency to drift is offset by friction with the road.

3. A bullet fired at 120 m/s (400 ft/sec) drifts 2.5 mm in 1 sec.

4. An artillery shell fired at 750 m/s (2500 ft/sec) drifts 60 m (200 ft) per 33 km (20 mi).

5. A missile launched from the north pole toward New York City at a rate of 1.6 km/s (1 mi/sec) lands near Chicago.

6. An airliner leaving Seattle for Washington, DC, along a great circle route and not corrected passes over South America near Lima, Peru.

Coriolis Effect

The rotation of the earth results in a process called the **Coriolis effect** (also referred to as the Coriolis force or Coriolis acceleration), which acts on any object moving free of the earth's surface, including aircraft, missiles, and long-range artillery (Table 5.2). The rotation of the earth causes winds in the northern hemisphere to turn in a clockwise direction, or to the right. In the southern hemisphere, the winds turn counterclockwise, or to the left (Figure 5.2). Mariners have long known and recorded the process. The French mathematician G. G. Coriolis (1792–1843) first explained the process, hence the name.

The rate of curvature imparted to the moving air is a function of the wind's velocity and the latitude. At the equator, the Coriolis effect is zero. It is maximum at the two poles. Although the Coriolis effect is relatively small, it is significant because air streams travel long distances over the earth.

Friction

Like Coriolis, **friction** can exert an influence only when air is in motion. It acts in opposition to air motion and results from contact with the surface, which induces turbulent flow. Friction works in opposition to the pressure gradient force to reduce wind velocity and the Coriolis effect. With the rotational effect reduced, the pressure gradient dominates. Thus, at the surface, winds blow from high to low pressure at an

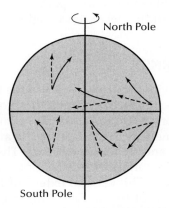

FIGURE 5.2 The Coriolis effect acts on the wind to deflect it to the right in the northern hemisphere and to the left in the southern hemisphere.

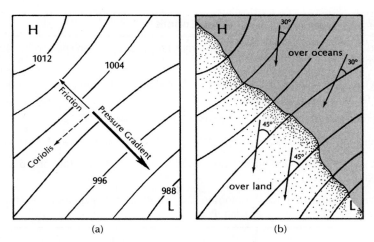

FIGURE 5.3   (a) The relative direction of the three forces that affect surface wind direction and velocity. (b) The mean angle of wind direction relative to the pressure gradient over land and the sea.

angle across the isobars. Over the oceans, where friction is less than over land, the winds cross the isobars at angles of 10° to 20° at 6 m above the surface. Oceanic winds blow at or about 65% of the velocity dictated by the pressure gradient. Over the land masses, where the surface is rougher and friction greater, the wind crosses the isobars at an average angle of 30° (Figure 5.3). Because of greater friction, winds blow with a velocity of about 40% of gradient velocity.

Rotational Forces

Rotating bodies are subject to the law of **conservation of momentum**. This law states that if no external force acts on a system, the total momentum of the system stays unchanged. Angular momentum is a function of the mass of the rotating body, the angular velocity (degrees per unit time), and the radius of curvature. The product of these elements stays constant when there is no external interference. A tetherball serves as a good illustration. As the ball goes around the pole and the radius shortens, the angular velocity increases to keep the momentum the same as it was initially. Another example of the conservation of momentum is a figure skater going into a spin. The moving skater turns into a slow spin with arms and perhaps a leg extended. When the skater pulls in his or her limbs while stretching upward, this action reduces the radius of the spin, and the skater turns faster and faster. Conservation of momentum is a part of the acceleration process in mesocyclones and tornadoes. As the converging air travels a shorter and shorter path, the velocity increases accordingly. Figure 5.4 provides a schematic model of conservation of momentum.

THE RESULTING PATTERNS

The processes outlined previously produce the winds that blow at all speeds, from the lightest hint to hurricane forces. A quantitative study of these processes is based upon fundamental equations relating to Newton's laws of motion. However, the forces do not act in the same way in all parts of the atmosphere. Clear differences are found between surface and upper level winds. The reason for this is shown in Figure 5.5. The turbulent flow induced is a result of friction produced by moving air in contact with a surface. The vertical extent of the irregular flow will vary depending on the roughness of the surface. However, for every surface and for wind at different speeds, it is possible to identify either geostrophic winds or friction layer winds.

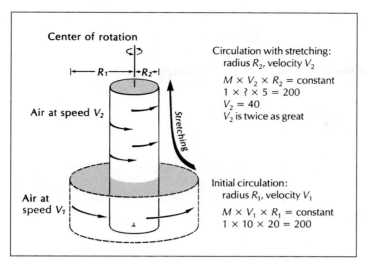

FIGURE 5.4 In a converging system such as a mesocyclone, vertical stretching reduces the radius of the rotating air and leads to an increase in wind velocity. In this example, the mass of the rotating air (*M*) is set to 1 for the sake of simplicity. It is the same for the initial rotating mass and for the stretched mass.

Geostrophic winds occur above the friction layer. Without the influence of friction, the main forces acting on moving air are pressure gradient and the Coriolis effect. Consider the parcel of air in Figure 5.6a. It will move initially from high to low pressure and, in the northern hemisphere, be turned to the right by Coriolis. Eventually, these two forces are balanced, and the geostrophic wind then moves parallel to the **isohypses**.

When depicted on constant pressure charts, as in Figure 5.6b, the flow of air parallel to the isohypses is apparent. Note that in these upper air charts, pressure is depicted by an elevation. Instead of showing equal pressure values using isobars, as on the surface map, these upper level charts show a single pressure surface. There is no need for isobars since the entire chart is a single pressure. In this case, the 500 mb surface is shown, and the isohypses connect places in the atmosphere with an

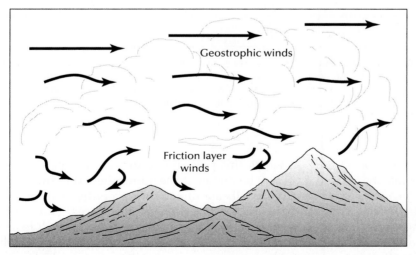

FIGURE 5.5 Moving air close to the surface is influenced by friction; above the friction layer, the lack of friction causes geostrophic flow. A friction layer and a geostrophic layer can be identified.

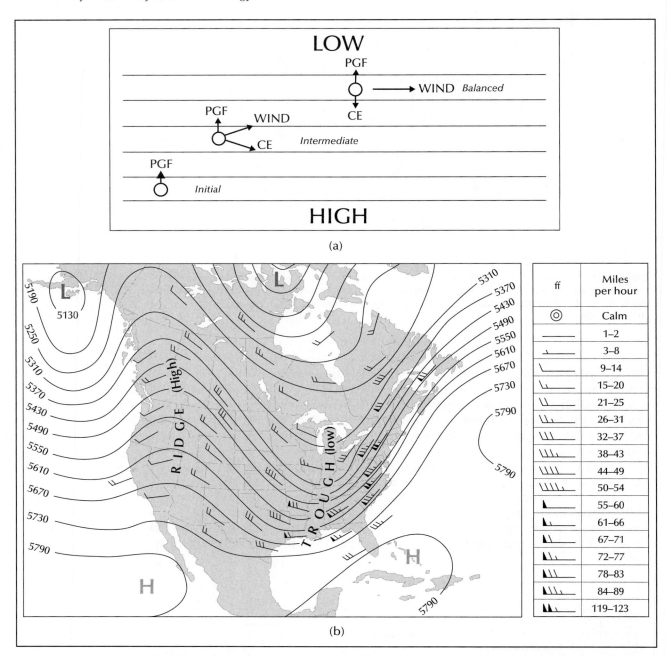

FIGURE 5.6   (a) The balance between pressure gradient force and Coriolis effect result in a geostrophic wind. (b) An upper air chart at the 500 mb level shows how winds generally blow parallel to the height contours. Note the high wind speeds. (From Lutgens F. K. and Tarbuck E. J., *The Atmosphere*, 10th ed., © 2007, Prentice Hall, Upper Saddle River, N.J.)

equal height, or altitude, at which this pressure occurs. For example, where warm air prevails, the 500 mb level will be higher in the atmosphere (Figure 5.7).

If pressure gradient and Coriolis are of equal magnitude, then the moving air should continue in a straight line for there are no net forces. However, around low-pressure systems, the air follows an almost circular path. In such instances, the pressure force toward the low is greater than the outward directed Coriolis effect. A similar effect occurs around high pressures. The resulting flows, which are a form of geostrophic winds, are called **gradient winds**.

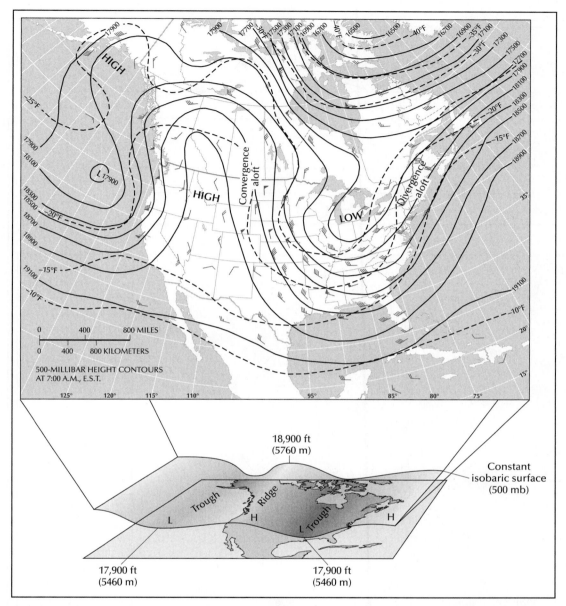

FIGURE 5.7 Pressure depiction on upper air charts is by the elevation of constant pressure levels. In this example, the height of a 500 mb surface is shown. (From Christopherson R. W., *Geosystems*, 7th ed., © 2009, Prentice Hall, Upper Saddle River, N.J.)

The lack of friction means that geostrophic and gradient wind can blow much faster than winds in the friction layer. The winds shown in Figure 5.6b are seen to commonly exceed 75 mph (120 kph). In places, much higher wind speeds are seen. These are identified as *jet streams*—corridors of high winds within flow of air above the friction layer.

Surface winds, because of friction, do not attain a balance between pressure gradient and Coriolis. As shown in Figure 5.8a, when air is moving, a balance is eventually established among the pressure gradient, Coriolis, and friction. The resultant winds blow across isobars rather than parallel to them (Figure 5.8b). It has been noted previously that the angle they cross the isobars varies depending on the surface.

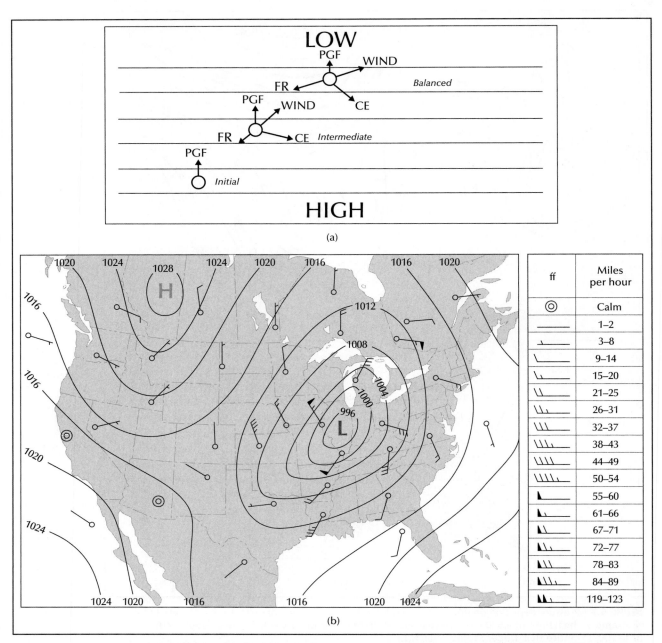

FIGURE 5.8 (a) The balance between pressure gradient force, Coriolis effect, and friction results in friction-layer winds. (b) A surface chart shows how winds relate to high- and low-pressure systems. Notice the low-wind speeds compared with those shown in Figure 5.6b. (From Lutgens F. K. and Tarbuck E. J., *The Atmosphere*, 10th ed., © 2007, Prentice Hall, Upper Saddle River, N.J.)

Not only is surface direction different from winds found aloft, but the rate at which they blow is appreciably lower. The 75 mph winds commonly found aloft are ranked as hurricane force winds at the surface. In fact, it is only under conditions where pressures well below normal are found that high winds occur at the surface. In these cases, as with hurricanes and tornadoes, the rapid circular motion of air is influenced by the rotation forces outlined earlier.

THE GENERAL CIRCULATION

The general circulation of the atmosphere is extremely important to Earth. It is the general circulation that carries water from the ocean over the continents to provide precipitation and move heat energy from the tropical regions toward the poles, warming the high latitudes. The general circulation is not static, but changes constantly. Next to the seasons, changes in the general circulation are the most significant changes that take place in our weather. Here we will demonstrate with a simple model how the general circulation works. Then building on the processes that control the general circulation, we examine how it changes.

Between latitudes of about 35° N and 35° S, incoming solar radiation exceeds Earth radiation to space (Figure 5.9). Toward the poles from 35°, there is a net loss of energy to space. For incoming solar radiation to exceed outgoing radiation in equatorial areas over a prolonged period, energy must be removed in some other manner than radiation. The mechanism for transporting this surplus energy poleward is the general circulation of the atmosphere and oceans. It is a mechanism that tends to equalize the distribution of energy over the surface of the earth. The atmosphere transfers about 60% of this energy and the oceans move the rest. This complex circulation system consists of several semipermanent areas of convergence and divergence and the air flow in and between them. The energy moves primarily as sensible heat and latent heat of water vapor.

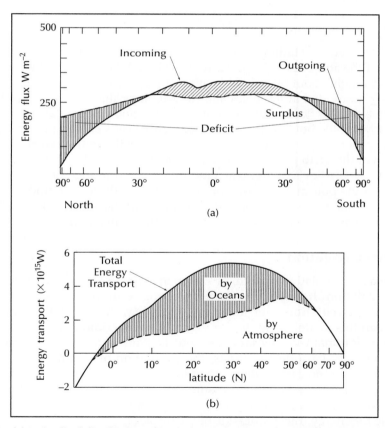

FIGURE 5.9   (a) Latitudinal distribution of incoming and outgoing radiation. Between latitudes of about 30° north and south, incoming solar radiation exceeds outgoing Earth radiation. Toward the poles from 30°, outgoing radiation exceeds incoming radiation. (b) For the distribution shown in (a) to exist, there is an equator-to-pole movement of energy by the ocean and atmosphere. The total amount of energy transported is shown by the solid line. The dashed line shows the amount of heat transported by the atmosphere. The shaded area is the amount transported by ocean currents. (Modified from Gill, 1982.)

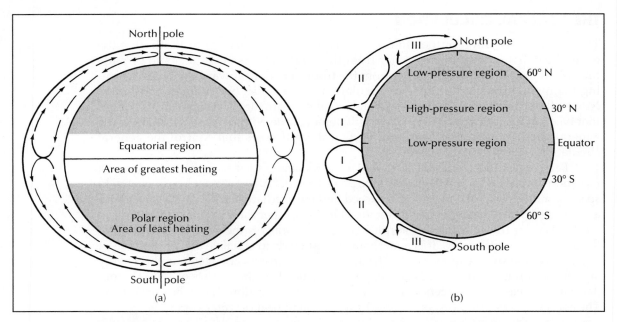

FIGURE 5.10   (a) The fundamental driving mechanisms of the general circulation. (b) A three-cell model that results from differential heating, the earth's rotation, and the fluid dynamics of the atmosphere.

In 1735, George Hadley proposed a cellular model to explain the primary circulation of the atmosphere (Figure 5.10a). He based the model on the basic pressure differences brought about by uneven heating of the earth's surface. Hadley postulated that cold air descended at the poles and flowed along the earth's surface toward the warmer equator. This flow would be countered by rising warm air at the equator and a flow aloft toward the poles so that two large **Hadley cells** existed in each hemisphere.

Observation of global wind patterns and later theoretical work showing that the rotation of the earth prevents the development of a single convective cell led to a modified view. Hadley's model was gradually altered to include a three-cell structure between the pole and the equator. In the revised **three-cell model**, there is still subsiding cold air at the poles and rising warm air at the equator (Figure 5.10b). In a rudimentary way, it illustrates the essentials of the primary circulation.

Meridional Circulation

A more detailed circulation pattern, as shown in Figure 5.11, is often used to show the general circulation. It actually represents a long-term average of the **meridional** (north–south) **circulation** over the globe. In reality, the circulation is much more complex than this model suggests. For example, the meridional transfer of air across the globe is dominant in low latitudes, but is some 60% weaker in midlatitudes. Although the low-latitude cell exists, the circulation of middle latitudes consists essentially of a west-to-east (zonal) flow of upper air that determines the position and location of moving high- and low-pressure systems at the surface. It is the circulation patterns associated with the traveling high- and low-pressure systems that is responsible for energy transfer outside of the tropics.

Zonal Circulation

The meridional depiction needs to be augmented by one showing **zonal circulation**. Figure 5.12 provides a model of the primary circulation of the northern hemisphere that stresses the difference between tropical and midlatitude circulation. The figure

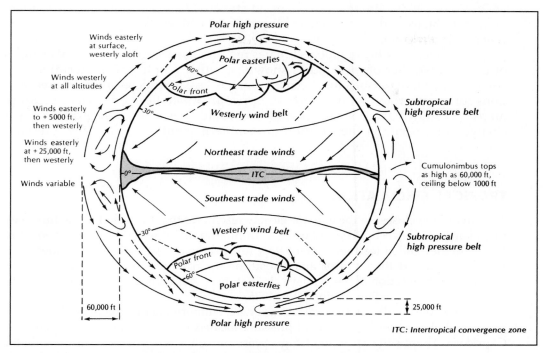

FIGURE 5.11   The vertical and longitudinal flows in the idealized general circulation.

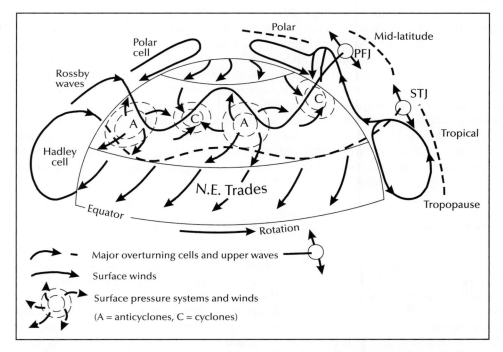

FIGURE 5.12   A general circulation model of the northern hemisphere showing the zonal flow dominant in middle and high latitudes. (After Hanwell, 1980.)

shows the tropics dominated by the Hadley cell and midlatitudes by upper air **Rossby waves**, so named for the meteorologist Carl-Gustav Rossby whose theoretical work provided the understanding of the mechanisms for their existence. The dominant air flow within the Rossby waves is west to east. Embedded in these upper air westerlies are corridors of rapidly moving air, or **jet streams**. Associated surface circulation, located in the westerly wind belt, are the cyclones and anticyclones that provide the changeable weather of the region.

An understanding of global climates requires assessment of seasonal changes of climate. Neither of the models shown, whether emphasizing meridional or zonal flows, takes seasons into account. Such is best considered using representations for summer and winter for both upper air westerlies and surface pressure maps. As shown later, such illustration provides clear evidence of the great changes that occur in the location of semipermanent high- and low-pressure systems.

TROPICAL CIRCULATION

The circulation closest to the equator closely resembles that proposed in the three-cell model and Hadley cells. The cell extends from around 25° to 30° to near the equator. Like a Hadley cell, it features surface flow toward the equator and counterflow aloft, with rising air at the equator and subsiding air near the Tropics of Cancer and Capricorn. This section of the primary circulation is called the Hadley because it operates effectively as Hadley outlined it in 1735.

Associated with this tropical cell are two semipermanent belts of surface divergence and above-average sea level pressure located near the Tropics of Cancer and Capricorn. These **semipermanent high pressure** zones on either side of the equator act as a source region for air flowing both north and south at the surface. As the air subsides, it heats adiabatically and is quite warm. Since the temperature is increasing and no moisture is being added, the relative humidity is quite low. These zones have high amounts of sunshine, clear skies, and low frequency of precipitation. In fact, there are major deserts associated with these areas of subsidence: the Sahara and Sonora deserts of the northern hemisphere and the Great Australian, Atacama, and Kalahari Deserts of the southern hemisphere.

If the air that flows out of these areas of subsidence has a trajectory over a land mass, it remains warm and dry and becomes a continental tropical air mass. The high pressure is due to the subsidence, which gives rise to divergence at the surface. These zones of subsidence and divergence result more from the motion of the atmosphere and the Coriolis effect than from thermal factors. Over the oceans, the surface winds are light and variable. In the days of wind-driven sailing vessels, these zones were known as the **horse latitudes**. When ships were becalmed, horses were thrown overboard to lighten the load.

Trade Winds

Lying between the subtropical belts of subsidence and the equatorial convergence zone is a region of surface winds flowing toward the equator known as the **trade winds**. These wind systems are most frequent in belts 10° in width and centered 15° on either side of the equator. They are the most regular wind systems found at the surface of the earth. The winds, which average 4 to 7 kph (2–4 mph), are most persistent over the eastern half of the oceans and are more dependable there than elsewhere. Both north and south from 15° these winds are less distinct. Their origin is in the subsidence of the subtropical high-pressure zones, and they flow into the equatorial low-pressure belt. The Coriolis effect gives the winds a westward turn, so they become northeasterly winds in the northern hemisphere and southeasterly in the southern hemisphere. The trade winds often have layers, with a surface layer that becomes increasingly moist and unstable as the air moves toward the equator. Above there is an overlying layer that is dry and stable.

The trades are known for providing the route across the Atlantic followed by the sailing ships of the age of exploration. These same winds play a major role in the

Pacific Ocean as well. The trade winds carried Thor Heyerdahl and his crew across the Pacific Ocean on the first modern crossing by raft:

> The wind did not become absolute still—we never experienced that throughout the voyage—and when it was feeble we hoisted every rag we had to collect what little there was. There was not one day on which we moved backward toward America, and our smallest distance in twenty-four hours was 9 sea miles, while our average run for the voyage as a whole was 42 1/2 sea miles in twenty-four hours. (Heyerdahl, 1950, p. 41)

The Intertropical Convergence Zone

Near the equator, the trade winds from both hemispheres converge to form a low-pressure trough where there is a gentle upward drift of air. This convergence zone is also called the **intertropical convergence zone (ITCZ)**. It is the area in which the trade winds from the north and south of the equator merge. The area of convergence is quite broad along the equator because of the extensive area of warm surface. Although this trough of low pressure is not very deep, it is quite consistent over the oceans and is deep enough to produce convergence most of the time. Where the converging trades have a trajectory over the ocean, the air contains large amounts of moisture, cloud cover is extensive, and precipitation is frequent.

There is less evaporation from the ocean along the equator than near the Tropics of Cancer and Capricorn. The reason is that equatorial air is quite humid, and the vapor pressure at the water surface is lower. Sea temperatures are not as high near the equator as they are farther out near the Tropics of Cancer and Capricorn. The addition of fresh water from precipitation and from streams draining adjacent land masses cools the ocean. Less solar radiation reaches the surface as a result of the more extensive cloud cover, which further reduces water temperatures.

In the center of the ITCZ over much of the Atlantic and Pacific Oceans are the **tropical easterlies**. These are stable winds of low velocity moving from east to west. Do not confuse them with the trade winds. These easterly winds are quite regular in direction. Occasionally, a change in circulation will bring quite unstable weather in the form of squalls and general rainstorms. The converging winds on occasion will rise some distance away from the center of the convergence zone. When this happens, quite stagnant conditions may result near the surface. Sailors of the 16th century called these stagnant conditions of low-velocity winds with ill-defined direction the **doldrums**. While traveling to the western hemisphere from Europe, sailing vessels became becalmed in this area, just as farther north, in the horse latitudes.

MIDLATITUDE CIRCULATION

The three-cell model of the atmosphere has inadequacies, and the real world differs most from the model in the midlatitudes. The model calls for a mean flow of air between the semipermanent high-pressure cells at 30° and the semipermanent low-pressure cells near 60°. In reality, the mean meridional flow in these latitudes is very weak. Surface winds are quite variable in both direction and velocity. The greatest frequency and highest average velocity show a west-to-east flow. Maximum westerly velocities occur near 35° N. Wind velocities at this latitude become more pronounced with height up to the tropopause. Thus, the primary flow in midlatitudes is zonal (west to east).

It is in the latitudes of 35° to 40° north and south that the maximum transfer of energy takes place. With a relatively weak meridional flow, there must be some other means of transporting the energy through the zone. In this region, there is a mixing zone between the warm tropical air and the cold polar air known as the **planetary frontal zone**. It is here that the midlatitude westerlies are strongest. The westerly flow increases up to about 12 km (7.5 mi). In the upper part of the troposphere and the

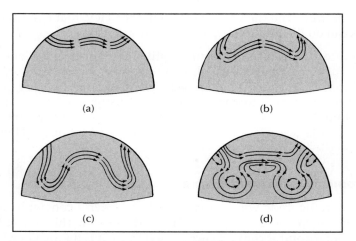

FIGURE 5.13   Model of the development of waves in the westerlies: (a) When the westerlies are zonal (flowing from west to east), there is minimal mixing of tropical and polar air. When waves form with air flow becoming more north–south, a great deal of cold and warm air is exchanged. (b) A wave has begun to form. (c) A wave exhibits large oscillations. Maximum exchange of warm and cold air occurs at this time. (d) As the system begins to stabilize, masses of cold air become isolated toward lower latitudes.

lower stratosphere, long meandering Rossby waves develop in the planetary frontal zone (Figure 5.13). These waves have an amplitude and wavelength that vary up to 6000 km (3730 mi). They are always in a state of flux, sometimes being standing waves and sometimes moving with the westerlies.

When the westerlies begin to loop widely, there is a greater north–south, or meridional, flow, although the net direction of flow is from west to east. Meridional flow causes exchange of warm and cold air. Warm air, and hence heat, is carried northward, and cold air is carried southward. When there is high meridional flow, temperatures may fluctuate rapidly and widely on a week-to-week basis. In the first week of February 1991, temperatures went from record lows to record highs over much of northern United States and southern Canada.

Figure 5.13d illustrates a situation in which the meandering is large enough that cells of cold or warm air become isolated from the main air flow. Sometimes these large masses of air stagnate and control the weather of the area in which they are located for up to several weeks. These are called **blocking systems**. These masses of air can bring either very warm or very cold conditions. Blocking highs divert moisture keeping storms away from the region and enhance the probability of drought occurring.

The westerlies shift back and forth between a zonal pattern and a meridional pattern at irregular intervals, which makes long-range forecasting difficult. Near the surface, large eddies or vortices develop that are 1000 to 2000 km (62 to 1240 mi) across and have lifetimes of several days. These eddies are the cyclones and anticyclones of the midlatitudes. They are rotational in character and thus carry cold air from the poles toward the equator and warm air from the subtropics toward the poles. These eddies carry energy toward the poles, but the net flow of air northward is very small. It is basically an exchange process of warm for cold air. Thus, there is a high-energy transfer with a low net flow of air toward the poles. It is also the high frequency of these cyclonic cells that forms the subpolar low-pressure zone in latitudes 50° to 60°. This convergence zone differs markedly from the equatorial convergence zone. The air flowing into it has very different temperature and moisture conditions.

The **upper air westerly** flows dominate the circulation of the midlatitudes. The temperature gradient between the equator and poles determines the strength and location of the subtropical jet stream. As polar regions warm in summer, the gradient is least and the strength of the westerlies diminishes. The jet migrates toward the pole. When a steep gradient exists, the westerly circulation is strong and the jet migrates closer to the tropics. The limit of the midlatitudes is along the axis of the semipermanent high-pressure cell located near 30°. From here to the equator, the Hadley circulation operates. The surface circulation depends on the flow of upper level westerlies.

POLAR CIRCULATION

Over the north and south polar regions, thermally induced high-pressure systems exist, although knowledge of them is limited. There is a predominance of high-pressure systems and divergent air flow over polar areas most of the year. The size of the area affects changes, expanding during the winter months. Outward from the polar highs are belts of weak easterly winds that are more pronounced in the summer in the northern hemisphere.

Pressure gradients are weak in the tundra, so this is not a stormy region. In fact, it is one of the least stormy areas of Earth's surface. High-pressure systems dominate with cold, still, dry air prevailing, particularly in the winter. The North American Arctic is even less stormy than other areas. Because of the mountainous character of Alaska, the cyclones of the Pacific Ocean do not move over the Arctic coastline. Some of the storms penetrate far into the arctic, but few do compared with other areas. Spring and fall are the seasons with the most active weather. It is the same in most high-latitude locations. Winds are light in velocity and variable in direction. The average velocity is less than 15 kph and less than 10 kph in the Canadian tundra.

In the region surrounding the Antarctic continent, there is an unbroken sea extending completely around the earth. In the winter, there is a much greater frequency of storms with high winds blowing from a variety of directions. This is the part of the world ocean that was named the "roaring forties," "the furious fifties," and the "shrieking sixties" by the sailors of the 16th century. Sailors on board vessels trying to sail around Cape Horn at the tip of South America named the winds. The finding of the Strait of Magellan made the passage around South America considerably less trying but by no means easy.

Winds are a predominant factor in the weather of the **polar deserts**. Westerly winds are most common at the surface to around 65°, where they give way to low-level easterly winds that extend to about 75°. Cyclonic storms develop over the oceans in the westerlies and move around the Antarctic from west to east. They normally do not penetrate far inland, and they account for much of the precipitation and weather along the coast. Winds of a high enough velocity to move snow and produce blizzard conditions occur at Byrd Station about 65% of the time. They are of high enough velocity to produce zero visibility about 30% of the time. A slight increase in wind velocity brings a large increase in blowing snow. The amount of snow moved by the wind varies as the third power of the velocity. It is unfortunate for those working in the Antarctic, but the most accessible locations are also the windiest. Mawson's base at Commonwealth Bay experiences winds that are above gale force (45 kph or 28 mph) more than 340 days a year. At Cape Denison, the mean wind velocity is 69 kph (43 mph). During July of 1913, it averaged 89 kph (55 mph).

SEASONAL CHANGES IN THE GLOBAL PATTERN

Some major elements of the general circulation shift location from season to season. The tropical Hadley cell shifts north and south with the seasons. This is a result of the shift in the thermal equator. As described earlier, the latitude that receives the most sunshine during the year changes from about 30° N to about

30° S latitude. This change in the heat equator causes the Hadley cell to move north and south as well. Most areas of the earth between the equator and the thirtieth parallel experience seasonal changes in wind direction and precipitation amounts. This is especially true over the continents since the land surfaces heat and cool more rapidly than does the ocean. Figure 5.14 shows the semipermanent highs and lows and the dominant winds that occur at the earth's surface. Note, however, that the actual system is not the same as the ideal shown in Figure 5.10. The earth is not a smooth surface, and the distribution of the land masses in the world ocean alters the winds.

The major semipermanent pressure zones tend to shift north and south with the seasons. Consider, for example, the location of the ITCZ. This convergence zone is mainly south of the equator in January (Figure 5.14a), but is north of the equator in July (Figure 5.14b). The subtropical highs and subpolar lows also shift location. This seasonal migration of the pressure belts and associated winds is a direct result of the changing distribution of temperature that occurs with the seasons. Along the equator, the ITCZ moves north and south during the year. This results in the convergence of warm moist air streams from both sides of the equator, leading to frequent and abundant precipitation.

Monsoons

In addition to the north–south shift in the general circulation, there is also an east–west shift in the relative position of the subtropical highs and subpolar lows. In midlatitudes, there is a large contrast in insolation between winter and summer. Because of the relatively low specific heat of Earth materials compared with water, the continents heat and cool more rapidly with the seasons. The heating of the land masses in summer weakens the subtropical highs, and in some cases breaks them down altogether and replaces them with a low.

The change in pressure systems brings about a marked reversal in wind and moisture conditions with the season. During the summer months, convergence predominates and high humidity and precipitation are general over the continents. In the winter months, when high-pressure systems predominate, divergence over the land is more frequent than convergence. As a result of the latter phenomenon, humidity and precipitation are both lower in winter than in summer. This seasonal shift of pressure, with all its ramifications, is a **monsoon**. The largest land areas in midlatitudes are in the northern hemisphere. Here is where the monsoons are most pronounced. Large parts of Asia and Africa are subject to this seasonal shift. It need be noted that the monsoon climates are also subject to interannual variation that have extended effects on the region's inhabitants.

The Asian Monsoon

The seasonal shift in pressure, wind, and humidity is greatest over Asia. During the winter, subsiding dry air covers much of the continent. This results in a dry season that is comparatively much drier than the North American winters. The subsidence produces offshore winds over much of coastal Asia. As these winds move out over the oceans, they evaporate large amounts of water from the sea surface and are sources of moisture for the offshore islands. During the summer months, the low-pressure system and convergence are strong, and the onshore flow of moisture is large (Figure 5.15). Moisture-laden winds blow north from the Indian Ocean and the Arabian Sea. They converge over the northern plains of India and the slopes of the Himalayas. One of the rainiest regions of the world is found in the Assam hills. Cherrapunji, India, is an extreme example of seasonal rainfall. Here the annual precipitation has reached 25 m (82.5 ft) and most falls during a four month period. Resulting precipitation on the coastal areas and oceanic sides of offshore islands is heavy.

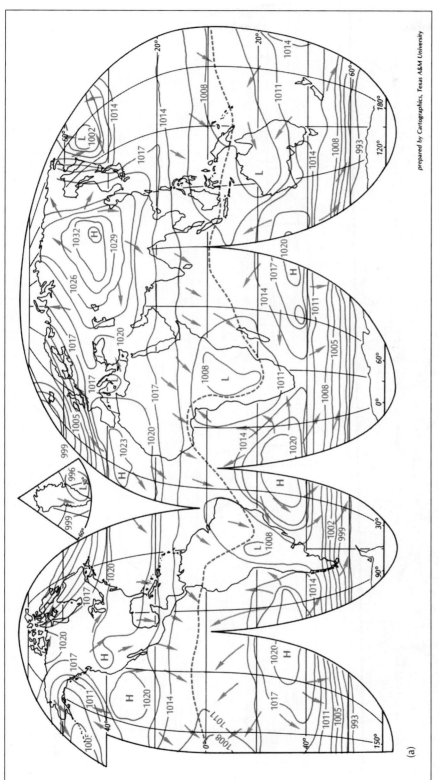

FIGURE 5.14 Mean sea level atmospheric pressure in January (a) and in July (b). Pressure is in millibars, and arrows indicate prevailing air flow.

prepared by Cartographics, Texas A&M University

(continues on next page)

(a)

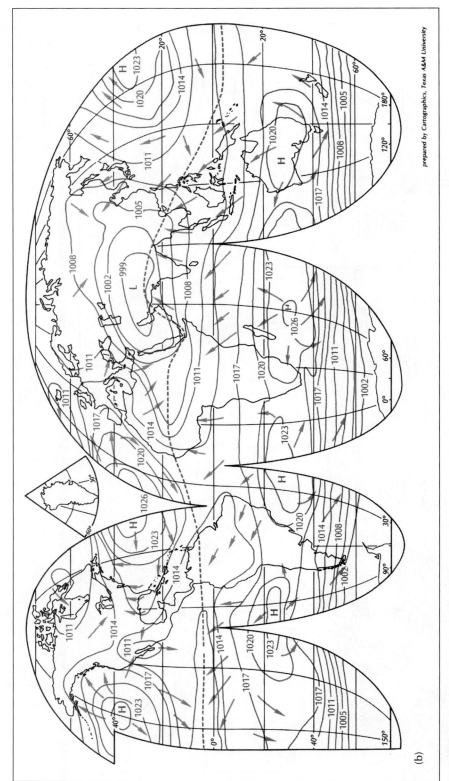

FIGURE 5.14 *(continued)*

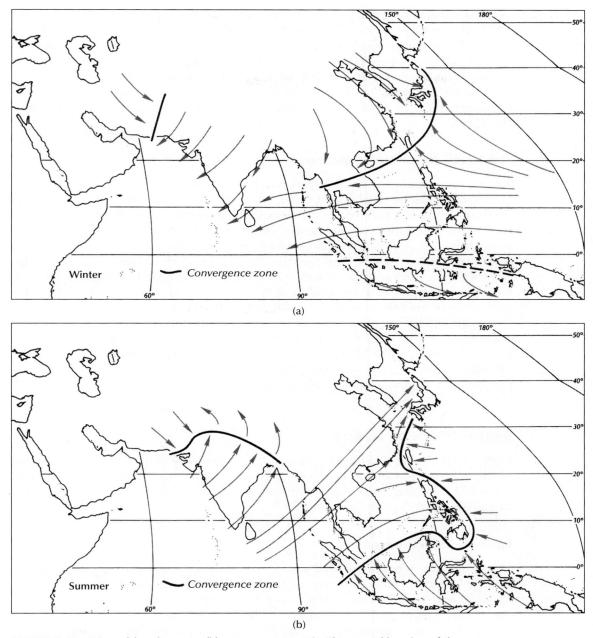

FIGURE 5.15 Winter (a) and summer (b) monsoons over Asia. The normal location of the convergence zone shifts substantially, so that most of the area has distinct rainy and dry seasons.

The monsoon is the product of many factors working together. One is that the ITCZ migrates to between 25° and 30°. A second is that the land mass interior records a large change in pressure that enforces the migration of the ITCZ. It is also likely that the Himalaya Mountains are a very significant topographic barrier, which splits the circulation over Asia. In the winter, the jet stream is over the mountains. The polar high expands and covers much of north-central Asia. Divergence from this strong system north of the Himalayas produces extremely strong offshore surface winds. South of the Himalayas and the jet stream, the anticyclonic flow is much weaker, although offshore flow is most common. In the summer as the temperature differential over the continent diminishes, the jet stream breaks down. The subtropical high and ITCZ shift rapidly northward and bring the sudden shift of winds onshore over the Indian subcontinent. North of the Himalayas, a weaker convergent system develops with attendant onshore flow of air. The precipitation is primarily convectional rainfall randomly distributed within the flowing moist air. Some traveling cyclones associated with the ITCZ add further rainfall. The peak season of rainfall moves northward over the subcontinent as the zone of convergence moves northward toward the slopes of the Himalayas.

The North American Monsoon

The monsoon applies to North America as it does to Asia. In summer, the subtropical high shifts north over the Pacific coast, and there are virtual desert conditions over much of southern California. Summer is the dry season for all of the United States west of the front ranges of the Rockies. On the eastern side of the continent, the low-pressure trough is more persistent, and convergence is frequent over this part of the continent. Onshore winds from the Atlantic high-pressure system centered over the Azores Islands carry considerable moisture. The convergence and lifting over the land areas produce quite a bit of rainfall, and there is a summer maximum of precipitation over much of the eastern part of the country. In the winter, the subtropical high over the Pacific and the subpolar low move toward the equator as the solar equator moves southward. As the subpolar zone of convergence moves southward, precipitation increases along the West Coast. To the east of the mountains, winter high-pressure systems are more frequent, and the Great Plains experience a season of low precipitation, although not absolute drought.

PERIODIC LOCAL WINDS

Daily variations in atmospheric pressure develop in many parts of the world and lead to distinctive wind patterns. These variations in pressure result from moving air caused by changing surface temperatures through the day, and they result in daily changes in wind direction and velocity.

Land and Sea Breezes

One of the daily winds associated with temperature differences of the ground surface is the land and sea breeze occurring along coastal areas and shorelines of large lakes. The land and sea breeze are a function of the change in temperature and pressure of the air over land in contrast to that over water (Figure 5.16).

In the hours after sunrise, land temperatures rise rapidly, while water temperatures rise slowly. When the overlying air above the shoreline heats, it expands and the density decreases. The air rises over the land surface and cold air flows in from the water to replace it. This flow of air landward is the sea breeze, which usually begins several hours after sunrise and peaks in the afternoon when temperatures are highest. This air will penetrate inland up to 30 km or more along an ocean coast and to a lesser extent around lakes. The cell is quite a shallow one, ranging up to several kilometers deep under the best conditions. The sea breeze is gusty and brings a decrease in temperature and an increase in humidity. The complementary land breeze is less well developed than the sea breeze. It occurs when the land surface cools at night and the water surface remains warm. The temperature and pressure relationships reverse with the higher pressure over the land. At night, however, the contrast between water and land is not as large as it is in the daylight hours. The water surface does not actually gain heat at night, but retains the heat acquired during the daylight period and thus remains relatively warm compared with the land. There is usually a lag in the development of the

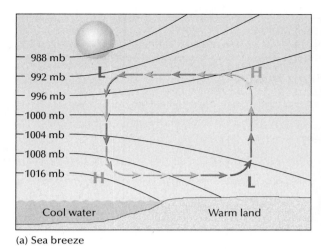

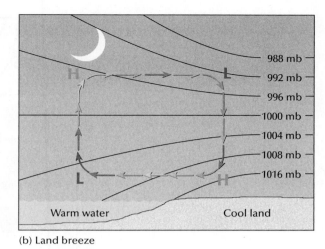

(a) Sea breeze   (b) Land breeze

FIGURE 5.16   Illustration of a sea breeze and a land breeze. (From Lutgens F. K. and Tarbuck E. J., *The Atmosphere*, 10th ed., © 2007, Prentice Hall, Upper Saddle River, N.J.)

breeze due to the general inertia of the system. The sea breeze does not set in until well after the temperature differences develop.

Richard Henry Dana, Jr., described the effects of land and sea breezes on commercial shipping in 1840 in the days of sail in the following passage from *Two Years Before the Mast*:

> The brig *Catalina* came in from San Diego, and being bound up to windward, we both got under weigh at the same time, for a trial of speed up to Santa Barbara, a distance of about eighty miles. We hove up and got under sail about eleven o'clock at night, with a light land breeze, which died away toward morning, leaving us becalmed only a few miles from our anchoring-place. The *Catalina*, being a small vessel, of less than half our size, put out sweeps and got a boat ahead, and pulled out to sea, during the night, so that she had the sea-breeze earlier and stronger than we did, and we had the mortification of seeing her standing up the coast, with a fine breeze, the sea all ruffled around her, while we were becalmed, inshore.

Figure 5.17 shows the effect of a lake breeze on the temperature at Chicago, Illinois, compared with that of Joliet, Illinois. Until about 3 P.M., temperatures were increasing in a similar manner in both cities. The lake breeze set in at Chicago at about 3 P.M., and there was a marked drop in temperature of some 3.8°C (7°F) in a little more than an hour. The temperature at Joliet, 56 km (34 mi) southwest of Chicago, does not show the decline. The lake breeze was not strong enough to reach as far inland as Joliet. The wind shift producing the drop in temperature was nearly 180°, switching from west to east. Later in the evening when the lake breeze died, there was a rather sharp jump in the temperature in Chicago.

Mountain and Valley Breezes

Similar in formation to the land and sea breezes are the mountain and valley breezes characteristic of some highland areas (Figure 5.18). In the daytime, the valley and slopes of mountains heat from the sun. The air near the surface heats, expands, and rises up the sides of the mountains. This breeze, called the valley breeze for the place of origin, is a warm wind and a daytime or late afternoon

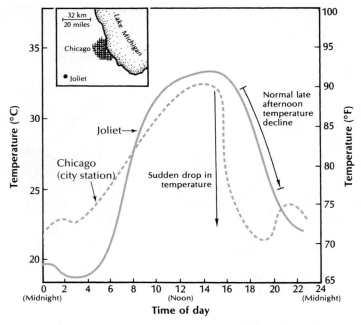

FIGURE 5.17 The effect of a lake breeze is shown by the difference in temperature over a 24-hour period between Chicago, on the lake, and Joliet, an inland location.

phenomenon. The clouds often seen forming over the hills in the afternoon are the result of condensation taking place as the air rises to cooler heights over the mountains. At night the valley walls cool. As the air at the surface cools, it flows down the slope due to greater density. This is the mountain breeze. In some mountain regions subject to frost, the valley slopes are the preferred places for orchards. The air moving up and down the valley slopes reduces the chance of the stagnant conditions conducive to frost formation.

Föhn Winds

As air descends, it is compressed and warmed. This property has given rise to a host of locally named winds that are collectively grouped using the European term **Föhn winds**. They sometimes result from synoptic patterns in concert with local topography, as in the case of the Santa Ana of California. This wind results from high pressure located over the Rocky Mountain area leading to winds moving from high elevations toward the coast. As the wind descends, it is warmed and becomes

FIGURE 5.18 The mountain and valley breeze. The mountain breeze is a nighttime breeze; the valley breeze is a daytime event.

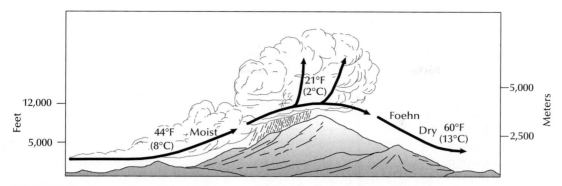

FIGURE 5.19 Cross-section of a mountain area showing the formation of a Föhn (Foehn) or Chinook wind.

increasingly dry. The wind can reach high velocities when channeled through valleys, such as the Santa Ana Valley. In the case of the hot, dry Santa Ana wind, desiccation of vegetation can increase the propensity for fires and the wind is classed as a hazard. In the case of another compressional wind, the Chinook, the warm dry wind is considered an asset.

The Chinook wind occurs on the eastern side of the Rockies when the synoptic situation causes winds to flow down the slope. As shown in Figure 5.19, the heating of the descending air can be considerable. In late winter, it can cause an early melting of snow in the high plains. This enables an earlier start to some agricultural activities.

Summary

The atmosphere is never still. It moves whenever and wherever there are pressure differences in the atmosphere. Once air moves along a pressure gradient, several processes affect direction and velocity. The Coriolis effect deflects the wind to the right or left depending on the hemisphere. Friction reduces the effects of the Coriolis force and determines how directly air will flow down the pressure gradient. Because different forces act at different levels of the atmosphere, friction layer and geostrophic winds are identified.

The general circulation is important in transporting heat energy from the equator toward the poles. This heat transfer reduces the imbalance in energy between the tropics and the polar regions. The general circulation also transports water from ocean to land. The winds follow a pattern resulting from the temperature difference over the earth, the Coriolis force, and the differential heating of land and water.

In equatorial regions, the winds are controlled by the subtropical highs and the intertropical convergence zone. In midlatitudes the airflow is between the subtropical high and subpolar low. The convergence of air into the subpolar low is of sharply contrasting temperature and moisture. This results in the development of the midlatitude lows, which produce much of the variable weather in midlatitudes. Over the poles subsidence predominates producing a flow of cold dry air toward the equator.

The seasonal imbalance in energy between the northern and southern hemispheres causes the planetary circulation system to shift north and south. As the solar equator moves north toward the Tropic of Cancer, the general circulation moves northward. Because of the extensive amount of land in the northern hemisphere, there is a land-to-sea shift in relative position of the subtropical high and subpolar low. This produces a seasonal reversal of pressure and wind direction known as the monsoon.

There are wind systems that blow regularly on a daily pattern. Examples include the land and sea breezes, the mountain and valley breezes, and winds warmed as a result of compression. Throughout the world, there are wind systems that have local importance. They influence many economic activities and have a major effect on human comfort. Some of these winds are peculiar to a single place, and others occur in similar circumstances at different places.

Key Terms

Blocking system, *96*

Conservation of momentum, *86*

Coriolis effect, *85*

Doldrums, *95*

Föhn wind, *104*

Friction, *85*

Geostrophic wind, *87*

Gradient wind, *88*

Hadley cell, *92*

Horse latitudes, *94*

High pressure cell, *83*

Intertropical convergence zone (ITCZ), *95*

Isohypse, *87*

Jet streams, *94*

Laws of motion, *83*

Low pressure cell, *83*

Meridional circulation, *92*

Monsoons, *98*

Planetary frontal zone, *95*

Polar desert, *97*

Pressure gradient, *84*

Ridge, *83*

Rossby waves, *93*

Semipermanent high pressure, *94*

Three-cell model, *92*

Trade winds, *94*

Tropical easterlies, *95*

Trough, *83*

Upper air westerly, *96*

Zonal circulation, *92*

Review Questions

1. What is pressure gradient, and how does it influence air motion?
2. What is the Coriolis effect?
3. How and why do friction layer winds differ from geostrophic winds?
4. What are Hadley cells?
5. Explain the location of the trade winds and why they were so named.
6. What are the major characteristics of the ITCZ?
7. Explain the importance of shifts in the thermal equator.
8. What causes a sea breeze?
9. Explain the cause of the Santa Ana wind.

6

Atmosphere–Ocean Interactions

The general circulation (described in the previous chapter) is not static but changes over time. The general circulation changes from day to day, week to week, and month to month. There are regular changes that take place in conjunction with the seasons. In some areas, the seasons are defined in terms of changes in wind directions and changes in precipitation. People who live in midlatitudes tend to think of the seasons in terms of temperature. No two years of weather are ever alike because of these changes. It is necessary to examine the weather record for only the past 50 years or so to find extreme short-term shifts in the weather. There are exceptionally cold or warm years and unusually wet or dry years. There are clusters of cold or warm years and wet or dry years. The problem with these changes is that much of human activity has adjusted to the usual pattern of seasonal changes. Whenever the circulation changes very far from the normal, it brings about all kinds of problems for living organisms. These anomalies result in benefits for some areas, but they entail widespread costs for others. Hazards such as floods, droughts, cold waves, and heat waves are examples of such deviations in the general circulation and in seasonal patterns.

Many processes and events cause the planetary circulation to change. Some of these events and processes are external to the planet, such as changing solar activity. Other processes that affect weather result from changes on the planet. Changes in the dust content of the atmosphere or changes in ocean currents are examples. Some events cause changes in the weather to occur rather quickly. The resulting weather changes may last for months or in some cases for several years. These changes in the weather between one year and the next do not herald a change in climate, and they are a natural part of the global climate.

OCEAN CURRENTS

The major wind systems that blow over Earth's surface exert friction on the surface of the ocean. This friction results in the movement of water in the general direction of the wind. The result of the prevailing winds of the general circulation is a series of ocean currents that follow the same pattern as the general circulation. These currents have a great influence on climate. On a global scale, there are semipermanent high-pressure systems over each of the major ocean basins. As a result of the wind systems around these high-pressure systems, there are circular patterns of ocean currents. These large ocean gyres move much slower than the wind systems that produce them, although they can move fast enough to be a hazard to small craft. The faster currents might move several kilometers per hour, whereas the average is more likely to be several kilometers per day. These are surface currents and do not go very deep into the oceans—typically no more than about 300 m (1000 ft) (Figure 6.1).

The surface currents with which most inhabitants of North America are familiar are those off our east and west coasts. These currents are part of the larger circulation of the North Atlantic and North Pacific. A detailed examination of the currents in the North Atlantic Ocean serves as an illustration of where these currents flow and their

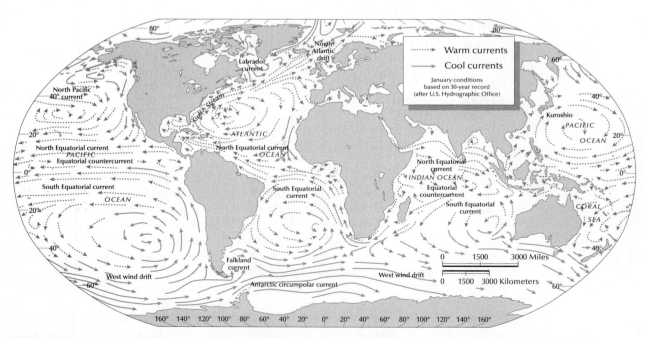

FIGURE 6.1 Primary circulation of ocean currents. Currents flowing toward the poles are relatively warm and carry large amounts of energy into midlatitude and polar regions. Currents flowing toward the equator are relatively cool. (From Christopherson R. W., *Geosystems*, 7th ed., © 2009, Prentice Hall, Upper Saddle River, N.J.)

effect on weather and climate. North of the equator in the Atlantic Ocean is a current that moves from east to west. Here the trade winds drag the surface water westward in a slow moving stream called the North Equatorial Current. The current splits on reaching South America, with most of the westward moving water flowing northwest along the coast into the Caribbean Sea and northward into the Atlantic off the North American coast (Figure 6.2).

Tropical regions of Earth receive more than the average amount of solar radiation. Polar regions are the major areas of Earth radiation to space. There is a continuous flow of heat from tropical to polar regions. This process cools equatorial regions and warms polar regions. This greatly reduces the imbalance in radiation between the two areas. Between the island of Cuba and Florida, there is a strong current of warm water flowing out of the Gulf of Mexico and into the Atlantic Ocean. This warm water from the Gulf of Mexico merges with the northward-flowing water from the Caribbean Sea near the West Indies and forms the Gulf Stream.

The Gulf Stream, along with the other poleward flowing currents, transports a tremendous amount of heat from tropical regions toward the poles. Although approximately 60% of the poleward movement of heat is accomplished by the winds of the general circulation, an estimated 40% of the heat is moved by ocean currents. The Gulf Stream is an extremely large current that transports enormous quantities of warm water northward off the east coast of the United States. This current is a major factor that influences weather along the east coast. The current provides a lot of latent heat energy and water vapor to the atmosphere that play a part in storm activity.

Hurricanes often track northward along the Gulf Stream due to the heat available to maintain the storms. Because of the size of these storms, even if they remain offshore over the current, they can have a major impact on coastal areas of the eastern United States. In the winter, the warm water of the Gulf Stream often provides enough heat for midlatitude lows to form and then move north along the coast. Some of these lows become well-developed **nor'easters**, which inflict major damage on beaches and

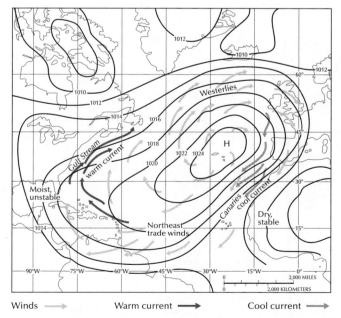

FIGURE 6.2 Model of atmospheric and oceanic circulation in the Atlantic Ocean. The surface ocean currents follow the same pattern as the atmospheric circulation. Note the desert conditions along the west coast of Africa due to the stabilizing influence of cool water offshore. (From Christopherson R. W., *Geosystems*, 7th ed., © 2009, Prentice Hall, Upper Saddle River, N.J.)

coastal property in the northeastern United States. How far northward the warm water flows and how close to shore the current flows vary with the seasons and from year to year.

The Gulf Stream is one of a number of currents that flow along the western side of ocean basins. In the North Pacific Ocean is the Kuroshio Current. Its origin is near the equator, and it flows northward toward the islands that make up Japan. The Kuroshio Current then turns eastward into the interior of the Pacific Ocean. These warm currents are not limited to the northern hemisphere. They are found in the Indian and South Atlantic Oceans as well. These currents flow away from the equator as do the Gulf Stream and Kuroshio Currents of the northern hemisphere. The Agulhas Current flows southward along the east coast of Africa. All of these currents flow fairly fast, move very large volumes of water, and they all affect regional climate and the global climate system.

In the general vicinity of New England, the westerly winds direct the Gulf Stream away from North America. It spreads out as it goes eastward and is known as the North Atlantic Drift. The current remains warm as it flows toward northern Europe, which makes winter conditions there unusually mild for those latitudes. Some of the warm water flows north between Iceland and the British Isles and along the Scandinavian Peninsula (Figure 6.3).

Part of the North Atlantic Drift splits and flows southward along the west coast of Spain and Portugal as the Canary Current, named for the Canary Islands. The water in this current is relatively cool as it flows south after having crossed the North Atlantic. It is cool only relative to the water temperature in mid-ocean. The equivalent current off western North America is the California Current, which flows southward roughly parallel to the west coast of North America.

Because the water making up this current comes from the North Pacific, it is quite cool. All along the west coast of the United States, the nearshore water is quite cool. As the current flows south from Washington to California, it might be expected that the water would warm as it receives more solar radiation. However, upwelling

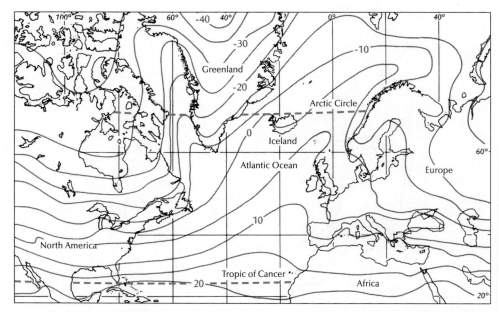

FIGURE 6.3 The Gulf Stream produces much warmer temperatures around the North Atlantic Ocean in winter than would be expected for the latitude. Note that the heat carried by the current produces warming beyond the Arctic Circle. Also note that the cool current off western Europe does not appreciably affect temperatures over the region. The isotherms are in degrees Celsius for the month of January.

cold water keeps the surface water rather cold all the way south past the Mexican border. This upwelling provides a cold surface over which the onshore winds cross before reaching land. The cold surface tends to stabilize the onshore-moving air, increasing the aridity along some coasts. This is particularly the case along southern California and Baja California. This cool current and upwelling keep summer water temperatures off the California beaches fairly cool and much cooler than the water at the same latitudes off the east coast. Temperatures in the California Current are typically below 21°C (70°F), as surfers and bathers are very aware. Another region that is exceptionally dry as a result of cool stabilizing water offshore is along the coast of Chile. Here precipitation seldom falls except during times of El Niño.

Changes within the planetary circulation systems of the atmosphere and ocean cause changes in weather patterns. Both the atmosphere and ocean are fluids, moving freely in response to internal and external influences. Neither stays in the same state for very long. They are each a dynamic system and together form a very complex, ever-shifting system. There is no normal condition. By far the most important process at work in the combined system is the heat exchange between the ocean and atmosphere. Seventy-one percent of the earth's surface is water. Hence, the ocean surface controls atmospheric circulation. The ocean absorbs far more solar radiation than does the atmosphere and has a much higher capacity to store heat. The ocean is the major source for heat in the atmosphere, so small changes in the temperature of the ocean can cause major changes in atmospheric heating. This in turn alters the general circulation of the atmosphere.

THE THERMOHALINE CIRCULATION

While the ocean currents previously mentioned are driven by the wind and tides, the **thermohaline circulation** is that part of the global ocean circulation that is caused by differences in the temperature (thermo) and salinity (haline) of seawater. The temperature and salinity of seawater determine its density. Cold seawater is more dense and heavier than warm water, and water with a high salinity is more dense than fresh water with the same temperature. These differences in seawater density caused by temperature and salinity are the driving force behind the thermohaline circulation. However, the wind driven currents play a major role in the process.

For example, as the Gulf Stream flows toward the north from the equatorial Atlantic Ocean, this circulation transports warm and saline water from the tropics to the north. As the water reaches higher latitudes, it cools and releases heat to the atmosphere, which warms the North Atlantic region of Europe. After releasing its heat, the current is further cooled by the cold polar wind. This wind enhances evaporation and evaporative cooling causing the water to become even colder. The water also becomes more saline as only fresh water is removed in the evaporation process. Eventually this cold, salty, dense water sinks into the deep ocean and then flows southward along the deep ocean basins at speeds that are slower than surface water currents to fill the basins of the polar seas. After reaching the Southern Ocean, this cold flow of water returns to the surface as a result of upwelling. Once the water is overturned and warmed, it merges with the established surface currents and begins its northerly trek again in this lengthy process sometimes referred to as the ocean conveyor belt.

There is some concern that the recent melting of ice sheets and glaciers along the North Atlantic could contribute large amounts of fresh water to the thermohaline circulation, causing it to become less dense and incapable of sinking. If this were to happen, the ocean conveyor belt could weaken or shut down. There is evidence that this happened as recently as 11,000 years ago. An ice dam suddenly collapsed, allowing a huge glacial lake to pour into the North Atlantic. The reduced salinity of the seawater shut down the conveyor belt for 1000 years, plunging Europe into a minor ice age. While a sudden shutdown in the near future is unlikely, there are signs that the thermohaline circulation is beginning to weaken.

THE WALKER CIRCULATION

In the Pacific Ocean, the Hadley cells operate similarly to the general circulation of the Atlantic. The trade winds blow from the subtropical highs toward the equator. They blow from the northeast in the northern hemisphere and from the southeast in the southern hemisphere. This results in a westward drift along the equator. In the upper troposphere, there is a counterflow of air from west to east.

Over the Pacific Ocean, there is a periodic zonal, or east–west, shift in the heart of the ITCZ. This seasonal shift is due to a shift in the area of ocean surface which has a temperature of more than 27.5°C (81°F). There is also an upper limit to the temperature of the sea, which is a little above 30°C (86°F). It does not get much above this upper limit because evaporation cools the surface. That is, as the water gets warmer, evaporation increases. In the southern hemisphere, the southeast trade winds blowing westward across the Pacific Ocean move the surface water westward. This westward movement of surface water results in sea level being 1 m (3.28 ft) or more higher on the western side of the Pacific Ocean than on the eastern side. The water also warms as it moves westward.

Along the coast of South America, cold nutrient-rich water from below the surface upwells and replaces the westward drift of warm water at the surface. The temperature of the cooler water from below might be as low as 20°C (68°F). As this cooler water moves westward in its long trek across the wide Pacific, it is warmed by absorbing solar radiation. By the time it reaches Micronesia, it has warmed to 24°C (75°F) or more.

On the opposite side of the basin, the layer of warm water at the top of the eastern Pacific Ocean is fairly thin, usually less than 100 m (660 ft) deep. At the bottom of this layer of warm water is a fairly sharp boundary called the *thermocline*. Below the thermocline, the water is a lot colder. The westward drift of warm water alters the depth of the thermocline south of the equator. In the western Pacific, the thermocline drops to a depth of 200 m. At the eastern edge of the Pacific Ocean, the thermocline rises almost to the surface. To counterbalance the westward drift of water at the surface, there is an eastward flow of water along the thermocline. This is the Cromwell Current, or the Equatorial Undercurrent. This is a strong current that in mid-ocean reaches a velocity of 1.1 m/s. Because the winds along the west coast of South America are weak and the cold water offshore stabilizes the lower atmosphere, there is fairly low precipitation along the coast. On the western side of the Pacific Ocean, there is much more precipitation since there is onshore flow of air and the air is more moist having traveled over the ocean.

In 1904, Sir Gilbert Walker became director general of observatories in India. He studied the monsoon system of Asia, partly as a result of the severe famines of 1877 and 1899. By the 1930s, he showed that there was a cyclical, interannual variation in the atmosphere over the southwest Pacific Ocean, which he called the **Southern Oscillation**. This oscillation brings major changes in pressure, winds, and precipitation over the southwest Pacific Ocean and the Indian Ocean. By the 1960s, enough data were available for scientists to conclude that the Southern Oscillation extends across the Pacific Ocean. This circulation system is the **Walker circulation**. It includes the event called El Niño and the *anos de abundancia*.

El Niño

Off the west coast of South America, an event takes place that the local people call **El Niño**. It appears as a warm current of water that replaces the normally cold upwelling water off the coast of South America. Fishermen of Spanish descendents named this event. It implies "the Little Boy" after the Christ Child, as the event appears most often in late December. El Niño has occurred frequently since first reported in 1541 (Table 6.1).

Every few cycles, El Niño is stronger than usual. It extends farther south and is exceptionally warm. It produces rain over the coastal desert. The rains bring a period

TABLE 6.1 El Niño and La Niña Events from 1960 to 2008

El Niño Events	La Niña Events
1963–1964	1964–1965
1965–1966	1970–1972
1969–1970	1973–1974
1972–1973	1974–1976
1976–1977	1984–1985
1977–1978	1988–1989
1979–1980	1995–1996
1982–1983	1998–2001
1986–1988	2007–2008
1991–1992	
1994–1995	
1997–1998	
2002–2003	
2004–2005	
2006–2007	

The events are defined as those where temperatures in the eastern Pacific Ocean were warmer or colder by +/−0.5°C.

of profound plant growth. These years are *anos de abundancia* (years of abundance). The desert provides abundant grazing for herds of sheep and goats.

In its greater context, El Niño is the collapse of the Walker circulation (Figure 6.4). A temporary circulation known as the **El Niño–Southern Oscillation (ENSO)** replaces it. Interactions between the ocean surface and the atmosphere are responsible for the shifts in the Walker circulation. Changes in sea-surface temperatures produce changes in atmospheric pressure and winds. Changes in atmospheric circulation change sea-surface temperatures. Once the process starts, the ENSO is self-perpetuating.

In low latitudes, surface temperatures of the ocean average around 27°C (80°F). Marked changes in sea-surface temperatures in the Pacific Ocean accompany the Southern Oscillation. These surface temperature changes take place mostly within 7.5° of the equator. Temperature increases of more than 3°C (5.4°F) take place in the warm phase, or ENSO. There are large shifts in the location of the warm pool. During El Niño, the area of warm water expands eastward, and the area of convergence and precipitation drifts eastward (Figure 6.5). Rainfall increases over the eastern Pacific Ocean and along the west coast of South America. Pressure differences decrease, and the trade winds die down. Pressure differences between Darwin, Australia, and the island of Tahiti show the Southern Oscillation. During El Niño, pressure increases at Darwin, and rainfall decreases. Pressure decreases at Tahiti, and rainfall increases eastward toward the international dateline. Sea-surface temperatures also increase toward the dateline. In the western Pacific, rainfall decreases and pressure increases. The ENSO weakens when the supply of warm water moves away from the equator, and the oscillation lasts anywhere from 14 to 22 months.

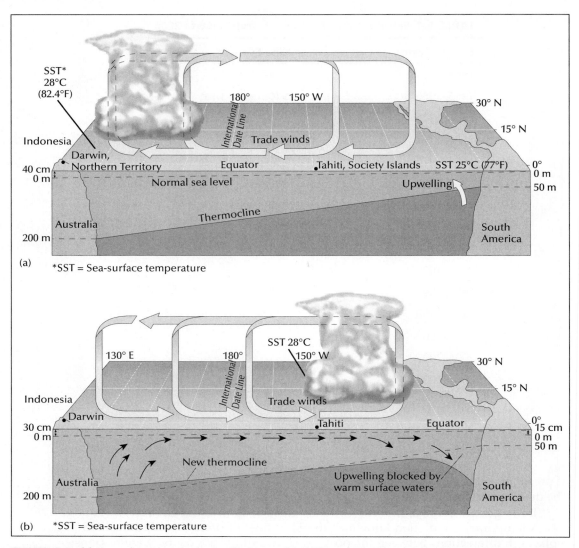

FIGURE 6.4 (a) Normal patterns in the Pacific Ocean. (b) El Niño wind and weather patterns across the Pacific Ocean. (From Christopherson R. W., *Geosystems*, 7th ed., © 2009, Prentice Hall, Upper Saddle River, N.J.)

The El Niño Episode of 1982–1983

Atmospheric pressures began to rise over the Indian Ocean in the summer of 1981. Sometimes the first sign of El Niño is in the early spring when there is a slight warming of the sea surface along the northwest coast of South America. In the spring and early summer of 1982, water temperatures stayed near normal. Across the ocean, the usually heavy rains of summer did not develop in Australia and Micronesia. Shortly after the summer solstice, the wind system reversed itself over the western Pacific. Still, water temperatures off South America remained normal. In early September, the United Nations sponsored a conference on forecasting El Niño. The conference did not forecast the coming El Niño.

With more uniform sea-surface temperatures across the Pacific, the trade winds weaken or even reverse. When the trade winds reverse and blow toward the east, a surge of warm water, known as the Kelvin wave, is produced. There is normally a warm current moving southward along the coast of South America. El Niño develops when this current goes much farther south than usual, reaching Ecuador and Peru.

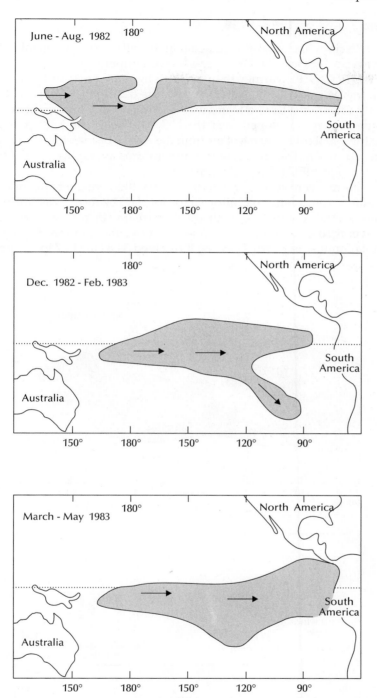

FIGURE 6.5   Shift in the location of the surface current and warm water pool during the El Niño of 1982–1983.

On September 25, 1982, sea-surface temperatures near the village of Paita, Peru, rose 4°C (7.2°F) in one day. By the first of December, the rise had reached 6.5°C (11.7°F). In some parts of the Pacific Ocean, surface temperatures rose as high as 31°C (88°F) and were as much as 8°C (14°F) above normal. Eventually, there was a swath of unusually warm water 13,000 km (8000 mi) long, straddling the equator. By the end of December 1982, the thermocline off the coast of South America pushed downward 150 m (495 ft). One result of this was that the eastward flowing Cromwell Current almost came to a halt.

The El Niño Event of 1997–1998

The southern Pacific circulation shifted again in the fall of 1997 and heralded the arrival of another event. By October 1997, sea-surface temperatures had risen to become approximately 5°C (9°F) warmer than average in some areas of the eastern Pacific Ocean. This was as warm as the water temperature during the strong El Niño event of 1982–1983. Weather in the United States was far from normal during the winter. A large high-pressure system developed over the U.S. Midwest. The high-pressure system blocked out the usual cold streams of air from the Arctic. Bismarck, North Dakota, experienced a record warm December, with temperatures averaging −2°C (28.5°F), well above the normal of −9°C (15.3°F) (Figure 6.6).

Winter storms were severe. In February 1998, the winter jet stream was crossing the United States from southern California to Florida. The result was storm after storm crossing the southern states beneath the jet stream. In the winter rainy season, California averaged double the normal amount of precipitation. Los Angeles set a record for total rainfall in February when it received 34.8 cm (13.7 in.) of precipitation.

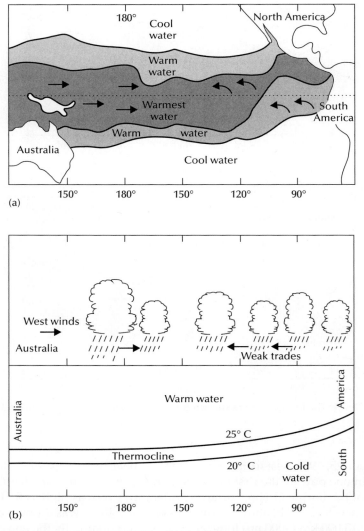

FIGURE 6.6 (a) Circulation during an El Niño. (b) The thermocline moves downward in the eastern Pacific Ocean. Cloud cover and convectional rainfall spread east across the Pacific. The warm pool of water shifts location. The surface winds slow or reverse direction, changing the surface temperature structure of the equatorial Pacific.

In California, floods and mudslides caused a number of fatalities and did extensive property damage. Cities all across the south experienced above-normal rainfall. Baton Rouge, Louisiana, doubled its long-term average. Tampa, Florida, received 12.9 in. more than normal. Florida also experienced a rash of tornadoes.

El Niño and Northern Hemisphere Weather

The Southern Oscillation directly affects atmospheric temperatures. In the immediate area of the Southern Oscillation, annual temperatures change by as much as 0.5°C (0.9°F). The ENSO affects atmospheric temperatures over the land masses within 20° of the equator. Temperature anomalies reach a maximum late in the calendar year and extend into the spring. For this reason, calendar year data tend to obliterate the effects of El Niño. The effect of the Southern Oscillation shows up much better in winter (October–March) data. Using winter data, when the Southern Oscillation is strongest, interannual variations of 0.3°C (0.5°F) appear over land.

Some studies show that the Southern Oscillation is connected to temperatures in the northern hemisphere. The Southern Oscillation affects temperatures in the northern hemisphere during the winter by as much as 0.2°C (0.36°F). The affected land areas are northwestern North America and southeastern United States. Large areas of the North Pacific Ocean develop an increase in temperature. There is a large area of warmer weather over northwestern North America and a small area of cooler than normal temperature around the Gulf of Mexico.

Counterpoint: La Niña

In the summer of 1987, the temperature in the Pacific Ocean along the equator dropped 4°C (7.2°F). This brought unusually cool weather to the eastern Pacific. This sudden cooling of the equatorial water is called **La Niña**, Spanish for "the girl." The name was applied to the phenomenon for the first time in 1986. La Niña exaggerates the normal conditions. During La Niña, the trade winds are stronger, the water off western South America is colder, and water in the western Pacific near the equator is warmer than normal. In the western Pacific, surface pressures are lower, and heavy rainfall occurs (Figure 6.7).

The strengthening of the Walker circulation causes the coastal deserts of Peru and Chile to become even drier—although that might not seem possible. On the western edge of the circulation, Southeast Asia gets even more summer precipitation than usual, causing massive floods in places such as Bangladesh.

THE IMPACTS OF EL NIÑO

The ENSO event of 1982–1983 was of unusual severity; many consider it to be the most destructive El Niño event in the past 100 years, and possibly the most extreme episode yet documented. Ecuador and Peru sustained the highest losses of life and economic damage.

The El Niño does bring *anos de abundancia* to the Peruvian portion of the Atacama Desert. In Peru, the benefits to some areas are more than offset by high costs in other parts of the nation. The costs of El Niño far exceed the benefits. The changes in the general circulation cause widespread social, economic, and environmental disruption. The event of 1982–1983 claimed some 2000 lives. The same heavy rains that cause the coastal desert to bloom created massive landslides and floods in the Andes Mountains. In May 1983, Guayaquil, Peru, received almost 20 times the average rainfall for that month. The flow in some rivers increased to 1000 times the mean flow. The winds caused $400 million of damage in Ecuador alone.

In the western Pacific Ocean, the effects of the shift in the ocean–atmosphere system was equally destructive. Six typhoons hit the island of Tahiti during the

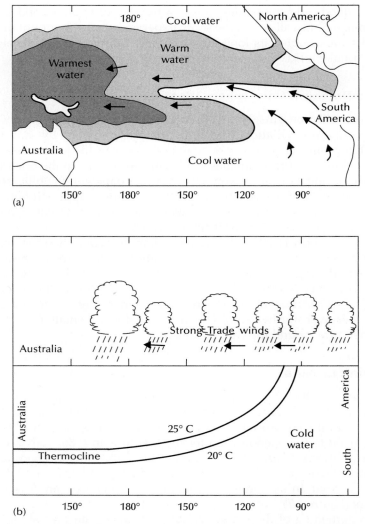

FIGURE 6.7 (a) Ocean temperatures and wind directions during La Niña. (b) The thermocline reaches the surface far west of South America. The warm pool of water is limited to the western Pacific Ocean. This restricts cloud cover and precipitation to the far western Pacific Ocean.

season—after decades without a single occurrence. Eastern Australia suffered severe drought. The southern hemisphere subtropical high-pressure system dominates the weather in Australia. In the summer, this high usually weakens to permit low-pressure systems to move onshore, bringing rain. In the 1982–1983 season, few low-pressure systems reached the continent.

There were severe weather events in many parts of the world, and many are attributed to El Niño. Table 6.2 lists some of the events. Because they occurred at the same time as El Niño is not enough reason to attribute them all to the phenomenon. For example, the unusually warm weather of the eastern United States during the winter of 1983–1984 is attributed to El Niño. However, during the El Niño in 1976–1977, the winter was one of extreme cold in the eastern United States.

Agricultural Losses

The greatest costs associated with El Niño are those related to agriculture. El Niño has frequently produced **drought** over widespread areas. In 1982–1983 and again in

---

TABLE 6.2 Worldwide Events Attributed to the El Niño Event of 1982–1983

- Above-normal temperatures in Alaska and northwestern Canada

- The warmest winter in 25 years in the eastern United States

- Drought in southeastern Africa

- Drought in Australia and New Zealand

- Drought in Southeast Asia, particularly Sri Lanka, India, Indonesia, and the Philippines

- Drought in Mexico

- Floods in Louisiana and Cuba

- Excessive beach erosion in California

- A 200 mm (8 in.) rise in sea level off California

- Death of coral reefs in the Pacific Ocean

- Slowing of the rate of Earth's rotation in late January

- Reduction in the salmon harvest off western North America

- Encephalitis outbreaks in the eastern United States

- Increased incidence of bubonic plague in southwestern United States

- High mortality of seabirds on the Farallon Islands

1997–1998, drought occurred in Australia, Indonesia, Mexico, the Philippines, and South Africa, causing damage estimates in the billions of dollars.

Disease Outbreaks

Both floods and drought that occur tend to result in the proliferation and spread of disease organisms. Stagnant pools of water form under both sets of conditions. These stagnant pools of water are an excellent habitat for mosquitoes that carry malaria, dengue fever, yellow fever, and encephalitis. Malaria outbreaks occurred in Columbia, Peru, India, and Sri Lanka during the 1982–1983 event. El Niño might have produced a local outbreak of hantavirus in the American Southwest. The heavy rainfall led to a burst of plant growth in the arid region. This in turn led to a rapid increase in deer mice that transmit hantavirus to humans. In areas of excessive rain, waterborne diseases such as dysentery and cholera erupt.

Wildlife Losses

While El Niño brings abundance to the desert, it creates havoc in the marine industries. Off the coast of Peru and Ecuador, there is a major anchovy fishing ground. The anchovy flourish in the cold and upwelling water, which is rich in nutrients. The nutrients provide food for the plankton, which in turn provide food for the anchovies. During El Niño, the upwelling continues but is restricted in the warm upper layer some 125 to 150 m deep (410-490 ft). There are fewer nutrients in the warm water, the supply of plankton drops, and the anchovies either leave the area or die in the alien environment.

Warm water and storms are extremely difficult on coral reefs. The 1982–1983 event did extremely severe damage to reefs off Columbia, Costa Rica, Panama, and the Galapagos Islands. Reefs off the Galapagos Islands suffered losses from 50% to 97%. In the 1997–1998 event, coral reefs around the Caribbean Sea were badly damaged from Mexico south through Belize, Panama, and Costa Rica.

FORECASTING EL NIÑO

Like many other cyclical phenomena, El Niño occurs every few years. The frequency appears to be on the increase. Since El Niño events have been well documented, they have averaged one about every 7 years. However, in the past few decades, they have occurred more frequently. There is one every 4 to 5 years, but they have varied from every other year to 10 years apart. In September 1982, a committee sponsored by the United Nations and meeting in Peru to discuss El Niño could not forecast the event that began later in the month. Since the event of 1982–1983, considerable progress has been made, especially in terms of early detection.

There are several problems in forecasting changes in the Southern Oscillation. First, it is not a periodic phenomenon. It does not have a fixed interval or a fixed amplitude. Each event begins and develops differently. Some events first appear off South America, with the appearance of warm water, and then the system develops westward. Other events appear first in the western part of the tropical Pacific Ocean. The 1982–1983 event began this way, with changes in the winds. This led to the eastward expansion of the pool of warm water and precipitation, which resulted in still further breakdown in the trade winds. Then warm water progressed still farther eastward, and the system went on to completion.

Part of the reason that forecasting the event is difficult is because it is not yet known what causes the system to form. There are currently several hypotheses. One suggests that it is the result of huge amounts of heat released on the seafloor as a result of magma pouring out onto the seafloor. Another suggestion is that ENSO is a result of high snowfall over Asia the previous winter. This hypothesis suggests that when there is a lot of snow on the Eurasian land mass in a given winter, there will be much more snowmelt, and a higher volume of meltwater reduces the normal summer heating of the land mass. Unfortunately, it will probably take at least several more ENSO events before any hypothesis can be adequately tested.

THE PACIFIC DECADAL OSCILLATION

The **Pacific Decadal Oscillation (PDO)** results in a climatic pattern similar to that of the ENSO. However, while a typical ENSO might last a year or more, the PDO can persist for decades, and unlike an ENSO event, the PDO tends to take place primarily in the northern Pacific. There appears to be a strong correlation between sea-surface temperatures and the PDO in the northern Pacific. The California Current has a warm phase and a cool phase. Records indicate that there was a major cool phase from 1947 to 1976, which took place between two warm phases from 1925 to 1946 and from 1977 to 1999 (Figure 6.8). During the cool phase, surface water temperatures are cooler than

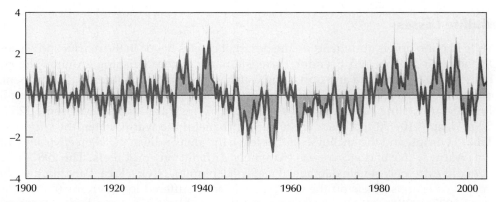

FIGURE 6.8   The Pacific Decadal Oscillation. (From Aguado E. & Burt J. E., *Understanding Weather & Climate*, 4th ed., © 2007, Prentice Hall, Upper Saddle River, N.J.).

normal along the coast of western North America. Across the ocean stretching from Japan well into the middle of the northern Pacific, sea-surface temperatures are warmer than usual. Because of these cooler temperatures along the west coast, the winters are abnormally wet and cold in the Pacific Northwest. During the warm phase of the PDO, the situation is reversed. Surface water temperatures are warmer than usual along the North American west coast, while from Japan into the central northern Pacific, sea-surface temperatures are cooler than usual. These warmer waters off the west coast result in winters that tend to be warmer, with less precipitation than normal, across the Pacific Northwest.

Studies suggest that during the cool PDO phase, record numbers of salmon can be observed in the Pacific Northwest, while to the north in Alaska salmon numbers decline. Conversely, during the warm phase, biological activity blossoms along the coast of Alaska, while further south along the west coast of North America fishery productivity decreases. Although the impacts of the warm and cool phases are well documented, the specific cause of the PDO and any links to similar events such as the ENSO are still unclear.

INTERANNUAL VARIATION IN THE MONSOONS

Much of Asia, the Middle East, and equatorial Africa are subject to the effects of an atmospheric circulation system known as the monsoon. Figure 6.9 delineates the region of Africa and Asia affected by the monsoon. The monsoon is a very pronounced seasonal shift in atmospheric pressure, which in turn brings about a marked change in wind direction, atmospheric humidity, and precipitation.

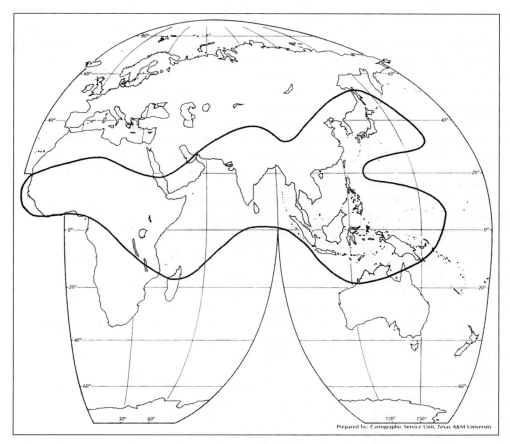

FIGURE 6.9 The monsoon region of Africa and Asia. Within this extensive area, seasonal reversals of wind direction are common and pronounced.

During the summer months, convergence predominates over Asia, and high humidity along with precipitation are the general conditions over the continent. In summer, the low-pressure system and convergence of air are strong. The summer monsoon winds blow across the subcontinent of India from southwest to northeast from April to October. The winds reaching the west coast of India have traveled over the ocean and have a very high moisture content. As the winds move onshore, they produce a lot of precipitation. At Bombay, the rains last from June until the end of September and total over 2 m (6.6 ft). The winds pass over the Western Ghats across the plains of Bengal toward the Himalayan Mountains. As a result of the direction of the winds and the topography of the Indian subcontinent, the annual rainfall varies tremendously. The Western Ghats divide India into two distinct climatic zones, as they run nearly perpendicular to the direction of the summer winds. There is a narrow, very wet climatic zone on the western side of the Ghats and a much wider and drier zone on the eastern side of the mountains.

During the winter months, when high pressure dominates the Asian land mass, divergence of surface air is more frequent than convergence. Divergence from this strong system north of the Himalayas produces extremely strong offshore surface winds. South of the Himalayas and south of the jet stream, the anticyclonic flow is much weaker, although offshore flow dominates. When divergence is present, subsiding dry air covers much of the continent, resulting in lower humidity and less precipitation. The subsidence produces offshore winds along much of coastal Asia. As the air streams out over the ocean, it evaporates large amounts of water from the sea surface and thus becomes a source of moisture for the offshore islands and along some coastal areas. There is a secondary rainy season in parts of India with the offshore flow of air, and there is a second harvest season in March and April.

The largest share of the rainfall in northern India occurs during the summer onshore monsoon. The Asian monsoon is quite variable from year to year regarding the time of year when the onshore and offshore flow begins and ends and the duration and intensity of the rainy season. The total amount of rainfall varies as a result. The agricultural economy of northern India and Pakistan are cruelly subject to the whims of this variable system. It is this summer monsoon that is so important for most of India. The primary growing season for rain-fed crops is during the summer rainy season, with the main harvest in December.

The demand and supply of food have been in a delicate balance for the human species throughout history. When the food supply has increased, there has been a gain in population; when food has been scarce, there has been some sort of trauma inflicted on the populace. Starvation results from too little food intake. During the long period of the hunting and gathering societies, starvation was probably near at hand for individuals, family groups, and tribes. The development of agriculture allowed the world population to expand rapidly. The basis for the supply of food—namely, agriculture—became more directly dependent on the weather. **Famine** is a phenomenon that affects whole populations of people over a broad area. It did not become a part of human experience until after agriculture began. However, as agriculture expanded, so did the frequency of famines. The number of times that famine has spread over the land is large. All histories of peoples and nations record famines. There are repeated references to famine in the Bible and other ancient histories.

There are many causes of famine, but one of the major ones is drought. Most catastrophic famines result from drought. Drought affects the quality and quantity of crop yields and the food supply for domestic animals. In severe drought, there may be a major loss of domestic animals. The loss of the milk products or meat furthers the effect of the drought. Major famines have occurred throughout the Asian continent from the time agriculture spread over the continent. India, China, Russia, and the countries of the Middle East have all suffered many times from drought-related famine. A famine occurred during the life of Abraham (about 2247 B.C.). Another massive famine occurred in Egypt before the exodus of the Israelites.

Drought and famine are common in India and China. The oldest record of famine in India goes back to 400 B.C. and in China to 108 B.C. From the time of the earliest noted droughts, there have been nearly continuous episodes of drought and famine in some parts of Asia.

Summary

Weather changes from day to day, week to week, and year to year. Many factors cause weather to change over time. Changes in weather that vary significantly from what is expected often bring disasters or economic stress. There is increasing knowledge that atmospheric circulation is very dependent on changes in the ocean surface, including changes in ocean currents. There is a global pattern of surface ocean currents that is similar to that of the general circulation of the atmosphere. In each of the major ocean basins, there is a circular series of currents that approximates the circulation around the semipermanent high-pressure systems. Such a system of currents exists in the North Atlantic Ocean. Part of that circulation along the east coast of the United States is a warm current known as the Gulf Stream. This stream of warm water affects the climate of both the eastern United States and western Europe. A larger ocean current, the thermohaline circulation, includes both surface currents along with deep ocean currents. This ocean conveyer is particularly sensitive to changes in density caused by changes in salinity, temperature, or both. This planetary-scale circulation in turn affects global heat exchange.

Changes in ocean currents, such as during an El Niño event, might bring substantial changes in weather and climate over large areas of Earth's surface. Perhaps the greatest impact of weather changes from year to year is when precipitation over land areas falls below normal. In subsistence or near-subsistence economies, the result is sometimes famine. Famine has plagued the human species from time immemorial. As the human population grows, ever greater numbers of people are affected.

Key Terms

Drought, *118*

El Niño, *112*

El Niño–Southern Oscillation (ENSO), *113*

Famine, *122*

La Niña, *117*

Nor'easter, *109*

Pacific Decadal Oscillation (PDO), *120*

Southern Oscillation, *112*

Thermohaline circulation, *111*

Walker circulation, *112*

Review Questions

1. Where does El Niño traditionally appear first and have the most drastic effect on climate?
2. In what part of the world is the Walker circulation found, and to what does it refer?
3. How does La Niña differ from El Niño?
4. How does El Niño affect North America?
5. How does La Niña affect North America?
6. The monsoons of Asia are a dominant climatic element. How does variation in the monsoons affect the region?
7. What sector of the world economy suffers the greatest amount of financial damage from El Niño?
8. How does the irregular shift in the Walker circulation affect the climate of western South America?

Air Mass and Synoptic Climatology

This chapter begins by looking at one of the basic causes of regional climatic differences, the role played by air masses. An **air mass** is a large horizontally homogeneous body of air that may cover thousands of square kilometers and extend upward for thousands of meters. Its uniformity is principally one of temperature and humidity. The study of air masses is integrally related to synoptic climatology. *Synoptic climatology* may be defined as the study of climates in relation to atmospheric circulations; it emphasizes the relationships between circulation, weather types, and climatic regional differences. The term originated with military operations in the 1940s, although some of its methods have been used since the beginning of the 20th century. The approaches used in synoptic climatology depend upon scale, ranging from the global through continental and regional to local scales.

This chapter includes a brief discussion of satellites. Since the first weather satellite was launched in April 1960, images from space have been of high significance in meteorological studies. For satellite climatology to develop, enough time had to pass in order to have enough imagery for the longer term study. This has been achieved, and the availability of remotely sensed images has allowed the development of satellite climatology.

AIR MASSES

Air masses derive their properties from the surface over which they originate and are thus identified and classified by the area of the earth over which they form—the source regions. Two basic categories of air masses are recognized on the basis of temperature and two on

TABLE 7.1 Classification of Air Masses

Name of Mass	Place of Origin	Properties	Symbol
Continental polar	Subpolar continental areas	Low temperatures (increasing with equatorward movement), low humidity, remaining constant	cP
Maritime polar	Subpolar and arctic oceanic areas	Low temperatures, increasing with movement, higher humidity	mP
Continental tropical	Subtropical high-pressure land areas	High temperatures, low moisture content	cT
Maritime Tropical	Southern borders of oceanic subtropical, high-pressure areas	Moderately high temperatures, high relative and specific humidity	mT
Equatorial	Equatorial and tropical seas	High temperature and humidity	E

moisture properties. Air masses are classified as tropical (T) if the source is in low latitudes and polar (P) if the source is in high latitudes. Air masses originating over land, and therefore relatively dry, are labeled continental (c); those originating over the oceans, and hence moist, are called maritime (m). This classification identifies four individual kinds of air masses: maritime tropical (mT), maritime polar (mP), continental tropical (cT), and continental polar (cP). Two additional categories are sometimes used in reference to the extremes of continental polar air masses and maritime tropical air masses. Continental Arctic (cA) indicates exceptionally cold and dry air; equatorial (E) indicates very warm moist air. The basic classification of air masses is given in Table 7.1, and a map of source regions is shown in Figure 7.1.

The general circulation of the atmosphere requires that energy exchanges occur over the globe. As a result, air mass movement from the source region acts as a mechanism for energy transfer in the atmosphere. The basic properties of the air mass will be modified the farther it moves from the source region. To indicate the manner in which air masses are modified, a third letter is sometimes added to the identification symbols. If the air mass is warmer than the land over which it is moving, then the letter *w* is added. Thus, an air mass designated mTw would indicate air that has its origin over the subtropical ocean and that is warmer than the surface over which it is moving. It may well be a warm, moist air mass moving onto land in the winter. If an air mass is colder than the surface it passes over, the letter *k* is added. Thus, a cPk air mass might be one passing from a Canadian source region and moving over warmer areas in the United States.

One of the best examples of air mass modification occurs when cP air masses pass over the Great Lakes in winter. Although the lake water is cold, it is warm relative to the air, and evaporation supplies moisture to the air mass. Once the air leaves the lakes to pass onto the warmer land on the eastern or southern shores, it becomes unstable and produces flurries. Greater snowfall may occur over the higher ground of the northern Appalachians. This lee-of-the-lake effect causes a remarkable gradient in snowfall amounts. In moving, the mass of air is not passing into a vacuum but rather replaces an existing air mass. The leading edge of an air mass will thus be a site of conflict between air of different properties. This is where **fronts** form.

FRONTS AND MIDDLE-LATITUDE CYCLONES

During World War I (1914–1918), remarkable advances were made in weather research in Norway. Cut off by the war from weather information of other countries, the Scandinavian countries established extensive station networks. This network of data enabled the meteorologists at Bergen in Norway to study the storm systems frequently encountered in northwestern Europe. From their observations came models detailing the structure of the **middle-latitude cyclone**.

The vocabulary selected by the scientists to describe their model reflects the war background of the time. The Norwegians developed the Polar Frontal Model, with the term *front* used to identify a zone of transition between air of different properties; the

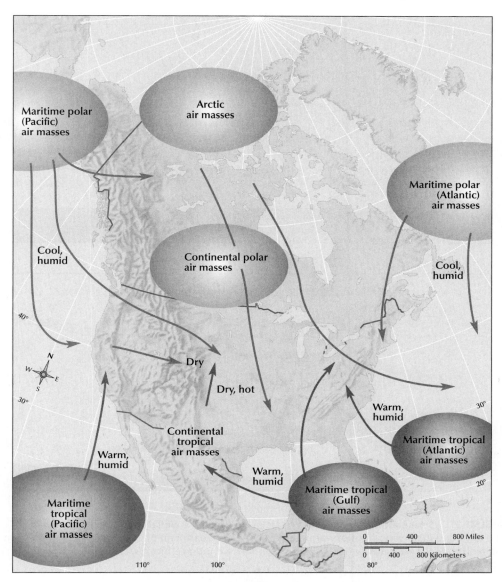

FIGURE 7.1 Air mass source regions for North America. (From Aguado E. & Burt J. E., *Understanding Weather & Climate*, 4th ed., © 2007, Prentice Hall, Upper Saddle River, N.J.)

analogy was to a front dividing the fighting armies. Various types of fronts were identified, as illustrated in Figure 7.2.

A warm front occurs when a warm air mass replaces colder air. Clearly, because warm air is less dense than cold, the warm air rises above the cold along a front that, in cross-section, has a slope of about 1:100 (1 unit of length vertically for every 100 of horizontal distance). This means that high clouds forming along the front may occur some 500 to 700 mi. (800 km–1300 km) ahead of the location of the warm front at the surface. Generally, stable air ascends along the front, resulting in the formation of stratiform clouds. As the cross-section shows, clouds may range from cirrus (many miles ahead of the front) to nimbostratus (where the front intersects the surface). Widespread rainfall often occurs during the passage of warm fronts. The majority of warm fronts contain stable air. However, if the air in the warm air mass is highly unstable, then cumuliform clouds will occur along the front. These may be embedded in stratiform clouds, creating problems for small aircraft pilots.

A cold front occurs when a cold air mass overtakes and replaces warmer air. Cold fronts are much better defined and generally move faster than warm fronts. The

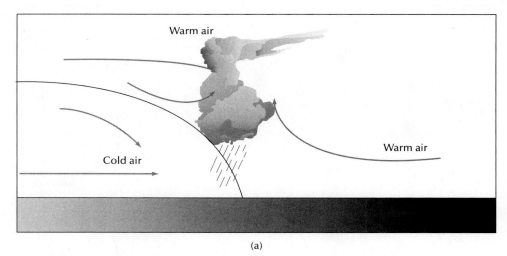

(a)

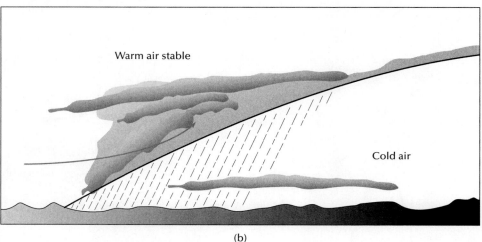

(b)

FIGURE 7.2   Schematic diagrams of (a) a fast-moving cold front with unstable warm air ahead of the front and (b) a warm front of warm stable air. (After Department of Commerce, Federal Aviation Administration, *Aviation Weather*, © 1965, Government Printing Office, Washington, DC, pp. 78 and 82.)

front is much steeper, sometimes having a slope of 1:50. The rapid ascent of warm air along the front results in cumuliform clouds with the possibility of severe weather. In the northern hemisphere, strong cold fronts are usually oriented in a northeast to southwest direction and move east or southeast.

At times, the forces are such that the frontal boundary neither advances nor retreats. This stationary front creates conditions similar to those along a warm front and may persist for a number of days, causing low-ceiling weather over wide areas.

The analysis of these fronts, together with an understanding of traveling high- and low-pressure systems, led to the identification of the middle latitude cyclone. A cyclone is a low-pressure area with an organized circulation pattern of winds. In middle latitudes, cyclones develop at air mass boundaries and are characterized at the surface by the formation of fronts. The surface history of the development of a middle-latitude cyclone is shown in Figure 7.3. Four stages are identified:

1. The initial stage, in which a V-shaped pattern occurs and initiates the formation of the central low pressure and the two fronts (Figure 7.3a–c)
2. The open stage, in which the cold and warm fronts have formed a distinctive warm sector (Figure 7.3d and e)
3. The occluded stage, in which one front overrides another (Figure 7.3f)
4. The dissolving stage, at the end of the life cycle (Figure 7.3g)

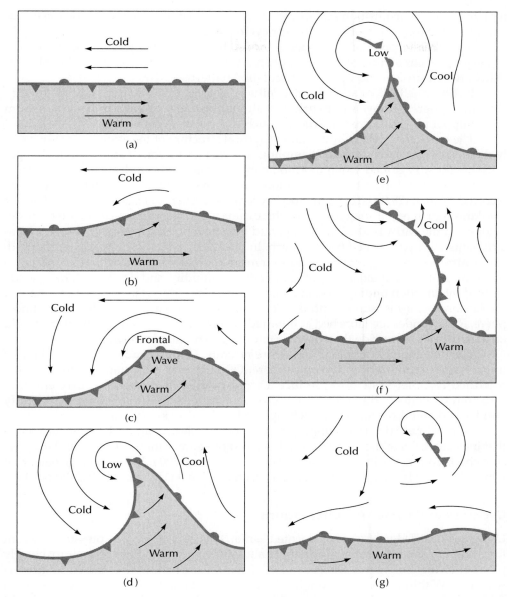

FIGURE 7.3 Schematic depiction of the life cycle of a frontal cyclone. See text for discussion. (After Department of Commerce, Federal Aviation Administration, *Aviation Weather*, © 1965, Government Printing Office, Washington, DC, p. 84.)

Note that this classical representation has been modified over the years, particularly in relation to the flow of air in the system. A recent view considers the conveyer belt model in which warm, cold, and dry air streams are conveyed into identified portions of the cyclone. Similarly, the idea of an occlusion resulting from one front overtaking the other is still accepted, but it has been found that there are alternatives to this. In some cases, elongation of the low-pressure area at the junction of the fronts can cause the occlusion. In other cases, the cold front may actually slide along the warm front to cause a similar effect.

Cyclogenesis

The identification of fronts and the concept of the structure of the middle-latitude cyclone were of major significance in the development of the atmospheric sciences. However, the explanation of why and where (e.g., a V shape interrupted the isobaric pattern along a front) simply could not be explained with only surface knowledge. It remained for the acquisition and interpretation of upper air data to provide the explanation for the formation of cyclones—or **cyclogenesis**.

Recall that aspects of the upper air circulation were dealt with in the discussion of the general circulation of the atmosphere. At this point, emphasis is on the relationship between surface and upper air circulation patterns.

In 1941, the U.S. Weather Service initiated a program to take soundings through the atmosphere. Today, a network of stations routinely obtains information from which upper air charts are constructed. Such was not always the case.

The conditions existing through the vertical structure of the atmosphere are largely derived by measurements and sampling. The first indication of atmospheric changes with height was gained when thermometers were carried first by kites and later by balloons. In 1804, the first manned balloons measured conditions; by 1862, an altitude record of 9000 m (29,000 ft) was attained by James Glaisher, an English scientist. A major breakthrough in monitoring the atmosphere came when recording instruments were developed in the early 1900s. Carried upward by a balloon that eventually burst, the instruments were parachuted back to Earth. In the 1920s, the radiosonde was introduced. This instrument package carried a radio transmitter that sent back to Earth continuous measurements, called *soundings*, of temperature, humidity, and pressure. A radiosonde tracked by direction finding ground equipment was introduced in World War II; named the *rawinsonde*, it measures wind direction and speed at various levels of the atmosphere. Radiosondes are launched at hundreds of ground stations throughout the world on a daily basis to provide the upper air information required for synoptic analysis.

The development of radar and, more recently, **Doppler radar** has been of major importance to monitoring the atmosphere. Of particular importance is the use of a modified Doppler system to produce a wind profiler. By simultaneously emitting three radar beams at different angles, changes in the backscatter beams can identify wind speed and direction at 72 levels up to a height of 10 km (6 mi).

As previously described, upper air conditions are shown on constant pressure charts that indicate the height contour at which a given pressure is found. Three standard pressure charts commonly in use are the 850 mb, 700 mb, and 500 mb. The average heights at which these levels occur are 5000 ft, 10,000 ft, and 18,000 ft (1525 m, 3050 m, and 5490 m).

Cyclones and Upper Air Circulations

Using the knowledge of winds presented earlier, the balance between pressure gradient force and Coriolis means that much upper air flow is geostrophic, moving approximately parallel to the isohypses. Observations of winds aloft show that the circulation is wavelike. Within the waves, troughs and ridges can be identified. These waves, with their ridges and troughs, migrate eastward across the United States in a complex of many interacting waves of variable lengths and amplitude.

Figure 7.4 illustrates how a wave may develop over the United States in the period of perhaps 4 days. The pattern passes from an west–east flow (**zonal flow**) to deep amplitude waves (**meridional flow**) and the eventual formation of a deepening trough, which may become cut off from the main west–east flow. The cutoff eventually disappears and a low-amplitude wave pattern is again established.

Winds in the upper air are strongest where the contours are closest together. The strongest bands within the westerly flow are jet streams, the cores of which are above regions of strong horizontal temperature gradients. The relationship between temperature and pressure aloft is shown in a cross-section through the atmosphere from warm to cooler latitudes (Figure 7.5a). Initially, an equal surface pressure exists. Yet, because of the density differences between warm and cold air, constant pressure surfaces are higher above warmer latitudes than cold. Using the geostrophic relationship, winds aloft will move from warmer to cooler areas and be deflected by Coriolis. As a result, westerly winds will occur.

Suppose that a front exists and the temperature/pressure gradient is modified (Figure 7.5b). At that junction, both gradients will be greatest, and geostrophic winds

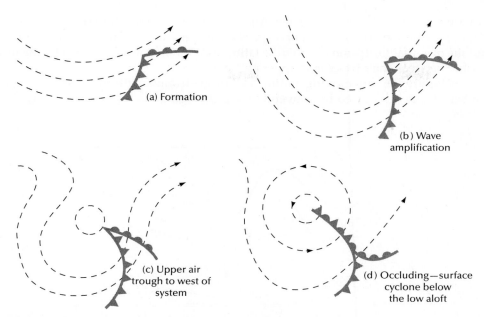

FIGURE 7.4   Development of a middle-latitude cyclone in relation to an upper air trough. System develops below a zone of divergence in the flow aloft (a). The cyclone deepens [(b) and (c)]. As a low forms in the upper air flow, the system occludes (d) and dissolves.

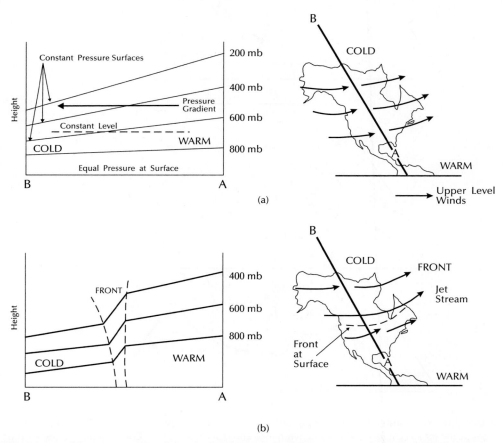

FIGURE 7.5   (a) A cross-section through the atmosphere from warm to cold latitudes shows that constant pressure surfaces are higher in warmer than cooler areas. This creates a pressure gradient toward the cooler air and, after Coriolis deflection, westerly winds. (b) If there is a strong temperature contrast (a front), then winds will blow faster to form a jet stream.

above the front will be much faster. A jet stream will be identified with a frontal boundary situation. The creation of a specific surface low-pressure area along a front (and the eventual formation of a middle latitude cyclone) is also explained by the relationship between the surface and upper air flows.

In Figure 7.6, air moving within the ridge–trough pattern must undergo a number of changes caused by converging and diverging air. As the air converges on

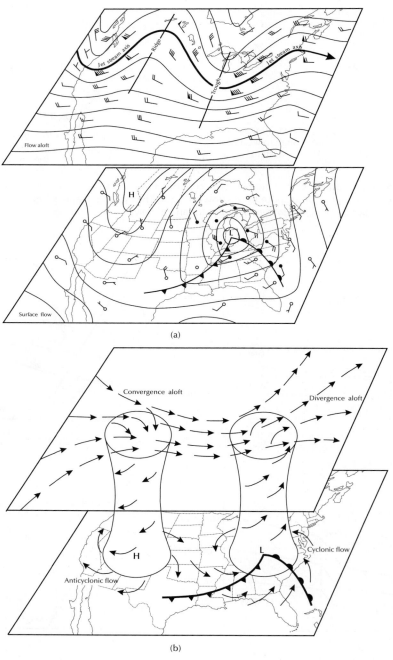

FIGURE 7.6   Relationship between upper air flows and surface pressure systems. (a) Troughs and ridges at the jet stream level produce convergence and divergence of air. Convergence, to the west of the trough, induces descending air and divergence at the surface; divergence aloft, to the east of the trough, induces rising air and convergence at the surface. (b) The resulting patterns of convergence and divergence result in high- and low-pressure systems at the surface. (From Lutgens F. K. & Tarbuck E. J., *The Atmosphere*, 10th ed., © 2007, Pearson Prentice Hall, Upper Saddle River, N.J.)

the western portion of the trough, air will slow down and create high pressure at the surface. On the eastern limb, air spreads out and speeds up to cause divergence. This **divergence** aloft is accompanied by **convergence** at the surface, with the resulting formation of low pressure. Thus, the favored position for the formation of a low-pressure area along a frontal boundary can be identified.

The relationships between upper air and surface conditions clearly rely on the location of the upper air waves and jet streams. Such relationships have been dealt with at length in many studies in synoptic climatology, and these studies have been aided by the development of weather satellites.

SYNOPTIC CLIMATIC STUDIES

Synoptic climatology is the study of local and regional climates in terms of properties and motion of the overlying atmosphere. The term was first used in the 1940s in reference to analysis of past weather situations to assess the frequency of sets of conditions. Meteorologists use the term *synoptic* to denote instantaneous weather conditions shown on a synoptic weather chart. By extension, synoptic charts are a major source for the study of synoptic climatology. To this source is added satellite imagery.

The basic procedures involved in synoptic climatic studies are the determination of circulation types and statistical assessment of conditions in relation to the identified patterns. Clearly, an important step in an analysis is the classification of atmospheric properties and processes on the basis of synoptic patterns. The basic division is in terms of scale, with subsets relating to whether the methods employed are subjective or objective. Subjective studies may use numerous daily weather map sequences to identify a synoptic pattern associated with a distinctive climatic event and hence classify the types that occur. For example, to identify the type of circulation regime that gives rise to stagnant air masses, the researcher might need to classify conditions associated with blocking systems. Objective classification takes advantage of the availability of digital data banks and the use of computers and statistical packages to use correlation analysis, factor analysis, and various clustering techniques. A classification of conditions relating air masses to pollution levels is illustrative of this approach. A good idea of the nature of synoptic studies may be derived by considering the approaches used at the global, continental, regional, and local scales.

The heart of upper air global circulation is based on the presence of planetary waves. These are very long waves that appear to sinuously wrap around the earth. At any given time, one, two, or three waves may exist, with their troughs and ridges located in preferred positions. These positions are partly determined by the unequal heating of continents and oceans and the changes that occur from season to season. Periodically, the jet stream associated with these waves may split to form meridional flow north and south of a high-pressure zone. This is a blocking high—a weather system that leads to extreme weather conditions, ranging from floods to drought, in large parts of North America. Note that a blocking high is not the same as the subtropical high, the latter occurring too far equatorward to split the flow of a jet stream.

The Hemispheric Scale

At the global or hemispheric scale, research often focuses on the patterns of circulation regimes. This may concern seasonal changes of pressure fields or analysis of circulation regimes. Thus, the role of zonal and meridional flow or the nature of blocking systems may be related to climates of large areas. Some objective schemes may calculate indices relating to typical and atypical circulations. Such analysis is often related to a specific area to become a continental study, as in Figure 7.7, which provides a classic example of a bitter-cold winter that occurred in the United States. The normal circulation

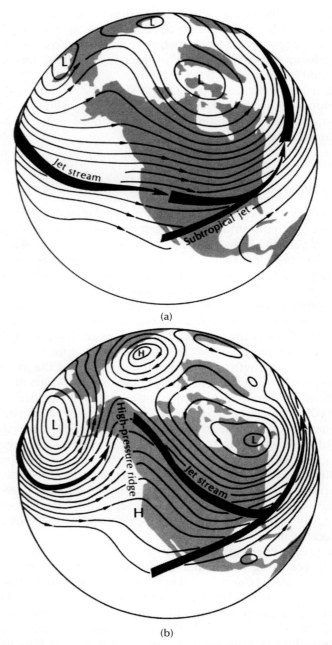

(a)

(b)

FIGURE 7.7   Schematic diagrams illustrating upper air circulations in winter: (a) the normal pattern; and (b) circulation that results in a severe winter over much of the United States.

of upper air flows associated with the Rossby regime is shown in Figure 7.7a. Compare these normal conditions with those showing an extreme winter season (Figure 7.7b). Remarkable differences are seen and can be assessed by comparing:

- The amplitude of the wave pattern
- The location of the high-pressure ridge over the Pacific Ocean
- The southerly extent of the low-pressure trough that is normally off the northern Canadian islands
- The intense high pressure in the polar area
- The deepening of the low-pressure system off the Aleutians

In all, the pattern of the frigid year is quite different from the normal; the result is that winters in many places in the United States vary appreciably from what is normally expected. A comparison of the two figures permits an estimate to be made of the ways in which various locations feel the effects of different circulation patterns. Of immediate note are:

- The high-pressure ridge that extends over the eastern Pacific Ocean causes the jet stream path to be far north of its normal position. This means that the West Coast of the United States does not fall in the path of frontal systems that normally bring rain from the Pacific. The area is dry.
- The same ridging means that the warm, moist air would extend to Alaska. Usually dominated by cold air at this time, the regions would experience exceptionally mild conditions.
- The exaggerated jet stream axis would be maintained by the movement around the low pressure over Labrador, and cold Arctic air would stream across eastern United States.

These conditions, hypothesized from the observed circulation patterns, actually occurred. For example, the temperature departure during January 1977 was as much as 10°C (18°F) below normal temperatures in the Midwest. As anticipated, large parts of the West received rainfall less than 50% of normal.

In recent years, much research has been undertaken to understand which patterns and anomalies occur, and many of them are rooted in **teleconnections**—the influence of local climate conditions by events happening in other world areas. As already noted, U.S. climates are greatly influenced by El Niño–Southern Oscillation (ENSO) events occurring in the Pacific Ocean—quite remote from the continental area. Other teleconnection correlations being studied include the Pacific Decadal Oscillation (PDO), which deals with sea-surface temperature patterns over periods of 20 to 30 years, and the North Atlantic Oscillation (NAO), which deals with cool and warm phases in the North Atlantic Ocean. This is sometimes called the Arctic Oscillation (AO). Although these and other teleconnections are being investigated, none thus far appears to have as much influence as ENSO events on the climate of North America.

Many regional synoptic catalogs have been derived. Perhaps the best known are the Grosswetterlage (large-scale weather pattern), which involves the classified synoptic patterns over a large region. A daily catalog of Grosswetter is published in Germany to show conditions in central Europe. Some regional-scale studies provide air mass and frontal classifications and cyclone tracks that influence an identified region and relate them to upper air flows. The formation and movement of middle latitude cyclones act as mechanisms for the transfer of heat and moisture over the globe. The movement of the storms in the United States, especially those that influence the central and eastern United States, tend to follow four main tracks. Figure 7.8 shows the cyclogenesis area by name and common tracks of the storms.

Both the Alberta and the Colorado lows form in the lee of the Rockies. The Alberta system brings light snow and rain to the northern tier states—a contrast to the Colorado low. This storm system is often regenerated in southern Oklahoma when moisture from the Gulf is drawn into the circulation. In winter, these storms can produce blizzards, whereas they often produce severe thunderstorm activity in the Midwest. The Gulf low originates in the Gulf of Mexico and moves toward the northeast to bring copious rain to the East Coast from Virginia to New England. The Cape Hatteras low greatly influences weather in the Mid-Atlantic states, bringing abundant rain and snow; New Englanders refer to these events as *nor'easters*.

Local-scale synoptic studies may define weather types according to some locally observed weather elements and group them accordingly. Frequently, locally observed weather events are intercorrelated and statistically condensed. This enables large

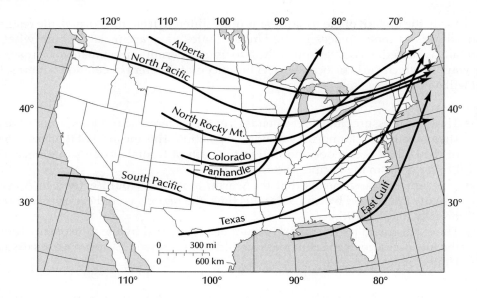

FIGURE 7.8   The main storm tracks across the United States. (From Aguado E. & Burt J. E., *Understanding Weather and Climate,* 4th ed., © 2007, Pearson Prentice Hall, Upper Saddle River, N.J. Fig. 10.16, 'p 299'.)

amounts of data to be related to an identified type of weather or to the occurrence of weather-related impacts on people and the environment. Local snowstorms, dust storms, and high pollution events are typical topics of study.

SATELLITE CLIMATOLOGY

World War II spurred the use of rocketry; by 1947, a modified V-2 rocket took the first successful photographs of Earth's clouds from an altitude of about 100 mi (160 km). Thereafter, with more rocket use, it became evident to atmospheric scientists that viewing weather systems over wide areas from above could play a significant part in understanding the atmosphere. Since that time, large numbers of satellites have been placed in orbit. Many of these are collecting environmental data.

Since the launch of the first specialized weather satellite, Tiros 1, in 1960, images of Earth and its cloud cover have become a common feature of TV weather programs. The changes that have occurred since the launching of the early satellites concern the orbit and sophistication of the sensors aboard the satellites. Thus, the types of data obtained from weather satellites are largely governed by their orbit and sensing instruments. Two fundamental orbits are used: the high-altitude, geostationary orbit and the low-level, polar orbit. In discussing these, it is possible to provide endless streams of acronyms, and the following accounts make a conscious effort not to do so.

Orbits and Sensors

Weather satellites are often classified on the basis of the type of orbit. Two fundamental orbits are used: the high-altitude, geostationary orbit and the low-level polar orbit.

A **geosynchronous satellite** is placed in orbit such that the satellite circles Earth at the same rate as the equatorial spin of Earth so that it stays in the same point above

Earth. To maintain the same rotational period as Earth, a satellite is in a fixed position at 35,786 km (22,236 mi) above Earth's equator. An orbit permits a full disk image of the globe to be derived every half hour or so. Because of its high altitude, the satellite scan views a huge area (the viewing area of a satellite is called a *footprint*). An advantage of the geosynchronous orbit is that it requires no tracking; that is, just like your home TV satellite signal, it is not necessary to constantly adjust the receiver to the satellite position.

The first series of geosynchronous weather satellites, the ATS series, were in equatorial orbit over the equator south of Hawaii. A new series, the GOES satellite systems, first launched in 1975, were the first operational geostationary satellites to provide high-resolution picture transmission (HRPT). The United States currently employs two geosynchronous weather satellites. GOES-11 is located over the eastern Pacific Ocean, and GOES-12 is located over the Amazon Basin. A number of other countries also maintain geosynchronous weather satellites.

Now in planning is the next generation of GOES satellites. They are currently scheduled for launch in 2015. There will be one each over the Amazon Valley and the Pacific Ocean. GOES-R includes two new satellites to replace existing 11 and 12. It will provide more frequent images and images with higher resolution. They will be able to scan the United States every five minutes. It will be able to map hurricanes every 30 seconds. Each will contain a lightning mapper, which will be able to record 70 to 90% of lightning strikes including cloud to cloud bolts. This will greatly aid in estimating the severity of storms and in tornado forecasting. A full Earth image is possible every 5 minutes compared to every 30 minutes at present.

Polar orbiting satellites circle the earth from north to south (or south to north) every couple of hours. These satellites are placed in orbit at much lower altitudes than the geosynchronous weather satellites. They orbit between 800 and 1500 km (496.8–931.5 mi) above the surface. The TIROS (for TV Infrared Observing Satellite) series was the first of the meteorological polar orbiting satellites. The National Oceanic and Atmospheric Administration (NOAA) satellites have orbits that pass very close to the poles on each rotation of the earth. The NOAA 15, for example, is in synchronous orbit at an altitude of 833 km (516.5 mi). Both visible and infrared radiation are sensed and sent to Earth during the day, whereas at night both channels are in the infrared. Great improvements over time have been introduced, and currently the United States uses the satellites designated NOAA followed by a letter of the alphabet, which changes to a number after successful launch (e.g., NOAA-K became NOAA 15). Figure 7.9 provides various satellite views from different orbiting systems.

Satellites carry a wide variety of sensing instruments but essentially may be classed as imaging and sounding systems. Imaging systems are either (1) those that give a series of instantaneous images such as the older vidicon systems or (2) scanning sensors (i.e., scanning radiometers) that build up images, line by line, in tracks. Sounding systems are those that measure emissions of radiation from different levels of the atmosphere to allow the construction of atmospheric profiles. The camera used on the TIROS series is the Advanced Very High Resolution Radiometer (AVHRR), which measures radiation brightness in selected wavelength bands. Each of these bands has a special purpose.

The GOES imager is designed to sense radiant and solar reflected energy from Earth. By imaging different wavelengths, as shown in Table 7.2, a variety of phenomena can be sensed. The GOES sounder takes readings of temperature, moisture (including clouds), and ozone through a column of the atmosphere using a radiometer that monitors energy from the visible wavelengths to the long wave, 15 microns. (See Chapter 2 for a reminder of details of the electromagnetic spectrum.) The NIMBUS series of satellites has been used as the basic research and development vehicle for these sensors.

(a)

(b)

FIGURE 7.9   Example of images derived from (a) a geostationary satellite (NASA) and (b) a polar orbiting satellite (NOAA).

TABLE 7.2 Spectral Bands on Advanced Very High Resolution Radiometer (AVHRR)

Band	Wavelength (μm)	Description	Examples of Use
1	0.58–0.68	Visible (green to red)	Cultural features; cities/farms
2	0.725–1.10	Near infrared	Land, water differences; penetrates haze
3	3.55–3.93	Infrared	Temperature differences; cloud heights
4	10.30–11.30	Infrared	Thermal mapping
5	11.50–12.50	Far infrared	Thermal mapping; water vapor correction

Limitations and Opportunities of Satellite Studies

Although important work is being completed in satellite climatology, its development is not as rapid as might be expected, for a number of reasons. Climatology, by its very nature, demands long runs of homogeneous data. Weather satellite data only recently reached the 30 to 35 years required for the confident evaluation of climatic norms. Frequent changes of satellite orbital patterns and sensors result in nonhomogeneous data sets. Methods of data processing have changed greatly since 1960, and there have been few coordinated international programs. This has resulted in extremely variable archived data. Despite this, satellite-based climatic studies have produced significant results.

Consider some of the sensors that have been used:

- ERB (Earth Radiation Budget) for monitoring energy budgets
- SAMS (Stratospheric and Mesospheric Sounder) for temperature profiles
- SBUV (Solar Backscatter, Ultraviolet Energy) for ozone profile retrieval and evaluation of ozone, terrestrial, and solar irradiances
- TOMS (Total Ozone Mapping Spectrometer) for ozone monitoring

Each of these has contributed to understanding specialized components of the atmosphere and has been especially useful in climatology. Many of the findings have been used in other sections of this book, and the following are but a few examples:

- Sea-surface temperature related to ENSO
- Evaluation of the solar constant for energy budget studies
- Global temperatures for comparison with surface temperatures and global warming/cooling
- The extent of ozone and its seasonal variations
- Properties and location of sea ice and polar ice caps
- Assessment of the variable gases in the atmosphere

As can be seen, satellite observations and measurement aid significantly in the understanding of the atmosphere.

Summary

Understanding of air masses, fronts, and middle-latitude cyclones represents major advances in the development of the atmospheric sciences. Although the surface patterns of fronts and frontal systems were well known, the understanding of the formation and life cycle of the middle-latitude cyclone could not be fully explained until upper air data became available.

Studies in synoptic climatology use both surface and upper air data. The scale of analysis varies from planetary to local, whereas the fairly recent availability of large computing capacity has allowed the rapid development of objective studies that complement the older subjective methods. Explanation of many synoptic events may be related to planetary wave analysis and teleconnections. ENSO events have been widely studied and publicized.

Satellite imagery results from both polar orbiting and geosynchronous platforms. The types of sensors aboard a satellite determine the types of imagery produced, with a wide range of electromagnetic radiation bands being used. Satellite climatology is becoming increasingly important as digital archives are growing.

Key Terms

Air mass, *125*
Convergence, *133*
Cyclogenesis, *130*
Divergence, *133*

Doppler radar, *130*
Front, *126*
Geosynchronous
    satellite, *136*

Meridional flow, *130*
Middle-latitude
    cyclone, *126*
Polar orbiting satellite, *137*

Synoptic, *133*
Teleconnection, *135*
Zonal flow, *130*

Review Questions

1. How are air masses classified?
2. Explain lee-of-the-lake snowfall.
3. Compare the major features of cold and warm fronts.
4. Describe the surface evolution of a middle-latitude cyclone.
5. Describe air flow around an upper air trough and ridge.
6. How have upper air data been acquired?
7. Explain the relationship between a jet stream and a frontal system.
8. Give an example of a subjective synoptic classification.
9. What are the two fundamental orbits of satellites?
10. What type of satellite provides your home TV signal? Why is this fortunate?

Climatology of Atmospheric Storms

C limatology traditionally dealt with the more stable aspects of weather, such as mean temperatures or mean amounts of precipitation over a particular month or year. But some events occur in our environment that are sporadic in space and time. They represent major deviations from normal conditions. They are frequently violent manifestations of natural processes that often result in disaster for the human species.

These events can be atmospheric, hydrologic, geologic, or biological, or they may come from space. They are also extremely difficult to predict. The duration of short-lived events varies over a wide range. Lightning strokes last milliseconds and are among the most instantaneous of phenomena. Floods may last for a few hours or for weeks. Many of

TABLE 8.1 Atmospheric Extreme Events

Event	Mean Duration
Lightning	Milliseconds
Tornadoes	Minutes
Hailstorms	Minutes
Thunderstorms	Minutes
Blizzards	Hours
Flash floods	Hours
Heat waves	Days
Cyclones	Days
Cold waves	Days
Nor'easters	Days
Regional floods	Weeks
Drought	Months

these extreme events are atmospheric and a part of the climate of a place. Table 8.1 lists many of the atmospheric hazards and the mean time span within which each phenomenon occurs. In some cases, there is considerable variation in duration, and so the mean life span of the event represents the most frequent duration.

There are a series of disturbances in the atmosphere called *severe storms*. These include thunderstorms, hailstorms, tropical cyclones, and tornadoes. These storms represent an atmospheric response to unequal distribution of energy in the atmosphere. They are thus an integral part of dynamic energy exchange rather than simple isolated events. The energy involved in such storms is prodigious. The concentration of this energy in hurricanes and tornadoes causes the loss of life and devastation associated with severe storms. Storms are mechanisms for global energy exchange, which makes them an important climatological element as well as a meteorological event. Analysis of storms provides equally important knowledge about their role in distributing energy.

Humanity has been exposed to these extreme events from the time the first of the species appeared on the planet. In prehistoric times, the number of people affected by any single event was small because there were few people, and they were scattered over one or more continents. Because of the sparse nature of the population, these events probably affected only individuals or families. Nearly all groups of people in historic times had legends of natural disasters. Most groups have legends about the creation and the great flood. The story of the biblical flood is replicated in folklore around the world. Written stories of natural disasters began to appear in different parts of the world as soon as writing was developed. One such written story, known as the Gilgamesh Epic, existed long before the Bible. From the Middle East, it contains the story of the deluge. The story may have existed long before the Gilgamesh Epic was written and handed down by word of mouth.

To explain natural disasters, in the past they were referred to as *acts of God* or attributed to the spirits. Now we know that these events result from natural processes that operate on the planet. In today's world, these events consistently make the headlines of newspapers and news magazines and occupy prime-time TV. When they affect humans adversely, we refer to them as *hazards*.

The impact of extreme events can take many forms and can be measured in terms of human stress. Both physical and mental health can be impaired by stress associated with extreme events. Some atmospheric events take lives instantly. Lives can also be lost in the aftermath of these events, due to injury, disease, or starvation. There may be long-lasting and far-reaching effects of these events that cannot be measured by fatalities alone. The ultimate extent of the impact depends on a variety of factors, including the magnitude of the event, the number of people affected, and the cultural stage of the people affected.

Although short-lived phenomena are similar in some respects, they differ widely in others. The size, frequency, and spatial and temporal characteristics of different phenomena vary. Some short-lived phenomena have a similar size with each occurrence. Tropical cyclones, although varying in size and intensity, have wind velocities and central pressures that are within a relatively narrow range. Other events vary widely in size. Earthquakes are an example of this class. The range of size that humans can feel varies immensely. There is the same kind of variation in flood magnitude. The longer the time period, the greater the chance of having a more extreme case of a short-lived phenomenon.

The frequency is a measure of how often an event of a given size might occur. The concept of frequency is far more important in the analysis of some events than others. Data show that the frequency of short-lived phenomena such as floods and earthquakes is increasing. It is virtually impossible to determine whether this is in fact the case because records of these events exist for only a limited time. Records of floods on the Nile River go back many centuries, but this historical data collection is an exception. Few nations began to compile data on extreme events until recent years. Only for giants of extreme events have records existed for more than a few decades.

Geographic distribution refers to the pattern of a phenomenon over the area in which it occurs. The areal extent ranges widely from a lightning bolt to a drought. Some phenomena have very well-defined limits, such as floods and avalanches. Others, such as hurricanes, are less well defined on the margins. The temporal pattern of short-lived phenomena also varies widely. Although most have a random element, there may be a daily, seasonal, or annual fluctuation in the probability of the event. There may also be a pattern or repetition that is irregular in occurrence, which makes forecasting even more difficult.

On a global basis, droughts, tropical cyclones, regional floods, and earthquakes are the four greatest hazards, measured in terms of casualties and economic cost. In the United States, the four top hazards are the same as on a global basis except that tornadoes replace droughts. These hazards rise above many others when considered over a period of many years. The major hazards in any given year or even for a few years might look different.

TROPICAL CYCLONES

The tropical cyclone is the typhoon of the Pacific Ocean and hurricane of the Atlantic. It is a **vortex**, or swirling mass of severe storms, that rotates counterclockwise in the northern hemisphere and clockwise in the southern hemisphere. The tropical cyclone gets its energy from the latent heat of condensation. The energy in an average hurricane may be equivalent to that of more than 10,000 atomic bombs the size of the one that destroyed Nagasaki. These storms range in size from a few kilometers to several hundred kilometers in diameter. In the middle is an eye that can be as large as 65 km (40 mi) across. The total area involved may be as much as 52,000 km^2.

The atmospheric pressure is nearly symmetrical about the center of a hurricane (Figure 8.1). The pressure may go as low as 870 mb, but this is rare. Such a drop in pressure represents a change of about 13% from normal. While such a drop can be relatively rapid, it is not necessarily an explosive drop because it might take place over a distance

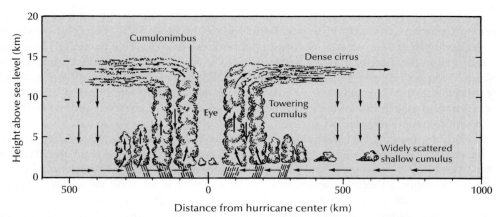

FIGURE 8.1 Vertical structure of a hurricane, showing the cloud pattern and wind flow.

of many kilometers and over a period of several days. The wind system associated with hurricanes is one of contrasts. In the eye of the storm, the winds are light and variable and of velocities not usually exceeding 25 kph (15 mph). The wind velocities increase rapidly away from the eye, reaching their highest velocities just outside the eye in the eye wall and at a height of about 0.8 km (0.5 mi).

For a storm to be classified as a hurricane, sustained winds must exceed 125 kph (74 mph). The maximum winds in most well-developed hurricanes reach 200 kph (125 mph) and in extreme cases may reach 300 kph (190 mph). Hurricane-velocity winds may extend over an area of 300 to 500 km (190–300 mi) in diameter, and gale-force winds of 65 kph (40 mph) may extend over an area of 600 to 800 km (375–500 mi) in diameter. Anemometers reinforced against hurricanes have measured such winds at velocities greater than 250 kph (150 mph). Although the winds within the storm area are of very high velocity, the storm moves at a relatively slow speed, averaging only 15 to 30 kph (10–20 mph). In Hurricane Wilma of 2005, the central pressure dropped to 882 mb. This is the lowest pressure ever recorded in a western hemisphere hurricane. It became a Category 5 hurricane and produced 281 kph (175 mph) winds before making landfall in southwest Florida as a Category 3 hurricane, with wind speeds of 203 kph (125 mph).

Some of the heaviest rains on record in low latitudes came from these tropical storms. Records of 500 mm (~19 in.) of rain over a 48-hour period are relatively common. One typhoon in the Philippines produced over 1600 mm (62 in.) of rain. Radar observations show that 500 to 800 km (300–500 mi) ahead of the storm, there are often fairly well-defined lines of thunderstorms. These storms can generate waves as high as 13 m (42 ft). Long sea swells as far as 1600 km (1000 mi) from the center are evidence of a storm's strength. Near the wall around the hurricane eye, wind often blows the tops off the waves, but in the eye, the waves are often very high.

Formation of Tropical Cyclones

There is no agreement yet as to why tropical cyclones form. The weather conditions required to produce them are known, and, once they have formed, the storms can be tracked. The precise set of factors that trigger tropical storms requires further research. The storms form only over large bodies of warm water when both the air and water temperatures are higher than normal. Thus, they only form in summer over tropical oceans. Temperatures of only a few degrees above normal are enough. Hurricanes form in an atmosphere of essentially uniform pressure and are not associated with atmospheric fronts. They appear to build on a wave of low pressure or on a minor disturbance in wind circulation. These atmospheric ripples may result from local differences in land and water heating or from instability in masses of tropical air.

The trade wind belt is typically a relatively shallow layer of warm, moist air above which is a deep layer of warmer, dry subsiding air. This forms the Trade Wind Inversion—a characteristic that limits the vertical development of clouds. The inversion is sometimes interrupted by a low-pressure trough, which allows thunderstorm development behind the wave. Increased convection and the normal pattern of high-altitude winds cause the trough to deepen so an isolated low-pressure system is formed. If the pressure continues to fall, winds accelerate and a tropical storm is born. The change in status from tropical storm to hurricane requires a mechanism to stimulate vertical air motion and convergence of air. Several possible trigger mechanisms exist, with an intruding high-altitude, low-pressure system being the most often cited cause. Derived from the remains of an upper tropospheric cyclone wave, these abandoned waves act in two ways to promote instability. First, there is divergence of the abandoned system; second, the low has a cold core so the lapse rate below is changed. Both the divergence and altered stability enhance surface pressure differences enough to generate a tropical cyclone. Once a hurricane begins to form, moist air from all sides converges toward the storm center. Condensation supplies the energy needed to develop the storm, and therefore a constant supply of water vapor is essential for the storm's continued existence.

The storm starts slowly at first, usually moving from east to west in low latitudes. As it gains strength, the speed increases, and its path curves gradually toward the pole. As long as the storm remains over warm water, it can grow in intensity. A storm can travel far north along the east coast of the United States because it follows the warm Gulf Stream. When a storm moves over the cold Labrador Current, it dissipates rapidly. If it moves over land, increased friction with the land surface and loss of the energy supply causes the storm to quickly dissipate.

Where and When Cyclones Occur

Tropical cyclones occur only in certain regions (Figure 8.2). They start over the tropical oceans between latitudes of 5° and 20°, and most form over the western sides of the oceans. They are rare along the equator due to the weakness of the Coriolis effect in that region. There are six general regions of occurrence: the Caribbean Sea and the Gulf of Mexico, the Northwest Pacific from the Philippines to the China Sea, the Pacific Ocean west of Mexico, the South Indian Ocean east of Madagascar, the North Indian Ocean in the Bay of Bengal, and the Arabian Sea. Over the central and western portions of the Pacific Ocean, an average of 20 tropical cyclones occur each year, mostly from June to October. In the eastern Pacific, southwest of Mexico, an average of three hurricanes form each season, but most rarely reach land. Some hurricanes occur outside these regions, but these are the areas of highest frequency.

Hurricanes also occur at particular times. The peak frequency corresponds with the period of highest sea temperature during the year and the time of the maximum displacement of the convergence zone. Thus, late summer and early fall are the seasons of maximum occurrence.

Storm Surge and Coastal Flooding

Losses from hurricanes are mainly due to the high water and accompanying waves. Winds produce heavy seas, with waves higher than 13 m (42 ft) reported. Long sea swells as far as 1600 km (1000 mi) from the storm center are evidence of both the strength of the storms and their far-reaching effects. Surges of water 2 to 3 m in height might form ahead of the storm, and the pounding of the waves on top of the surge

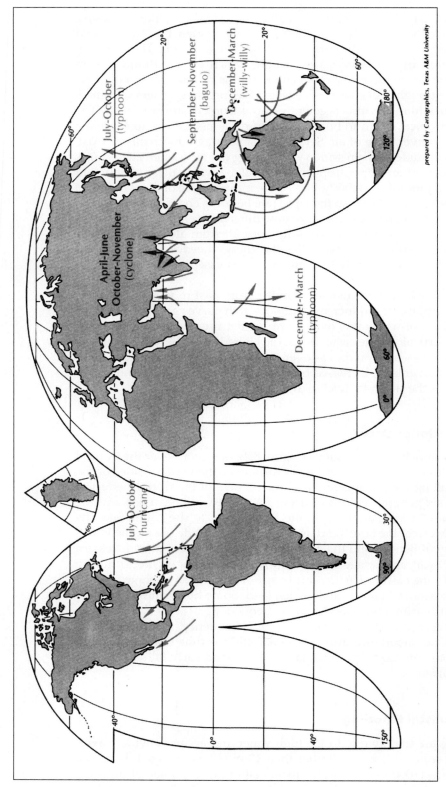

FIGURE 8.2 Areas of the world that experience hurricanes most frequently.

TABLE 8.2 Saffir-Simpson Hurricane Damage Scale

Scale Number	Central Pressure		Winds		Storm Surge		Damage
	mb	in.	mph	kts	ft	m	
1	980	28.94	74–95	64–82	4–5	1.5	*Minimal*: damage mainly to trees, shrubbery, and unanchored mobile homes
2	965–979	28.50–28.91	96–110	83–95	6–8	2.0–2.5	*Moderate*: some trees blown down; major damage to exposed mobile homes; some damage to roofs of buildings
3	945–964	27.91–28.47	111–130	96–113	9–12	2.5–4.0	*Extensive*: foliage removed from trees; large trees blown down; mobile homes destroyed; some structural damage to small buildings
4	920–944	27.17–27.88	131–155	114–135	13–18	4.0–5.5	*Extreme*: all signs blown down; extensive damage to roofs, windows, and doors; complete destruction of mobile homes; flooding inland as far as 10 km (6 mi); major damage to lower floors of structures near shore
5	920	27.17	155	135	18	5.5	*Catastrophic*: severe damage to windows and doors; extensive damage to roofs of homes and industrial buildings; small buildings overturned and blown away; major damage to lower floors of all structures less than 4.5 m (15 ft) above sea level within 500 m of shore

does most of the damage. A storm's arriving at the time of high tide compounds the problem. Additional flooding results from the high amounts of rainfall that usually accompany the storm.

When a hurricane forms, its size, intensity, and path determine the potential for damage. The size and intensity of tropical cyclones provide a basis for classifying them. A widely used scale is the **Saffir-Simpson Scale** (Table 8.2). This scale ranges from a value of 1 for the weakest storm to a value of 5 for the most severe. The scale is based on atmospheric pressure, wind speed, size of the waves generated, and relative damage the storm might cause.

North American Hurricanes

Major hurricanes are classified as 3, 4, and 5 on the Saffir-Simpson Scale. Between 1900 and 2008, more than 75 major hurricanes made landfall on the eastern United States and in the Gulf Coast. However, of the hurricanes influencing the United States since 1900, fewer than half were classified as major. The majority were Category 1 storms, and more than 50% of all hurricanes affecting the United States have occurred in September (Table 8.3).

To consider losses, it is necessary to distinguish between deadliest and costliest hurricanes. Fatalities in the United States have been declining in recent years as a

TABLE 8.3 Number of Hurricanes Affecting the United States (1900–2008) by Category, Month, Number of Casualties, and Cost

	Hurricanes by Category (Saffir-Simpson Scale Number)					
	1	**2**	**3**	**4**	**5**	**All**
United States	68	41	54	17	3	183
Florida	22	18	21	7	1	69
Texas	15	10	11	6	0	42
Louisiana	12	6	10	3	1	32
North Carolina	13	7	10	1	0	31

	Hurricanes by Month (Category 3 or Higher on Saffir-Simpson Scale)					
	June	**July**	**August**	**September**	**October**	**All**
United States	3	4	18	39	9	73
Florida	0	2	3	17	7	29
Texas	1	1	8	7	0	17
Louisiana	2	0	5	6	1	14
North Carolina	0	0	2	8	1	11

Hurricanes by Casualties			
Name (Location)	**Year**	**Category**	**Deaths**
Galveston, TX	1900	4	8000*
Lake Okeechobee, FL	1928	4	2500
Katrina, LA/MS/AL/FL	2005	3	1833
Florida Keys, FL	1935	5	408
Audrey, TX/LA	1957	4	390

Hurricanes by Cost			
Name	**Year**	**Category**	**Cost ($ billions)**
Katrina	2005	3	81.0
Andrew	1992	4	35.0
Wilma	2005	3	20.6
Charley	2004	4	14.0
Ivan	2004	3	13.0
Rita	2005	3	10.0
Hugo	1989	4	9.7

Data obtained from the National Hurricane Center, www.nhc.noaa.gov, adjusted to year 2000 dollars.

*May actually have been as high as 10,000 to 12,000.

TABLE 8.4 States with the Most Hurricane Landfalls (1900–2008)

Rank	State	Number of Hurricane Landfalls
1	Florida	68
2	Texas	42
3	North Carolina	33
4	Louisiana	32
5	South Carolina	17

result of better warnings. In contrast, dollar damages have been increasing due to continued construction on coastal lands.

Hurricane Hazard in the United States

In the United States, there has been a major movement of population into the coastal areas along the Gulf of Mexico and the Atlantic Ocean. The population along the coast has almost doubled in the years since 1930. In Florida alone, more than 5 million people live close enough to the sea to feel the effects of a hurricane surge. Nearly 90% of the Florida population is in urban areas, and a large proportion of these people are living in mobile homes. Coastal cities in Florida are among those that have a high chance of being struck by hurricanes (Table 8.4).

THUNDERSTORMS

Observers for the National Weather Service consider a thunderstorm to begin when thunder is heard or overhead lightning or hail is observed. Severe thunderstorms require three-quarter-inch hail and/or wind gusts of 50 kts. The storm is considered ended 15 minutes after the last thunderclap is heard. Note that this definition makes no mention of rainfall; in fact, in dry climates, thunderstorms often occur without measurable precipitation.

Regardless of whether precipitation occurs, thunderstorms form from an initial uplift of moist, unstable air and the release of sufficient latent heat to cause continued uplift. Thus, the keys to storm formation are a source of moist air and a mechanism to produce the required uplift. Given the moisture and convective heat sources, the greatest number of thunderstorms occurs in moist, tropical realms, especially in Africa. The second requirement, which is a mechanism to initiate cumulonimbus cloud development and, hence, thunderstorms, is obviously more varied. This mechanism can result from intense convective activity to forced uplift at fronts, along squall lines, or from terrain.

Isolated thunderstorms that occur in summer in the midlatitudes typify the air mass thunderstorm. They occur in a disorganized manner, often consisting of a single cell or several distinct cells less than 10 km (6 mi) wide. While air-mass thunderstorms sometimes occur because of differences in surface heating, other trigger effects often cause their development. Alternate mechanisms responsible for growth include converging winds and topography. An example of the former occurs in Florida, where convergence of the moist ocean air along both coasts produces frequent thunderstorms over the peninsula. The role of topography is apparent on the slopes of the Rockies. Air near the slopes is heated more intensely than air at similar levels over flat land, resulting in a distinct upslope movement and the potential for the formation of cumulonimbus clouds.

Single-cell thunderstorms are generally much less violent than those associated with the forced upward motion of air that occurs along cold fronts and squall lines. Some of the most severe thunderstorms are associated with squall lines. These occur mostly ahead of cold fronts and, unlike the isolated air mass type, are an integral part of large-scale circulation patterns. The schematic sequence given in Figure 8.3 provides the essential details of the storm. Warm, moist air at surface levels lies ahead of an advancing cold front. At the 850 mb level (about 1500 m), this air flows from a warm, southerly source. In contrast, at 500 mb (about 5500 m), a westerly stream of cool, dry air—with divergence—flows across the surface systems. The combination of the unstable surface air and the divergence aloft leads to extensive vertical development of clouds and a line of thunderstorms.

The differential flow of air at varying altitudes adds to the severity of a storm. The westerly high-level flow tilts the top of the storm clouds so that falling precipitation does not slow the updrafts, as it does at the mature stage of air mass thunderstorms, thus extending the storm's life and increasing the potential for hail to form. At times the size and severity of storms allow them to be classed as supercells. These are enormous storms whose updrafts and downdrafts are sufficiently balanced to enable cells to last many hours. Note that the squall line may be located hundreds of kilometers ahead of the cold front. In fact, squall lines in tropical climates do not require the presence of a front for their formation.

When a number of individual thunderstorms grow in size and organize into a large convective system, it is termed a *mesoscale convective complex*. These are often large enough to cover an entire state in the Midwest and can persist for more than 12 hours.

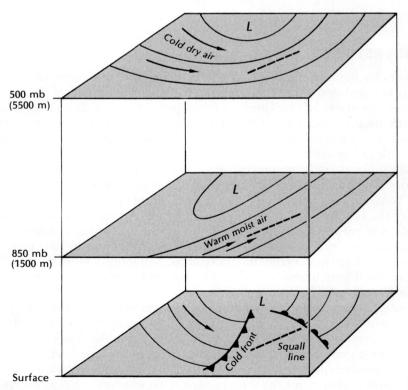

FIGURE 8.3 Schematic depiction of airflows, shown at three levels of the atmosphere, that give rise to severe weather.

Thunderstorm-Associated Hazards

Thunder is the resulting sound from the violent expansion of air close to a lightning bolt. Thus, lightning is an integral part of severe storms and is itself a distinctive hazard. Lightning is an electrical discharge resulting from separation of positive and negative charges within clouds and between clouds and the ground. When the differences in charge are great enough to overcome the insulating effect of air, a lightning flash results. While the reason for the separation of charges is still imperfectly understood, a number of plausible theories try to explain it. One of these suggests that soft hail (graupel) in the cloud becomes polarized, with negative charges at the top and positive charges at the bottom of the particles. At this stage, the graupel has no net charge, but collisions result in larger falling pellets acquiring a negative charge while smaller, positively charged graupel is carried aloft. The separation of charges leads to lightning inside the cloud, between neighboring clouds, or between the cloud and the positively charged surface below.

The distribution of lightning is obviously associated with the distribution of thunderstorms. However, deaths resulting from lightning in the United States do not correlate perfectly with this distribution because lightning-caused deaths are not concentrated in one area. This is probably due to the time of thunderstorm occurrence. A study of injuries and fatalities from lightning shows that 70% occur in the afternoon, and only 1% occur between midnight and 6:00 A.M. Thus, where the proportion of night storms is high, injuries are fewer because people generally are not exposed to them.

Hail also results from thunderstorm activity, and its formation is demonstrated in Figure 8.4. Essentially, hailstones form as a result of adding supercooled water to an initial nucleus. The eventual size of the stone depends on the length of time spent on its passage through the cloud. The height of the cloud is related to the intensity of

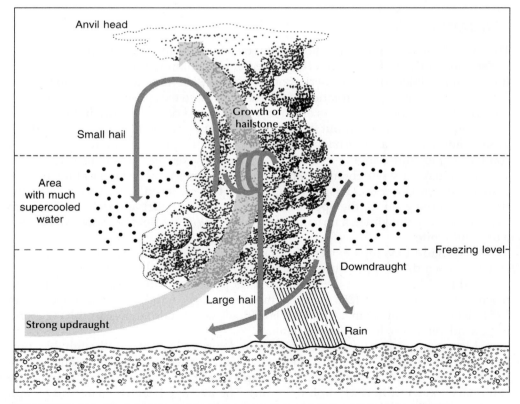

FIGURE 8.4 Simplified representation of the formation of hail in cumulonimbus clouds.

surface heating. The updrafts in the cloud carry hailstones to the glaciated regions causing the supercooled water coating to freeze and adding ice to the mass. As the stones fall, additional layers of liquid are added in the lower portions of the cloud, and the cycle repeats itself. Finally, gravity overtakes the power of the updrafts, and the hailstones fall from the cloud.

Hail exists in three forms. Graupel is usually less than 5 mm (0.2 in.) in diameter and has a crisp texture that causes it to be crushed easily when it strikes the ground. Graupel may serve as the nucleus for small hail, which is often mixed with rain. Because of the thin layer of exterior ice, a small hailstone remains intact on reaching the ground. Although graupel and small hail are about the same size, neither is large enough to cause much destruction. Destructive hail, called true hail or severe hail, can attain large sizes. The remarkable Coffeyville, Kansas, hailstone weighed 3.7 kg (8.2 lb).

Neither the size nor the number of stones associated with severe hail has been systematically measured for long periods throughout the United States. However, based on all occurrences of damaging hail, whether large or small, Texas, Oklahoma, Kansas, Nebraska, and Missouri (ranked in order of the most to least severe) experienced over one-half of the total severe hail occurrences in the 48 continental states over a period of 12 years. Studies have shown that the diurnal distribution of hailstorms peaks between 3:00 P.M. and 6:00 P.M. local standard time. Of the severe hail that falls in the continental United States, 40% occurs within these three hours.

Downbursts are another hazard associated with thunderstorms. These are localized downdrafts at the base of a cumulonimbus cloud that strike the ground and spread out horizontally. When the extent of the downburst is less than 4 km (2.5 mi), it is called a microburst. An intense microburst can have winds as high as 256 kph (160 mph). This short-lived wind can cause much damage at the surface and, because of the wind shear associated with it, is a major hazard to aircraft. In fact, some of the worst commercial aircraft disasters have been related to landing in downburst conditions.

TORNADOES

The tornado is the most intense vortex that occurs in the atmosphere. It is a converging spiral of air with wind speeds estimated at several hundred kilometers per hour. It is the most violent of atmospheric storms, but it is seldom larger than a kilometer in diameter. The direction of rotation is counterclockwise except in rare instances. The tornado depends on moisture as an energy supply and occurs mostly in moist tropical air. Funnel clouds form out of other clouds. They work downward toward the ground and classify as a tornado only if they reach the ground. The funnel cloud of condensed moisture hanging from the storm cloud makes the tornado readily visible. The cloud may vary widely in thickness and sometimes may be larger at the bottom than at the top. Most often it is gray in color due to condensed water vapor. Yet as the tip contacts the ground, the appearance changes due to dirt and debris picked up from the surface. In winter, a tornado may touch down on a field of snow and become a brilliant white.

A tornado develops extremely low atmospheric pressure in the center. The record known drop in pressure occurred in Minnesota in 1904, when the pressure dropped to 813 mb (against the mean sea level pressure of 1013 mb). Wind velocities reach as high as 400 kph (250 mph). In most tornadoes, wind velocities are less than 145 kph (90 mph). Direct measurement is difficult as the sudden pressure changes and high wind velocities destroy instruments.

Movement of the funnels over the ground surface is erratic. Although they normally move parallel to cold fronts, they occasionally move in circles and figure eights, and they may even stay in one spot. Their speed over the ground ranges from nearly stationary to as much as 110 kph (65 mph). The average is between 40 and 65 kph (25–40 mph). The surface path of most tornadoes is short and narrow, varying from a few meters to 2 km (10 ft–1.2 mi) in width. The path averages less than 40 km (25 mi) in length. Tornadoes stay on the ground an average of 15 to 20 minutes. However, in May 1977, a tornado traveled 570 km (340 mi) across Illinois and Indiana, existing for 7 hours and 20 minutes. They stay on the ground longest and travel the straightest over flat, open land. Hills, large structures, and other wind barriers alter their course.

Formation of Tornadoes

Tornadoes occur most often with thunderstorm activity. Ideal conditions for tornado formation are those found ahead of a cold front. Tornadoes form from the collision of a mass of warm, very moist air with cooler, drier air from polar regions. Extreme turbulence develops along the air mass boundary, and eddies occasionally develop into strong whirls through which warm air escapes upward. If the air is extremely unstable, the convergence intensifies, and the storm forms.

For tornadoes to form, several prerequisites are essential: (1) There must be a mass of very warm, moist air present at the surface; (2) there must be an unstable vertical temperature structure; and (3) there must be a mechanism present to start rotation.

The Great Plains is the foremost tornado region in the world, and in this region, a set of weather conditions often provides these three elements. Low-pressure centers develop east of the Rocky Mountains. They typically have a cold front extending to the south and a warm front extending east from the low. In the warm sector, there is a south–north flow of warm moist air from the Gulf of Mexico. Above this air is a stream of cold dry air from the west. This air comes from the Pacific Ocean as rather cool moist air. As it crosses the western mountain ranges, it loses its moisture.

The boundary between the warm moist air from the Gulf of Mexico and the dry air from the west is a zone of great turbulence and is termed the *dry line*. If the air pouring over the Rocky Mountains is warm enough, it will move out over the warm moist air from the Gulf. The result is that the dry line extends more horizontally rather than vertically. This provides a stable but potentially explosive condition. If there is some disturbance, either from the jet stream above or the surface below, there may be a violent exchange of air.

One such disturbance is heat from the ground surface. As the ground heats during the daylight hours, it steadily radiates more and more heat to the air above. The warm moist air moving northward heats during the daylight hours, and by late afternoon, the surface air gets quite hot. The air near the surface becomes hot enough to break through the dry line above. This results in the explosive development of thunderstorms. The upward rush of air reaches velocities as high as 165 kph (100 mph). Under favorable conditions, the storms will grow through the tropopause to heights of 18,000 m (60,000 ft). These **supercells** produce heavy rainfall and large hail. Violent updrafts develop in large thunderstorms and are one reason that all aircraft try to avoid flying through them.

For a tornado to develop, something must start the column rotating. The mechanism exists in the form of wind shear. **Wind shear** is a change in wind speed and/or direction with height. The upper-level winds are blowing across the path of the lower-level winds and at higher speeds. It is this wind shear aloft that starts the rising column of air rotating counterclockwise. As more air flows in and the storm stretches in height, the rate of rotation increases. When the center of the system begins to spin, it becomes a **mesocyclone** and may be as much as 10 km (6 mi) across (Figure 8.5). The rotation starts in the middle level of the storm and works downward. If the process continues, the mesocyclone grows vertically through the thunderstorm and intensifies. It is unfortunate in a sense that the core of the storm begins to rotate first. The rotation is not visible on the outside of the thunderstorm. In most thunderstorms, the supply of warm moist air gradually shuts off, and the storm dies as the energy dissipates. If the supply of moisture continues, the rotating core stretches both upward

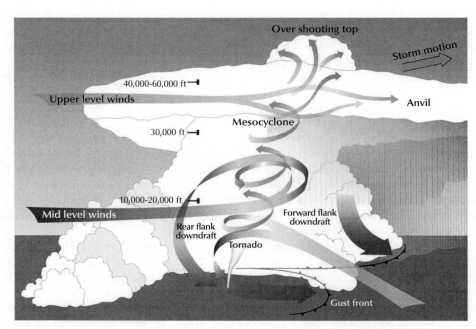

FIGURE 8.5 Diagram of the development of a tornado. (From the NOAA National Severe Storms Laboratory.)

and downward toward the ground. The strengthening results in higher wind velocities, a smaller vortex, and lower pressure.

When turbulence and vertical motion become excessive in a severe thunderstorm, mammatus clouds appear at the cloud base. These clouds occur only when there is extreme turbulence and are a good sign of tornado potential. Rotating clouds at the base of a thunderstorm are an indicator that a mesocyclone exists. As vorticity increases, the rotating clouds drop below the rest of the thunderstorm to form a wall cloud. Within this spinning mass of air, a funnel cloud may form and drop below the cloud. If a tornado forms, it comes out of this wall cloud. Hail is an indication of the severity of the thunderstorm and the high potential for a tornado forming. Heavy falls of hail accompany most, but not all, tornadoes.

Distribution of Tornadoes

Every state in the United States has experienced tornadoes, but their occurrence in regions north of 45° and west of the Rocky Mountains is infrequent. Most of them occur in the Great Plains, a region often called Tornado Alley. Nine states in the United States—Kansas, Iowa, Texas, Arkansas, Oklahoma, Missouri, Alabama, Mississippi, and Nebraska—report an average of more than five tornadoes per year. Figure 8.6 shows tornado hazard potential. The map provides the average of tornadoes for each 10,000 sq mi by state. By expressing the numbers per unit area, it becomes possible to compare the relative occurrence in states of different sizes. Table 8.5 provides data on the distribution of tornadoes.

The time that tornadoes most often occur in the United States is similar to that of thunderstorms. The daily pattern consists of a maximum concentration in a two-hour period between 4 P.M. and 6 P.M. About one-fourth of tornadoes occur from 4 P.M. to

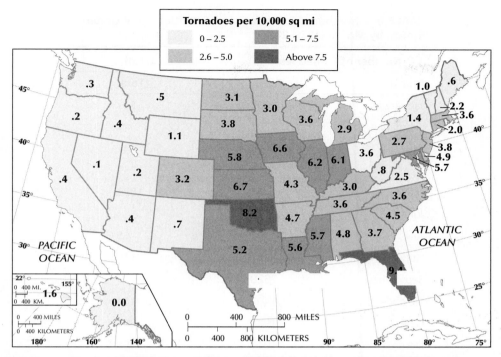

FIGURE 8.6 Annual average number of tornadoes per 10,000 sq mi by state, 1953–2004. (From the NOAA National Severe Storms Laboratory.)

6 P.M., and two-thirds of all tornadoes occur in the six-hour period from 2 P.M. to 8 P.M. This is the time when Earth radiation to the lower atmosphere is greatest. Tornadoes are least frequent around dawn, when the air near the surface has cooled and is relatively stable.

Although tornadoes occur in all months, the annual pattern contains a maximum from May to September (Table 8.6). The number of tornadoes that occur varies from year to year. There is also a seasonal pattern of tornado occurrence.

Hazards of Tornadoes

Concerning atmospheric events, tornadoes are among the biggest killers in the United States. From 1950 to 2005, there were nearly 47,000 reported tornadoes, killing more than 4900 people and injuring nearly 80,000 others. The low was 15 fatalities in 1986, and the high was 519 in 1953. Property damage from tornadoes on an annual basis is

TABLE 8.5 Number of Tornadoes, Ranked by State (1950–2005)

Rank	State	Number of Tornadoes
1	Texas	7484
2	Oklahoma	3235
3	Kansas	3227
4	Florida	2854
5	Nebraska	2417

TABLE 8.6 Number of Tornadoes and Fatalities in the United States, by Month (1950–2005)

Month	Number of Tornadoes Total	Deaths Total
January	1007	132
February	1207	374
March	2931	647
April	5969	1359
May	10,565	1058
June	9441	544
July	5336	65
August	3381	119
September	2528	101
October	1401	104
November	1921	261
December	1011	158

Data obtained from the NOAA Storm Prediction Center.

particularly large. From 1950 to 2005, property damage in the United States totaled more than $20 billion, averaging over $360 million per year. When a tornado is large and well developed, it destroys most ground objects touched by the lower end of the funnel. If the funnel lifts off the ground, surface destruction decreases rapidly. Considerable damage to roofs occurs because during construction, most roofs are not fastened securely to the frame.

Other factors that increase the hazard potential are the development of second homes and the increased use of mobile homes in North America. Second homes are often built in high-risk areas. These sites include beach homes, mountain cabins, and homes along rivers and around lakes. There has also been a major increase in the number of mobile homes sold in the United States. More than half of all mobile homes are outside standard metropolitan areas. Many mobile home parks are in locations where they do not have to meet building codes as stringent as they would have to meet within city limits. Mobile homes are more vulnerable to damage from wind than standard dwellings built on a foundation.

T. Theodore Fujita of the University of Chicago introduced a method for assessing the relative severity of tornadoes in the late 1960s. This scale is now known as the **Fujita Scale** (Table 8.7). The severity of the storm is determined only after the tornado is over. The Fujita Scale is therefore not a forecasting device, as is the Saffir-Simpson Scale. The Fujita Scale rating is based on the worst damage. This qualitative measure provides approximate wind speeds. There is an inverse relationship between the number of tornadoes in each class and the severity of the storm. Most of the storms are in the weakest class. Only 2% are in the most violent classes. However, this 2% is responsible for more than 60% of the deaths from tornadoes. Deaths and damage from tornadoes will continue to increase because of the increased density of population, increased property values, and lack of adequate warning systems.

TABLE 8.7 Fujita Scale of Damaging Wind

Category		mph	kts	Expected Damage
Weak	0	40–72	35–62	*Light*: tree branches broken, signs damaged
	1	73–112	63–97	*Moderate*: trees snapped, windows broken
Strong	2	113–157	98–136	*Considerable*: large trees uprooted, weak structures destroyed
	3	158–206	137–179	*Severe*: trees leveled, cars overturned, walls removed from buildings
Violent	4	207–260	180–226	*Devastating*: frame houses destroyed
	5	261–318	227–276	*Incredible*: structures the size of autos moved more than 100 m, steel-reinforced structures highly damaged

In 2007, a new scale called the Enhanced Fujita Scale, or the **EF Scale**, (Table 8.8) was developed. The EF Scale makes use of numerous indicators to assess tornado damage, such as damage to mobile homes, barns, and trees. Based on examination of these indicators, an estimate is made of probable wind speed, and the tornado is categorized as having a strength between EF0 and EF5.

The Tornado Outbreaks of 1925 and 1974

The most lethal tornado outbreak in North America occurred in the spring of 1925. Once thought to be a single tornado, it actually consisted of a series of perhaps seven tornadoes. They developed over Missouri and traveled northeast across Illinois and Indiana for a distance of 703 km (437 mi). In the path of the storms, more than 600 people lost their lives.

On April 3, 1974, a strong cold front moved across the eastern half of the United States. On April 4, an intense squall line formed ahead of the cold front. An extremely violent outbreak of severe thunderstorms and tornadoes accompanied the squall line. Over a period of 16 hours, 148 tornadoes and an unknown number of severe thunderstorms spread death and destruction. Tornadoes formed from Michigan south through Alabama and Georgia. There were 307 confirmed deaths, 6000 injuries, and property damage of more than $600 million. The storms moved along a combined path of 4180 km (2598 mi).

Tornado Forecasting

The main factor in the high death rate from tornadoes is the problem of prediction and detection. Tornadoes are so localized in extent and random in distribution that it is very difficult to forecast them. Once a tornado is located, radar can determine the

TABLE 8.8 EF Scale

Category		mph	kts
Weak	EF0	65–85	56–74
	EF1	86–110	75–95
Strong	EF2	111–135	96–117
	EF3	136–165	118–143
Violent	EF4	166–200	144–174
	EF5	> 200	> 174

general direction and rate of movement. Because of the random nature of tornadoes, the National Severe Storms Forecast Center in Kansas City, Missouri, issues tornado watches. When conditions are such that tornadoes can occur, the center issues a tornado watch for the area concerned. When a tornado is actually sighted or detected by radar, a tornado warning is broadcast by the nearest National Weather Service Office. This warning gives the location and direction of travel of the storm.

Radar: A Tool for Analysis and Forecasting

Radar provides a means of examining what is happening inside storm clouds such as mesocyclones. A radar transmitter sends out radiation that ranges from 1 to 20 cm in length and is much longer than visible light. When the waves strike an object, part of the beam scatters back to the radar antenna. Cloud particles are very small and are detected with very short radar waves. Longer radar waves penetrate cloud particles and reflect from the larger raindrop-size particles. The harder it rains, the more of the beam reflects. The radar provides a map of where precipitation is occurring and how intense the precipitation is. When a tornado takes place within a mesocyclone, a hook-shaped pattern of rainfall often appears. Although the presence of this hook in a mesocyclone is a good indicator of the possible presence of a tornado, it is not completely reliable. The hook-shaped pattern of rainfall may appear without a tornado. Or there may be a tornado and no hook-shaped pattern of precipitation. In addition, a tornado may be on the ground by the time the hook develops.

Doppler Radar and Tornado Forecasting

A relatively recent development in tornado research and forecasting is Doppler radar. **Doppler radar** makes use of a phenomenon known as the Doppler effect. The frequency of waves from an object coming toward a person is higher than that of an object going away from a person. A train whistle is a good example. As a train approaches, the pitch of the whistle is higher than when it passes and is going away. The frequency of radar waves reflected from rain varies with the direction the rain is moving. By converting the frequency of the radiation to a color on a computer screen, it is possible to determine the direction of wind within a mesocyclone. If a storm develops into a mesocyclone with winds forming a vortex, the pattern of wind velocities shows this on the radar image. A mesocyclone develops before it spawns a tornado, sometimes as much as 20 minutes before. Therefore, when the mesocyclone forms with its rotating winds, Doppler radar can provide up to 20 minutes of warning for the affected area. Because Doppler radar can establish the speed and direction of the storm, it is possible to make a fairly accurate forecast of the likely path of the tornado. Once a tornado forms, a distinct pattern of very rapidly changing wind direction appears on the radar screen.

FLOODS

The uneven distribution of precipitation over both time and space causes large inequalities over the earth's surface. Although some places are perpetually dry and others wet, sometimes the precipitation reaches extremes of too much or too little. Floods and droughts can create major hazards for the human population.

Technically, a *flood* may be defined as the condition in any stream or lake when it rises above bank full. Although arbitrary, this is a useful definition of a flood. All natural stream floods are due primarily to surface runoff, which may result from heavy rainfall, the melting of snow, or a combination of both. Floods caused by rain can result from either short periods of high-intensity rainfall (such as rates of 2.5 cm/h or 25–45 cm/day) or from prolonged periods of steady rains lasting for several days or weeks. Flood runoff from small watersheds usually results from different causes than floods on large drainage basins. Small watersheds are defined somewhat arbitrarily; the average is 25 km^2 or less. Watersheds of this size can be completely covered by a single convective storm, and most floods on small drainage basins are caused from cloudbursts. The rainfall is so intense that the stream channels cannot carry off the water as fast as it falls. The floods that isolated New Orleans and created havoc in other parts of Louisiana and Mississippi in April 1983 were caused by a series of thunderstorms over several days. The flood that destroyed parts of Rapid City, South Dakota, in 1972 was caused by an extremely intense storm that covered a very restricted area. The rainstorm lasted less than eight hours and produced flooding on only a few small watersheds.

On large drainage basins, extended precipitation from cyclonic storms or massive snowmelt is required to produce flooding. The Mississippi River floods of the spring of 1993 are a good example. The winter in the upper valley was very wet. The ground was saturated. When the spring melt occurred, almost all the water ran off the land surface. To add to the problem, a series of cyclonic storms moved across the drainage basin. The result was one of the highest floods in recent history on the lower Mississippi River. In June 2008, Illinois, Iowa, Indiana, Michigan, Minnesota, Missouri, Nebraska, and Wisconsin were affected by major flooding. Several weeks of rain raised the flow in nine rivers at or above the highest levels on record. The worst flooding occurred in Cedar Rapids on the Iowa River, where more than 24,000 people were forced to evacuate, and the river eventually crested 9 ft (2.7 m) above flood stage.

The time of occurrence of floods varies over the earth's surface. All around the Red Sea is a complete network of stream beds that are dry most of the time. During short rainy periods, they often flood. The Amazon River Basin receives large amounts of rain in all parts sometime during the year, and somewhere in the large basin there is usually a flooded area at any given time. The Ganges River of India is frequently flooded during the monsoon season because melting snow in the Himalaya Mountains and excessive rain combine to produce overflow.

DROUGHTS

Droughts that create a major problem for humans at this stage in our history are not just dry conditions but abnormally dry conditions; the absence of precipitation when it normally can be expected, and a demand exists for it; or much less precipitation than can be expected at a given time. In some locations where daily rainfall is the usual condition, a week without rain would be considered a drought. In parts of Libya, only a period of 2 or more years without rain would be considered a drought. Along the flood plain of the Nile River, rainfall is unimportant in determining drought. Prior to the construction of the Aswan High Dam, a drought was any year when the Nile failed to flood. The annual flood provided the soil moisture for agriculture along the river all the way from Khartoum, Sudan, to Alexandria, Egypt.

In the monsoon lands of the world, a rainy season producing half the normal precipitation may bring drought; in areas that normally have two rainy seasons, the failure of one would be considered a drought. Because precipitation and water supply vary greatly over the earth, the term *drought* has different meanings in different places. Thus, drought is a relative term, and the total rainfall in a place is not a suitable indication of its presence. It is for this same reason that it is so difficult to devise a widely applicable quantitative index for drought. Nearly every location has its own criteria for drought conditions.

Drought also has different meanings depending on user demands. There are often distinctions made among meteorological drought, agricultural drought, and hydrologic drought. Meteorological droughts are irregular intervals of time, most often of months or years in duration, when the water supply falls unusually far below that expected on the basis of the prevailing climate. Agricultural droughts exist when soil moisture is so depleted as to affect plant growth. Because agricultural systems vary widely, drought must be related to the water needs of the particular animals or crops in that particular system. There are degrees of agricultural drought that depend on whether only shallow-rooted plants are affected or whether deep-rooted plants are affected as well. Both growing season and dry season precipitation can affect crop yields. The Thornthwaite Method is often used to assess relative drought. An alternate quantitative index is the Palmer Drought Index. A slightly modified Palmer Drought Index is used to develop a generalized map of abnormally dry conditions for crops that is published as part of the Weekly Weather and Crop Bulletin.

The temporal and spatial scales of drought vary widely. The duration of droughts varies as much as the timing of their occurrence. How long a drought will last still cannot be predicted for the same reasons that other irregular oscillations cannot be predicted with accuracy. Drought simply ends when the rains come and the streams rise. At present, we are unable to predict when they will occur or how long they will last or to prevent them from occurring. All that is certain is that they are a part of the natural system, and they will occur again, perhaps with even greater duration and intensity than in the past. They may be very local in extent, covering only a few square miles, or they may be widespread, covering major sections of continents. Even in large-scale droughts, the intensity is likely to vary considerably.

Summary

There are many atmospheric events that in most cases are not typical of usual day-to-day weather. They range over a wide variety of weather conditions. Many, if not most, of these events may cause casualties or property damage. The extent of these losses varies greatly from one event to another. Some of these events occur in very limited areas, and others occur over much of Earth.

Drought is an event that may last for months or years and may cause massive starvation, death, and extensive property damage. Drought in developing countries may be extremely difficult to combat as both food and water may become scarce. Relief efforts are often hampered by the sheer scale of the event. Drought may cover a large part of a continent and affect millions of people. At this stage in history, it is not possible to forecast drought in many areas. The increasing knowledge of changes in atmospheric circulation allows some forecasting of probable drought conditions. An example is drought associated with El Niño.

At the other extreme are floods. Flash floods that occur on small watersheds are very difficult to forecast, and there is little warning time. As a result, fatalities often result from these storm-produced floods. Floods on large watersheds take a longer time to develop, and forecasting is more reliable, so that evacuation and other preparations are possible.

Tornadoes and hurricanes are two types of storms that affect the United States on a relatively frequent basis. Both are dangerous storms and often cause fatalities and extensive damage. Hurricanes affect coastal areas primarily, and the Gulf Coast and Atlantic Coast are the most frequently hit. Tornadoes have occurred in all 50 states; however, the frequency is much greater in the Great Plains than elsewhere. Each type of storm has a season of maximum occurrence. Forecasting of hurricanes is more reliable and earlier due to a number of factors, including their size and speed of travel.

Key Terms

Doppler radar, *158*	Fujita Scale, *156*	Saffir-Simpson Scale, *147*	Vortex, *143*
EF Scale, *157*	Mesocyclone, *153*	Supercell, *153*	Wind shear, *153*

Review Questions

1. Which natural hazards are most deadly in the United States?
2. On a global basis, the major hazards are not the same as in the United States. How and why do the two groups differ?
3. What are the requisite conditions for the formation of hurricanes?
4. What are the requisite conditions for the formation of tornadoes?
5. Why is Doppler radar better than standard radar for detecting tornadoes?
6. How does a hurricane surge differ from waves generated by hurricanes?
7. Drought causes large numbers of casualties on a global basis, but not in the United States. Why is this the case?
8. Regional floods result in casualties and extensive damage even in arid regions. How can this be?
9. The amount of damage caused by both tornadoes and hurricanes is increasing in the United States. Why is this the case?
10. Why is evacuation from tornadoes and floods on small watersheds not a practical means of reducing casualties?

PART II

Climate Change

CHAPTER

9

Natural Causes of Climatic Change

From all the available evidence, it is clear that the climate has changed over time. It is natural that the possible cause of such change has been the basis of much study. The question of what causes climates to change has not been completely answered—but not, however, because of a lack of theories. In the 100 years or so since the magnitude of climatic change was realized, it has been estimated that for every year that has passed, a new theory has been postulated. Not all of these have been satisfactory for some have failed to account for the two basic ingredients of a viable explanation. They must (1) explain the onset of ice ages through geologic time (i.e., account for long periods of warmth with cooler interruptions) and (2) account for the warming and cooling periods that occur within an ice age. Not all theories can do this, especially in quantitative terms.

The search for the explanation of climatic change has become increasingly important in recent years. Human activities have resulted in a modification of both the atmosphere and the earth's surface to such a degree that the changes are now an integral part of explaining climatic variations. Given this fact, this chapter is concerned with what may be termed the *natural causes* of climate change. This discussion includes theories that explain changes that occurred long before human habitation of the earth. The role of human activity in inadvertently modifying climate is examined in subsequent chapters. Obviously, human-induced changes and natural changes are not independent for natural changes are still taking place while people are modifying the system.

The basic reason for climatic changes on Earth is essentially very simple: Change is related to the flows of energy into and out of the system and the ways in which energy is exchanged within the earth–ocean–atmosphere system.

SHORT-TERM CHANGES

There are some processes that affect the energy balance for relatively short periods of time. In this case, we are defining *short-term* as changes in weather or climate for periods measured from individual to thousands of years. Of course, some of these changes could be classified as interannual variations, as discussed in Chapter 7. However, there is a tendency for these types of changes to persist for years.

Variation in Solar Irradiance

In most theories of climatic change, it is assumed that the output of energy from the sun is constant or nearly so. However, a fluctuation of less than 10% in output from the sun could explain all the climatic changes that have occurred on Earth. Thus theories of climatic change have used changes in the sun as their basis.

Two main approaches have been considered. The first visualizes an actual change in the radiating temperature of the sun over long time periods. The second considers shorter times and deals with cyclical phenomena—specifically sunspots.

The solar energy received at the earth's surface can change due to the amount of energy given off by the sun, changes in the transparency of the atmosphere, or changes in the distance between earth and the sun. There is little doubt that solar irradiance affects weather. The average temperature of the sun is near 5438°C (9820°F) but varies slightly, and hence the energy output varies. The actual variation in solar irradiance is small and difficult to measure. Most measurements are made from within the earth's atmosphere. The turbidity of the atmosphere affects the measurements. Until recently, measured differences in the energy were less than the accuracy limits of the instruments used to record the variations. Beginning in January 1977, the temperature of the sun's surface fell 11°C (20°F) in a single year. Kitt Peak National Observatory near Tucson, Arizona, measured the drop. This is the first time the observatory measured such a change, but measurements only began in 1975.

During the 1980s, the brightness of the sun faded. The irradiance of the sun declined 0.07% from 1981 to 1984. Satellites outside the earth's atmosphere made these measurements. A change as small as 0.1% for a decade or more might change the earth's climate in a measurable fashion. The net effect is less irradiance and less energy reaching Earth. Computer models show that a drop in solar irradiance from 1 to 2% would bring about conditions similar to those of the Little Ice Age. Snow and ice would spread over high latitudes in the northern hemisphere. A decline of 2% for 50 years would be enough to cause renewed glaciation. A drop of 5% should be adequate to bring about a major glaciation of the earth.

Sunspot Activity

Scientists have suggested for a long time that sunspots are responsible for changes in weather patterns and climatic cycles. Detailed analysis of the sun's outer surface, or **photosphere**, shows some 1400°C (2520°F) lower than surrounding areas. Intense magnetic fields are associated with them. Sunspots were seen and recorded as early as 28 B.C. in eastern Asia. They have been studied intensively since the invention of the telescope shortly after A.D. 1600.

The number of sunspots occurring at any one time varies from as few as 5 or 6 to as many as 100. Both long- and short-term fluctuations occur in sunspot activity. Two periods of major deviation from normal have existed in the past 1000 years. An unusually large number of sunspots formed in a 200-year period around 1180. Between 1645 and

1715, no sunspots occurred for years at a time. The total reported for the entire 70 years was less than the number that usually occur in a single year. There was minimal auroral activity, and the solar corona was less visible than usual. This period, known as the **Maunder minimum**, is the only such period in historic times. It was also a period of exceptional cold known as the Little Ice Age. The Maunder minimum appears in both historical records and studies of tree ring growth using carbon 14 dating methods. There is now a record of growth rates of trees dating back 8000 years. Each period of weak sunspot activity correlates with periods of cold, as shown by tree ring growth and glacial advances.

There are some signs that sunspots follow an 11-year cycle and that multiples of this cycle occur at 22 and 33 years. Still other cycles appear at periods as short as 5.5 years and as long as 90.4 years. The key to sunspot cycles is that magnetic fields on the surface of the sun reverse their magnetic polarity every 22 years—an interval twice the length of the apparent sunspot cycle.

Repeated studies trying to correlate rainfall with the fluctuation in sunspot cycles have not yet produced statistically significant results. As Tannehill (1947) stated over 50 years ago:

> Some drought years have come close to the top of the sunspots (1893 and 1917), some near the bottom (1901 and 1933), some with increasing spots (1925 and 1936), and some with decreasing spots (1910 and 1930). No matter how we select the years or how we group them, we see no obvious relation to sunspots.

Attempts to correlate weather with sunspots continue. There is controversy about the relationship between sunspots and weather because no clear physical connection between them has yet been established.

Variation in Atmospheric Dust

The amount of energy available at the earth's surface depends on the extent to which the energy is modified as it flows through the atmosphere. Changes in transparency of the atmosphere result from changes in dust content of the atmosphere, cloud cover, and ozone content of the upper atmosphere. A reduction in radiation absorbed by the earth by as little as 1% can produce a change in surface temperatures by as much as 1.2°C to 1.5°C (2° to 2.7°F).

The amount of fine ash injected into the atmosphere from major volcanic eruptions is sometimes very large. This dust absorbs and scatters a significant portion of solar radiation. Although most of the scattered and absorbed energy eventually reaches the ground, a small part goes back into space without affecting temperatures in the lower atmosphere. Maass and Schneider (1977) examined temperature records for 42 weather stations scattered over the earth, each with a temperature record at least 85 years long. They correlated these data with levels of atmospheric dust and concluded that stratospheric temperatures have increased as a result of the injection of volcanic aerosols. They also found there is a drop in annual temperatures at the earth's surface following major volcanic events. Some individual cases support these findings and others do not. Mt. Asoma in Japan erupted in 1783, and cold years followed from 1784 to 1786. The famous year without a summer in 1816 came the year after the massive eruption of Mt. Tambora in the Dutch East Indies. So much dust blew into the air that almost total darkness existed for three days at distances up to 500 km (310 mi) from the mountain. The explosion of the volcano on the island of Krakatoa in the East Indies blasted some 53 km^3 (12.3 mi^3) of solid debris more than 30 km (19 mi) into the atmosphere. Winds in the upper atmosphere distributed the dust over the planet. For the next three years, Montpelier Observatory in France recorded a 10% drop in the intensity of solar radiation. Unusually cold years from 1884 to 1886 accompanied the drop in insolation.

Mt. Katmai in Alaska erupted in 1912, ejecting 21 km^3 (5 mi^3) of rock. Observations at Mount Wilson, California, and at Bassour, Algeria, show a 20% drop in solar radiation during the following months. It is perhaps worth mentioning that unusually cool weather occurred for a month before Mt. Katmai erupted. Mt. Agung in Bali erupted in 1963, and observations at Mauna Loa, Hawaii, show that the receipt of direct solar energy dropped sharply by nearly 2% for a period afterward. Much of the scattered and absorbed radiation eventually reached the lower atmosphere, and total radiation reaching the surface dropped by only 0.5%.

Other volcanic eruptions, including some greater than those already mentioned, produced no cooling effects on the weather. For example, in 1835, Mount Cosequina in Nicaragua blew 49 km^3 (12 mi^3) of ash into the atmosphere, but no atmospheric after-effects were reported. Other notable eruptions have also occurred without any clear effect on weather.

The 1982 eruption of El Chichon in Mexico blew nearly 10 times as much ash and gas into the atmosphere as Mount St. Helens in 1980. It was the largest volume of rock ejected since the eruption of Mount Katmai in Alaska in 1912. The impact each of these two eruptions had on the atmosphere was very different. The blast from Mount St. Helens went laterally, and most of the gas and ash stayed in the troposphere and rapidly precipitated out. Perhaps the most significant aspect of El Chichon is that the main eruption was vertical. As a result, most of the sulfuric gases and sulfuric acid aerosols went into the stratosphere, where they remain for much longer periods. The effect of this cloud of debris on insolation is easier to determine than its effects on actual surface temperatures. The reduction in solar radiation at the surface could have reduced Earth temperatures by about 0.25°C (0.45°F) during 1983. However, no actual temperature drop was measured.

The major difference, then, between volcanoes that can influence world temperatures and those that have only a local effect is penetration of the stratosphere by eruptions that produce large amounts of sulfur dioxide. This gas joins with water vapor in the atmosphere to form tiny droplets of sulfuric acid, which can remain in the atmosphere for several years after an eruption. The sulfuric acid droplets reflect sunlight back to space, producing a cooling effect on surface temperatures.

A number of researchers have attempted to derive measures of the role of volcanoes in climatic change. One, the **Volcanic Explosivity Index (VEI)**, is based on the fact that very explosive volcanoes will penetrate the stratosphere and be more effective than those that do not. However, as noted, the impact of the eruption is also contingent on the amount of sulfur dioxide ejected. The **Dust Veil Index (DVI)**, as the name suggests, is a measure of turbidity in the atmosphere—a factor influenced by volcanic activity.

Although there is clear evidence to show that volcanic activity can have an impact on global climate, it is no easy task relating it to climatic changes of the past. Large eruptions such as Krakatoa (Indonesia, 1883), Katmai (Alaska, 1912), and Tambora (Indonesia, 1815) can be shown to have short-term effects on energy flows. To account for major glacial periods, it is necessary to assume a long-term residence of volcanic dust in the atmosphere. One hypothesis is that, during active times of Earth-building forces, continued volcanic activity over long time periods have an extended effect. Once glaciation is initiated, the role of volcanic dust would be secondary to changing surface albedos.

Human-Induced Changes in Earth's Surface

In the short period that people have inhabited the earth (in terms of geologic time), they have brought about massive changes in the environment. These changes have had a significant impact on the earth's climate. To examine the impacts, it is convenient to deal with changes that modify the earth's surface and those that directly affect the atmosphere.

Humans have been altering the environment since they first controlled fire, domesticated animals, and originated agriculture. Modifications began in an early

TABLE 9.1 Changes in Albedo That Occur with Land Use

Land Type Changed	Earth's Surface (%)	Change in Albedo
Savanna to desert	1.8	0.16 to 0.35
Temperate forest to field, grassland	1.6	0.12 to 0.15
Tropical forest to field, savanna	1.4	0.07 to 0.16
Salinization, field to salt flat	0.1	0.1 to 0.25

epoch, when the hunters and gatherers used fire to make hunting easier and to drive game during the hunt. Records from early explorers of Africa refer to massive fires, which probably represented the annual burning of grazing areas south of the Sahara. In their visits to the Americas, European explorers noted that Native Americans used fire to improve hunting grounds and catch game.

The result of these activities was the deforestation of large areas of the world. In the tropical realm, it may be that the savanna grasslands are a response to deforestation by fire; in temperate regions, grasslands in North America and eastern Europe, prairie and steppe, may be a partial response to burning of woodlands that once existed.

With the development of agriculture, deforestation became even more extensive. Extensive forests in China, the Mediterranean basin, western and central Europe, and North America were cleared for farming. The extent of the change is illustrated by the fact that 50% of central Europe has been converted from forest to farmland over the past 1000 years. But deforestation has not been the only extensive alteration. The misuse of marginal lands led to overuse and the eventual desertlike environment, and it started the process of desertification. Desertification now exists in India, Africa, South America, and Asia.

The advent of a technological society, such as that in which we now live, created further changes. Destruction of the environment in the quest for raw materials, creation of artificial lakes, generation of energy, expansion of farmlands, urbanization, and other processes have significantly changed the face of the earth.

The result of the cumulative changes is their modification on the energy interchange that occurs at the earth's surface. It was pointed out that the surface climate that occurs is a function of the energy that arrives at the surface and the way it is utilized. Of considerable importance in this respect is the amount of energy that is reflected from the surface—energy that does not enter the heat balance of the system. The amount of energy reflected depends on the albedo of the surface, and changes in surface cover over time have appreciably altered the albedo of large areas. Table 9.1 provides examples of how much albedo may have changed as a result of human impacts.

The eventual result of the long-term changes described here is a reduction of surface temperatures. The worldwide temperature decrease resulting from the change has been estimated at about 1°C (1.8°F). This value is open to question because albedo changes lead to other modifications (e.g., cloud cover and dust) that also play a role in determining the earth's average temperature. Despite this, it is clear that surface change has had a significant impact on Earth's climate as a whole and most certainly on the local areas where changes are the greatest.

LONG-TERM CLIMATIC CHANGES

Long-term climate changes may persist for as long as millions of years. These processes are thus extremely slow when we consider them in the context of a human life span.

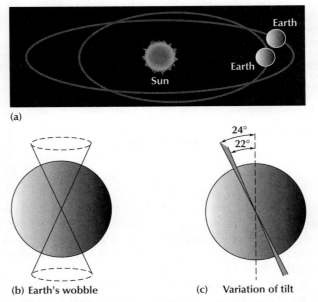

(a)

(b) Earth's wobble   (c)   Variation of tilt

FIGURE 9.1 Astronomical factors that may affect broad climatic cycles. (a) Earth's elliptical orbit varies widely during a 100,000-year cycle, stretching out to an extreme ellipse. (b) Earth's 26,000-year axial wobble. (c) Variation in Earth's axial tilt every 41,000 years. (From Christopherson R. W., *Geosystems: An Introduction to Physical Geography*, © 2009, Pearson Prentice Hall, Upper Saddle River, N.J., Figure 17.31, p. 563.)

Earth–Sun Relationships

Variations in the earth's motion around the sun explain diurnal and seasonal differences in the amount of solar energy arriving at the surface. However, the angle of the earth's axis and the distance from Earth to the sun vary over time.

THE OBLIQUITY OF THE ECLIPTIC   The term **obliquity** refers to the angle of the axis in relation to the plane in which the earth revolves around the sun. At the present time, the angle is 66.5°, which gives an obliquity angle of 23.5°. This angle is not constant; on a cycle of a period of about 41,000 years, the angle varies some 1.5° about a mean of 23.1° (Figure 9.1). The effects of changing obliquity can be quite large. An obliquity of 0° would lead to equal lengths of day and night over the globe and result in a lack of seasonal changes, which would cause well-defined climatic zonation. Another extreme is an angle of obliquity of 54°. Such an angle would produce great extremes in the lengths of summer and winter days and nights. For example, at the December solstice much of the northern hemisphere would have 24 hours of darkness. Extreme temperature differences would occur from summer to winter. Although the actual changes in the angle of obliquity are not as large as these examples, they are sufficient to cause distinctive changes in the distribution of Earth's climates.

EARTH'S ORBITAL ECCENTRICITY   The earth moves around the sun in an elliptical orbit; the **eccentricity** of the orbit is derived by comparing the path to that of a true circle. Currently, the orbit is relatively close to a circle: Its eccentricity is 0.017. Over the past million years, this value has changed from almost circular (e = 0.001) to an extreme value (e = 0.054). This change influences the amount of solar radiation intercepted by the earth and also modifies the dates at which the solstices and equinoxes occur. This factor is used to derive the precession of the equinoxes.

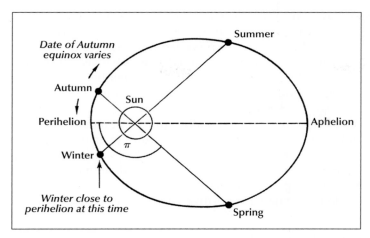

FIGURE 9.2 Precession of the equinoxes occurs as dates of solstices and equinoxes move along the orbit.

PRECESSION OF THE EQUINOXES   Because of varying Earth motions, the days on which the earth reaches its closest and most distant point from the sun (the **perihelion** and **aphelion**, respectively) change over time (Figure 9.2). At present, perihelion occurs during the northern hemisphere winter. Ten thousand years from now, the date of perihelion will pass to the northern hemisphere summer season. And then the lowest input of solar radiation will correspond to winter in the northern hemisphere, the land hemisphere, so more extreme cold temperatures will occur.

The dates at which these events occur can be calculated. It is possible to reconstruct the times at which the various cycles of change reinforce one another to produce very high or very low radiation values within a hemisphere. In fact, long before the advent of high-speed computers, the Yugoslavian scientist, Milutin Milankovitch, actually derived values going back thousands of years. Although his derived values provide the basis for understanding the relationships, more recent researchers have constructed models of the cycles and their relative weight in influencing climate.

Distribution of Continents

Modern geophysical research has provided a unifying theory to the old idea of continental drift—the concept of plate tectonics. There is now little doubt that the present positions of the world land masses are but a transitory location in the long-term evolution of the continents and oceans. Given this, one set of theories of climatic change deals with the relative location of continental land masses in relation to the position of the poles and the equator.

An interesting pattern results when the continental distribution during the last two great earth-cooling periods is reconstructed. Reconstructed maps for the Permo-Carboniferous glaciation (250 million years ago) and those of the most recent glaciation, the Pleistocene glaciation, show that in both cases, there is a concentration of land masses in the polar realms. The presence of large land masses in polar areas is much more conducive to glacial formation because land masses lack the heat storage and transfer mechanisms of oceans. It has been suggested that a primary requirement for the formation of great ice caps is the polar location of continents. Of course, there have been times when such a location occurred but no glaciation resulted. It seems that the relative location of the continents may well be an important factor in ice age occurrences, but actual ice formation must rest on other causal factors.

Intimately related to the idea of moving continents is the role of mountain building and continental uplift. As an explanation of this theory, one only has to think of the formation of snow and ice on Mt. Kilimanjaro, a mountain located astride the equator. With increasing height of land masses, the potential for ice formation is greatly increased.

Geologists have long noted the relationship between times of extensive mountain-building periods and ice ages. For example, the Permo-Carboniferous ice age and some more recent cold periods were preceded by extensive mountain-building periods. There is, however, a lag time of millions of years between the mountain building and the onset of the glaciation. To this criticism must be added the fact that some mountain-building periods—for example, the Caledonian period in Europe 370 to 450 million years ago—did not give rise to ice ages.

Despite these criticisms, it is generally agreed that mountain building, or at least the presence of high mountains, certainly contributes to an optimum condition for ice formation. The idea that mountain building can influence climatic change is reinforced by modern research that shows that mountain ranges do influence upper air circulation patterns by changing upper air vorticity. It is known that, on the leeward side of the mountains, a cyclonic circulation—positive vorticity—is induced in air flow.

Variation in the Oceans

Modern research in climatology is paying increasing attention to the role of the oceans in the climatic system. Variations in sea-surface temperatures have been linked to changing circulation patterns and weather anomalies. In terms of climatic change, the oceans have received attention, providing in some cases the basis of entire theories of climatic change. Some of the ways in which the oceans influence the prevailing climates on Earth include the following:

- *As a factor in the relative elevation of land*: A drop in sea level would increase the heights of the continents and enlarge land masses in area.
- *As a heat storage mechanism*: The oceans are less variable than the continents, and changes in the relative temperature of oceanic waters would influence world climates. Variations in the energy storage in the oceans can occur because of changes in salinity, evaporation rates, and relative solar penetration.
- *As a mobile medium*: Ocean water plays a significant role in the redistribution of energy over the earth's surface. Ocean currents transport large amounts of heat, and any changes in their relative extent would have extended results.

Although the oceans are highly significant in the nature of climate that occurs over the earth's surface, many researchers think that their role in climatic change may be a secondary effect. That is, the oceans react to other changes, such as a modified atmospheric circulation; they do not create them. This idea suggests that the atmosphere leads and the ocean follows.

Extraterrestrial Impacts on Climate

At intervals throughout the history of Earth, exceptionally large objects from space have struck the earth. We need only look at the surface of the moon to get some idea of what our planet would look like were it not for weathering and erosion, which continually erase the evidence of impacts. When these objects struck Earth, some altered the climate tremendously for short periods and in some cases for periods of thousands of years. One such example is that associated with the K-T boundary mass extinction. The boundary between the Cretaceous and Tertiary geologic periods occurred approximately 65 million years ago. Much evidence now points to a large object impacting the earth at that time.

This impact was either the sole cause or one of the causes for the elimination of over half of all species of organisms living on Earth at the time. It certainly altered the climate. The impact produced firestorms over the earth, destroying much of the vegetation. At the time of impact, the tremendous heat from the impact and the scattering of molten meteorites and gases ignited forests and grasslands over the entire planet.

These same fires would have removed much of the atmospheric oxygen and added large amounts of carbon dioxide.

The mass of material blown into the atmosphere as the object struck was perhaps 100 times the mass of the meteorite. A huge dust cloud was raised that blocked out the sun for months. The fires also contributed a huge cloud of soot to the atmosphere. The soot from the fires would have further reduced the amount of sunlight reaching the surface, halting or greatly reducing photosynthesis. The reduction in photosynthesis had a lethal effect on marine organisms. A high percentage of those existing at the time also became extinct. The sustained cloud of solid particles would have greatly reduced solar radiation to the ground and caused temperatures to drop.

Other drastic changes in atmospheric chemistry occurred. Precipitation would have turned into acid rain and snow, as the highly sulfurous particles combined with water particles in the atmosphere. Undoubtedly the stratospheric ozone layer would have disappeared for some time.

The combination of fire, lack of sunlight, and cold would have destroyed much of the vegetation that constituted the food supply of the larger animals, such as the dinosaurs. In fact, mainly the largest animals were eliminated. Most were probably killed by the initial blast and firestorm, but those left alive would have succumbed to the loss of food supply.

Following the months of cold, the temperature rebounded to levels higher than before the impact, adding to the stress on living organisms. The impact greatly disturbed the global system for some time, and much of Earth's biota simply was killed outright or could not adjust to the changes. The event affected seasonal variability of climate for nearly a million years The temperature cycles due to variation in solar output discussed earlier in this chapter were greatly strengthened by the event. The amplitude of the cycles jumped substantially following the event and gradually returned to their previous levels after hundreds of thousands of years. The greater seasonal variation in the global climate was most likely due to the greatly reduced seasonal variation in carbon dioxide. Currently, there is more carbon dioxide in winter, helping retain heat, than in summer. This is due to seasonal changes in photosynthesis. Because Earth's vegetation was largely destroyed during the event, there was little change thereafter in the seasonal level of carbon dioxide to dampen the seasonal variation in heat loss from the ground.

OTHER THEORIES OF CLIMATE CHANGE

The theories presented herein are a partial representation of those that have been suggested. Other researchers have introduced ideas ranging from the possible influence of the periodic passage of the earth through an interstellar dust cloud to variations in atmospheric water vapor caused by both natural and human activities. There is no single theory that can account for all the observed events; it is evident that earth's climates result from a spectrum of causal elements.

Summary

What causes climates to change is a fascinating topic. Many theories have been postulated, but no single one can totally satisfy all necessary requirements. Although there is little doubt that changes in Earth–sun relationships may be a basic cause of long-term climatic change on Earth, it appears that the effects must be considered in conjunction with other factors. The problem of explaining the change is further complicated because human activity is now a major factor in the modification of climate. To explain what is happening, the so-called natural causes must be considered together with the impacts of humans. This is well demonstrated by the extensive surface changes created by human activity and the addition of carbon dioxide to the atmosphere. In cities, the human impact is so great that a new set of climatic conditions is created.

Key Terms

Aphelion, *171*

Dust Veil Index (DVI), *168*

Eccentricity, *170*

Maunder minimum, *167*

Obliquity, *170*

Perihelion, *171*

Photosphere, *166*

Precession, *171*

Volcanic Explosivity
Index (VEI), *168*

Review Questions

1. At the present time, what is the angle of the obliquity of the ecliptic?

2. As the eccentricity of Earth's orbit about the sun increases, what effect does this have on the seasons?

3. The time of year when Earth is closest to the sun is now in the northern hemisphere winter. As perihelion moves to the northern hemisphere summer, what effect will this have on seasons in the northern hemisphere?

4. What relationship does the Maunder minimum have with the Little Ice Age?

5. Recent large volcanic eruptions appear to have affected global climate for a few years. How have these volcanic eruptions affected climate?

6. What is the mechanism by which volcanic eruptions affect climate?

7. In the past, when large amounts of land have been located near the poles, what have been the global climatic conditions?

8. Evidence of climatic changes is found in association with impacts of large objects from space. What has been the effect of these impacts on global climate?

9. What mechanisms appear to have operated to change global temperatures with the impact event at the K-T boundary?

10

Reconstruction of Past Climates

Using the instruments of today, minor changes or trends in weather are routinely monitored and analyzed. However, the period for which weather instruments have been available is but a tiny fraction of Earth's history. To understand current climates and predict future climates, it is essential that they be considered in the framework of climatic change over geologic time. To achieve this, the climates of the past must be reconstructed. This process reflects much painstaking, detective-like research because much of the evidence is based on the relationship between climate and other environmental processes and past life. Prior to presenting an account of how climate has varied, this chapter deals first with the way in which past climates are deduced.

The reconstruction of climates over time is a fascinating puzzle. Instrumental records of meaningful spatial extent have become available only in very recent times. Therefore, information about earlier climates requires the use of **proxy data**—observations of other variables that serve as a substitute or proxy for the actual climatic record.

TABLE 10.1 Paleoclimatic Data Sources and Their Characteristics

Data Source	Variable Measured	Potential Geographic Coverage	Period Open to Study (years BP)	Climate Inference
Ocean sediments (cores, accumulation rate of < 2 cm /1000 years)	Isotopic composition of planktonic fossils; benthic fossils; mineralogic composition	Global ocean	1,000,000+	Sea-surface temperature, global ice volume; bottom temperature and bottom-water flux; bottom-water chemistry
Ancient soils	Soil type	Lower latitudes and midlatitudes	1,000,000	Temperature, precipitation, drainage
Marine shorelines	Coastal features, reef growth	Stable coasts, oceanic islands	400,000	Sea level, ice volume
Ocean sediments (common deep-sea) cores, 2–5 cm/1000 years)	Ash and sand accumulation	Global ocean (outside red clay areas)	200,000	Wind direction
Ocean sediments (common deep-sea cores, 2–5 cm/1000 years)	Fossil plankton composition	Global ocean (outside red clay areas)	200,000	Sea-surface temperature, surface salinity, sea ice extent
Ocean sediments (common deep-sea cores, 2–5 cm/1000 years)	Isotopic composition of planktonic fossils; benthic fossils; mineralogic composition	Global ocean (above $CaCO_3$ compensation level)	200,000	Surface temperature, global ice volume; bottom temperature and bottom-water flux; bottom-water chemistry
Layered ice cores	Oxygen isotope	Antarctica; Greenland concentration (long cores)	100,000+	Temperature
Closed-basin lakes	Lake level	Lower latitudes and midlatitudes	50,000	Evaporation, runoff, precipitation, temperature
Mountain glaciers	Terminal positions	45° S to 70° N	50,000	Extent of mountain glaciers
Ice sheets	Terminal positions	Midlatitudes and high latitudes	25,000 to 1,000,000	Area of ice sheets
Bog or lake sediments	Pollen type and concentration; mineralogic composition	50° S to 70° N	10,000 to 200,000	Temperature, precipitation, soil moisture
Ocean sediments (rare cores, > 10 cm/1000 years)	Isotopic composition of planktonic fossils; benthic fossils; mineralogic composition	Along continental margins	10,000+	Surface temperature, global ice volume; bottom temperature and bottom-water flux; bottom-water chemistry
Layered ice cores	Oxygen-isotope concentration, thickness	Antarctica; Greenland (short cores)	10,000+	Temperature, accumulation
Layered lake sediments	Pollen type and concentration (annually layered core)	Midlatitude continents	10,000+	Temperature, precipitation, soil moisture
Tree rings	Ring width anomaly, density, isotopic composition	Midlatitude and high-latitude continents	1000 to 8000	Temperature, runoff, precipitation, soil moisture
Written records	Phenology, weather logs, sailing logs, etc.	Global	1000+	Varied
Archeological records	Varied	Global	10,000+	Varied

Source: After Kutzbach, 1975.

Proxies are paleoclimatological archives. In recent years, the development of highly sophisticated methods of analyzing and dating materials has led to great advances in reconstructing the past. Table 10.1 provides a list of paleoclimatic data sources, some of which are discussed in the following pages.

EVIDENCE FROM ICE

Study of the processes that modify Earth's surface provided the first indication that climates have varied over time, particularly in relation to the existence of ice ages. Perhaps the best-known early researcher in formulating these ideas was Swiss scientist Louis Agassiz, whose work began in the early 1800s.

Beginning his research in Switzerland, Agassiz noted that the valleys had a U-shape rather than the typical V-shape of river valleys. Other researchers had previously noted the same phenomenon and commented on the fact that the U-shaped valley contained a relatively small river that was out of proportion to the valley's size. The other researchers explained this occurrence by referring to the biblical flood so vividly recorded in the Old Testament. It was assumed that the huge valleys must have been carved by much larger streams when the flood occurred.

Agassiz was also impressed by the presence of large boulders set amid assorted finer sediments that had obviously been transported from an area that was quite distant from where they landed. These boulders, called *erratics*, aroused much curiosity, and their presence was again attributed to the great biblical flood.

Other geologists disagreed with this view of a catastrophic cause for the erratics. These geologists maintained that the forces that caused the U-shaped valleys and erratics were the same forces that now operated. They maintained that present processes provided the key to what happened in the past. Accordingly, they believed that the large erratics scattered over many areas of the world had been dumped by ice. Basing their ideas on observed facts, they concluded that the erratics had been deposited by icebergs that floated on an extensive sea that formerly covered large areas of Europe. Using the same reasoning, they decided that the assorted finer materials in which erratics occurred had been derived from melting icebergs, and they applied the term *drift* to these assorted sediments.

At first, Agassiz was convinced that the iceberg theory best explained the erratics, but in his work he met other scholars who had different ideas. Scientists such as Jean de Charpentier and Ignace Venetz had studied the landscape of the Swiss Alps and were convinced that the glaciers they saw had once been much more extensive. They believed the glaciers were responsible for the valley shapes, the drift, the erratics, and the parallel striations found on hard rock surfaces.

The accumulation of this type of evidence prompted Agassiz to formulate a comprehensive theory of extensive **glaciation**. He suggested that a great ice sheet had extended from the North Pole to the Mediterranean and that the moraines, striations, erratics, and drift seen in Switzerland had resulted from the action of glaciers. The idea of an ice age was born. Today it seems somewhat astonishing that the idea of major glaciation covering much of the earth became accepted only in the 19th century.

Glaciers and Glaciation

The results of ice erosion and deposition are characteristics of large parts of the northern hemisphere. Figure 10.1 shows a typical view of an area that has experienced mountain (Alpine) glaciation. The resulting features—the U-shaped valleys, hanging troughs, aretes, and tarns—are well known and clearly indicative of ice activity.

The results of the work of ice are an important interpretive device because glacial modification of the landscape provides a measure of deducing temperature and precipitation conditions. The advance or retreat of glaciers has been used in interpreting

FIGURE 10.1 The retreating terminus of Stephens Glacier, Alaska, with several of its tributaries. (Photograph by Jason Cheever/Shutterstock.)

climate change over historic periods. There is evidence that glaciers shrunk about 3000 B.C. and that Alpine snow level was at least 1000 feet higher than today. By 500 B.C., there was a marked advance followed by another recession. Between the 17th and 19th centuries, a general resurgence of ice was observed in the Alps and Scandinavia. The 20th century tended to be a time of glacial retreat across most of Europe.

Features associated with continental glaciation differ markedly from those of Alpine glaciation. Depositional features such as the position of terminal moraines and the erratic boulders when studied in detail allow reconstruction of the extent and movement of the ice. The erosional features, such as glacial striations, ice-gouged lakes, cirques, and aretes provide similar evidence. Using such findings, it is possible to reconstruct the climate conditions that existed in North America during ice advance and retreat. Figure 10.2 provides one interpretation.

Ice Sheets and Cores

In the great ice sheets of Greenland and Antarctica lies a wealth of climatic information. Ice sheets are formed layer by layer from the snowfall each year. With time, the snow is compressed into crystalline firn and then into hard glacial ice, often containing numerous bubbles of trapped air. By drilling into the ice, we can learn a great deal about the atmospheric environment in which original snow formed and was deposited. **Ice cores** are derived by driving hollow tubes into carefully selected ice sheet sites where the original ice deposit has been minimally disrupted. The cores are stored at below-freezing conditions for analysis. At the Greenland Ice Sheet Project 2 (GISP 2), a core more than 3050 m (10,000 ft) has been obtained. This reflects a record of some 110,000 years. At the Vostok drilling in Antarctica, a core of 3350 m (11,000 ft) represents some 420,000 years. The European Project for Ice Coring in Antarctica (EPICA) produced a core of 3190 m (10,500 ft) and a record of 720,000 years. Ice cores provide both direct and indirect (proxy) indications of past climates.

Isotopes of an element have the same number of electrons and protons but have a different number of neutrons. Both oxygen and hydrogen, the elements comprising water/ice, have isotopes. Oxygen commonly exists as oxygen 16 (^{16}O) and the rarer and heavier oxygen 18 (^{18}O). The two naturally occurring isotopes of hydrogen are

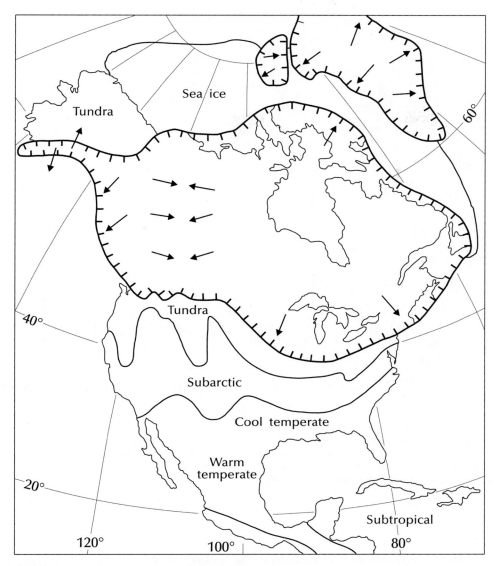

FIGURE 10.2   The maximum extent of the Pleistocene ice advance in North America.

hydrogen 1 (^{1}H) and hydrogen 2 (^{2}H). The latter is a heavier and rarer isotope and is called deuterium (D). ^{1}H^{16}O is the lighter and more common form of the water molecule. Separation of isotopes (fractionation) is a function of temperature. Water in the oceans contains primarily ^{1}H^{16}O, which evaporates more readily than the heavier ^{2}H^{18}O. The ratio may be measured accurately using a mass spectrometer, and results are reported in parts per thousand.

Snowfall each year, together with the constituents it contains, is buried by successive annual accumulations. Depending on the temperature of evaporation and the distance the vapor traveled before undergoing deposition and falling as snow, the ratio of the two isotopes will vary. This variation is retained in the layers of snow that are deposited. Eventually the weight of overlying snow causes lower layers to turn to ice. This becomes the enduring ice of Earth's ice fields.

Air deposited with the snow is trapped in gas bubbles (or inclusions) in the ice, which represent the composition of air at the time the snow first fell. From these inclusions, it is possible to determine the levels of greenhouse gases over time. Figure 10.3

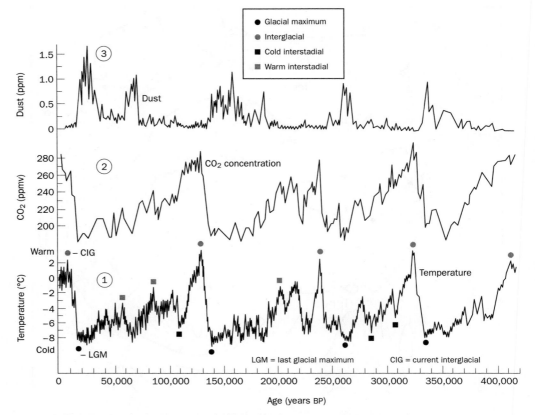

FIGURE 10.3 The trend of atmospheric carbon dioxide (CO_2), temperature, and dust concentration for the past 400,000+ years, as recorded in the Vostok, Antarctica, ice core. The atmospheric temperature at Vostok is plotted as a deviation from present-day mean temperature. (From O'Hare G., Sweeney J. & Wilby R., *Weather, Climate, and Climate Change: Human Perspectives*, © 2005, Pearson Education Limited, Upper Saddle River, N.J.)

provides the record of 420,000 years from the Vostok drilling, which, in this case, provides the varying concentrations of carbon dioxide (CO_2). The relationship between the isotopic data and the CO_2 data provides evidence of how global temperatures vary with greenhouse gas concentrations—an important factor in the evaluation of climate change. Air inclusions in ice cores are used not only to re-construct ancient atmospheric chemistry, but also to determine air pollution levels since the Industrial Revolution. Carbon dioxide, methane, and nitrous oxide are some constituents that have been monitored. All have increased over the past 200 years.

Ice contains oxygen, and oxygen has atoms whose atomic weights vary from the normal atom. Of these isotopes of oxygen, the two most common are oxygen 18, the heavy isotope, and the lighter oxygen 16. By analyzing the ratio of one isotope to another in the ice deposit, it is possible to determine the existing environmental temperature. When there is a higher ratio of oxygen 18 to oxygen 16, scientists can assume the climate was colder. From this it is possible to derive a pattern of temperature over time as represented by the layers of ice in the core. Other significant information is gained from the trapped air bubbles. Minute samples can be used to determine, for example, the level of carbon dioxide in the atmosphere. Layers of dust may give insight into the storminess of the atmosphere or the extent of volcanic activity.

The annual layers in ice cores provide considerable information about contaminants in the atmosphere for both ancient and more recent times. For example, radioactivity levels in the ice layers show that large peaks occurred at the time of the

Chernobyl disaster in Russia and the times of atomic testing that occurred in the 1950s and 1960s. Volcanic activity may be assessed by examining the annual layers of ice cores, which is akin to tree ring analysis. The analysis of the core provides the peaks of sulfate levels of volcanic origin. Precise dating of cataclysmic events can be determined. A marked peak occurred some 73,000 years ago, when an Indonesian volcano called Toba erupted. Other peaks provide a clear sequence of Icelandic volcanic effects.

The improvements in transferrable technology have led to ice cores being derived from places other than Greenland and Antarctica. Of considerable importance are the cores derived from the high mountains of Peru. They provide a guide to temperature changes in low latitudes. Results obtained are correlated with the temperature changes indicated in tropical marine sediments. Evidence suggests that tropical oceans were much cooler during the ice ages.

Periglacial Evidence

There is valuable geomorphic evidence regarding ice ages other than that provided by ice. **Periglacial areas**, areas close to but not covered by continental ice, experienced a climate quite different from that of today. Some areas that are now quite dry experienced much wetter climates as a result of modified circulation patterns. Many inland basins, for example, have been occupied by large lakes as a result of higher precipitation. Such pluvial (from the Latin for "rain") lakes, so named because they resulted from increased precipitation in earlier times, have been widely identified.

The western part of the United States, particularly in the Great Basin area, shows fine examples of the extent of such lakes. In this area, it is estimated that glacial lakes Bonneville and Lahontan were enormous. At its maximum, Lake Bonneville occupied some 50,000 km^2, or an area approximately the size of Lake Michigan. Evidence of the extent of this great lake is found in the present Bonneville Salt Flats and in the *strand lines*—the shore areas indicative of the former level of the lake. Such lakes would necessarily modify the drainage systems, and the formation of a lake and its eventual overflow might lead to a totally new drainage direction.

ANCIENT SEDIMENTS

Sediments deposited during geologic time offer evidence of the climatic environment in which they formed. One example is offered by **evaporite** (salt) deposits. Salt deposits are formed when, on a long-term basis, evaporation exceeds precipitation in an area where water flows in from other areas. Water evaporates to leave the salts that were formerly in solution as sediments. Eventually these may be buried and turned into sedimentary rock. The large evaporite beds of the western United States, Germany, and central Asia are similar deposits being formed at the present time.

Wind-deposited materials also provide a guide to prevailing winds in earlier times. During the last ice age, sand dunes formed around glacial lakes and along shorelines. The structure and location of the dunes have been used to estimate local wind conditions. Similarly, pockets of loess were deposited over wide areas. Loess is made up of silt picked up from the edge of the glacier by the wind and scattered over a broad area many miles from the glacier.

Of particular importance in climatic reconstruction are ocean sediments. These provide many kinds of evidence including fossil isotopic composition and plankton accumulation, which are described later in this chapter. In addition, the sediment contains volcanic ash, which provides information about crustal activity, and sand which in turn provides insight into prevailing wind directions during a past age.

SEA LEVEL CHANGES

Changes in ocean level can occur when the volume of water or the volume of the ocean basins increases or decreases. Many factors can cause either of these to occur, but most of them—such as the accumulation of sediments on the ocean floor or the extrusion of igneous rocks into the oceans—require many years. Rapid changes mostly result from ice alternately accumulating and melting. If the present Antarctic ice sheet were to melt, some researchers calculate that there would be enough additional water to cause a sea level rise of 60 m (200 ft). The removal of ice from a large land area would, however, be accompanied by an upward readjustment of the land. The ocean floors would sink farther into the crust under the weight of additional water. Even allowing for this, melting of the Antarctic ice would still cause a rise of 40 m (135 ft)—sufficient to flood most of the world's major ports.

The rise and fall of sea level during the Pleistocene is an important guide to glacial and interglacial periods. Submergence and emergence of coastal areas and marine terraces point to the amount of water tied up as ice. Using appropriate dating methods, such evidence provides a guide to glacial advance and retreat.

PAST LIFE

Studying species of organisms that lived in the past provides much information about the climate of the time. All plants and animals have a preferred set of physical conditions in which they live. Plant and animal fossils can be used in the reconstruction of past ecologic conditions, including the climates in which they lived.

The Faunal Evidence

Invertebrate fossils are fossils of creatures without backbones. They are widely used to establish the geologic sequence of rock, and they have proved valuable for reconstruction of past climates. However, caution should be exercised because it is possible to draw biased conclusions from incomplete evidence. This restriction may apply particularly to the invertebrates because many fossil species lived in fresh or salt water, and they were not directly affected by the atmospheric climate because the sea, lake, or river in which they lived acted as a buffer to direct exposure.

Although the physiology of fossil animals in relation to modern species is used to determine paleoclimates, much information has recently been gained from the chemistry of invertebrate organisms. Of particular note in this respect is the use of isotopes, especially as part of cooperative research.

CLIMAP (Climate: Long-Range Investigation, Mapping, and Prediction) was one of the important multidisciplinary projects funded to study and understand climates of the past. It investigated ocean-atmosphere changes to obtain fundamental knowledge of what produces climate variations and how changes might be predicted. Another project, COHMAP (Cooperative Holocene Mapping Project), had similar undertakings but concentrated on the climates of the past 10,000 years. One of the keys to the CLIMAP research was obtained from cores taken from layers of mud that cover ocean floors.

The layers that exist contain billions of microscopic skeletal remains of plankton. When the creatures die, their remains fall to the ocean depths and are incorporated in the mud layers. Because the tiny organisms are adapted to temperature, each species lives within a certain ocean climate. If that climate changes, they drift away to be replaced by plankton better adapted to the new conditions. Thus, the fossils found in the mud layers provide a record of temperature changes in the oceans—hence, in the atmosphere above the ocean. To obtain a sample of the undisturbed layers, cores are taken from the ocean floor. Analysis of the cores is a lengthy task since 20 to 50 species and up to 500 individuals exist in each few centimeters of core.

Dating the time at which the organism was deposited is achieved through *carbon 14* analysis. The basis of this technique is that skeletons contain both ordinary carbon and a minute trace of the isotope carbon 14. The proportion of carbon 14 to carbon 12 remains fixed while the organism is alive. After it dies, the carbon 14 begins to decay. By knowing the ratio of carbon 12 to carbon 14, one can determine the age of the shell.

Large research ships, such as Japan's *Chikyu*, are capable of drilling deep into the ocean floor, allowing the reconstruction of past climates dating back millions of years. This is accomplished by using oxygen isotopes to analyze the tiny shells of foraminifera that are deposited on the ocean floor. In this case, the oxygen is in the calcium carbonate ($CaCO_3$) of the shells. The shells are separated from the mud and silt and are dissolved in acid. The ratio of oxygen 18 to oxygen 16 is calculated to determine the temperature of the ocean when the foraminifera formed. As with ice cores, the presence of higher levels of the heavier oxygen 18 isotopes identifies climate periods when the oceans were much colder.

Fossils of animals with backbones—those of vertebrates—provide important clues to past climates. Much can be interpreted from the fossil distribution and their physiology. The great changes in vertebrate life over geologic time result in quite different interpretive methods. For example, the extinction of the dinosaurs at the end of the Cretaceous period promoted much discussion. One view is that progressive and increasing world aridity might have been the cause of their extinction. Alternatively, an asteroid could have been the forcing factor.

Although the physiology of vertebrates is probably the most widely used method for interpreting the ecologic conditions under which they lived, important evidence is also offered by the way in which they are fossilized. For example, some are found right side up in a standing position. To explain their death, it might be assumed that they became bogged down in a swamp environment. By relating such evidence to surrounding deposits and other indicator fossils, it becomes possible to reconstruct the environments in which they lived. The fact that many fossil remains are found close together indicates the death of animals through a catastrophe. Obviously, the catastrophe can take various forms—freezing or drought, for example—but by correlation to other past climatic indicators, the nature of a particular event might be deduced.

The Floral Evidence

Plant distribution provides an important guide to the distribution of climate at the present time; and the same is true of paleoclimates. Identification of vegetation patterns and their changes over time is widely used to interpret past climates. Often the evidence is used in relation to other environmental features. For example, it has already been noted that the extent of mountain glaciers varies. A change will be reflected in mountain vegetation, particularly the elevation of the tree line on the mountain. This feature can be traced over time and has been used to evaluate climatic trends in selected areas.

The physiology of plants, like that of animals, provides much information. The development of drip leaves in plants is indicative of their existence under very moist conditions. The fossil remains of plants with thick, fleshy leaves are probably indicative of arid or semiarid climates. The interpretation of the climate of the Carboniferous period, a time of prolific vegetation that gave rise to great thicknesses of coal, provides examples of this use of plant physiology. Many of the fossil plants in the Carboniferous appear to be related to horsetail ferns and club mosses, both representative of a marsh or swamp environment. Such an interpretation is endorsed by fossils of plants with layered roots, such as those found in modern bog plants, and by fossils with many minor structures which appear to indicate that some of the plants actually floated on water.

Trees of the Carboniferous lacked a development of growth rings. This characteristic is indicative of a climate without marked seasonal differences; and the dominance of trees over herbaceous plants would further indicate a swamp environment. In all, the representative vegetation suggests a warm, moist climate that favored a luxurious, if wet, plant cover. Similar evidence has been used to help reconstruct the climates of late Paleozoic and Mesozoic era rocks. More recent deposits can be interpreted by pollen analysis. For much more recent plant distributions, tree ring analysis is used.

The study of pollen grains, or spores, is termed **palynology**. Its success depends on the fact that many plants produce pollen grains in great numbers (e.g., a single green sorrel may produce 393 million grains; a single plant of rye, 21 million grains) and that pollen is widely distributed in the area in which plants are found. Most importantly, the outer wall of the pollen grain is one of the most durable organic substances known. Even when heated to high temperatures or treated with acid, it is not visibly changed. This is important because pollen possesses morphological characteristics that allow identification of groups above the species level.

For pollen to be of value in interpreting the past distribution of vegetation, and hence inferring the climate that occurred, it is necessary to obtain a layered sequence of the pollen. As shown in Figure 10.4, this often occurs in ancient lakes or peat bogs where seasonal pollen deposits were covered by sediments. Cores taken show a sequenced pattern. Pollens from the cores are identified, and a frequency distribution of plant types is derived. Thus, a high proportion of spruce pollen in the lower core might give way to oak pollen at higher levels. This variation would indicate that a vegetation change had occurred over time and that the difference could be related to a passage from cool to warmer climatic conditions.

Even a relatively rudimentary classification of pollen type (e.g., pollens from trees compared with pollens from other plants) could provide a rough guide to changing climatic conditions. The shift from pollen associated with the nontree climates of the cold tundra to tree climates might indicate an amelioration of climate conditions. In-depth statistical counts obviously provide more detail. Much work in palynology has been completed in Scandinavia, where the first palynological stratigraphy method was devised.

Despite the important progress using this method, it does have shortcomings. A vegetation cover attains maturity only after a fairly lengthy period of time, and it is quite feasible that the vegetation established through pollen analysis represents a successional stage that is not totally representative of the prevailing climate. In some areas, the vegetation cover is mixed; thus, it becomes difficult to establish any dominant type that can be related to climate. It has been pointed out that from Neolithic times, people have extensively altered the forest cover, and human-induced changes might give misleading results.

Tree ring analysis, or **dendrochronology**, was pioneered by A.E. Douglas and his colleagues at the University of Arizona. Initial studies were used in an attempt to relate seasonal growth of trees to sunspot cycles, and a great deal of significant work

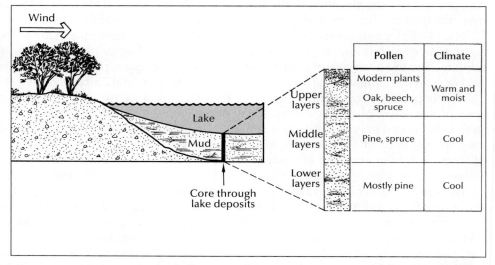

FIGURE 10.4 Simplified diagram showing the method of reconstructing past climates using pollen analysis.

was completed. In the quest for ancient living trees, the bristlecone pine (*Pinus aristata*) was found to be 4000 years old. Analysis of such ancient trees permitted reconstruction of climates of the American southwest during various settlement periods.

Tree ring analysis depends on the fact that growth rings record significant events that happened during the life history of the tree. Growth rings are formed in the xylem wood of trees. Early in the season, the xylem cells are smaller and darker. The abrupt change from light- to dark-colored rings delineates the annual increments of growth. Study of these rings, their size, and variations provides information about the varying environmental conditions to which the tree was subjected. The method is, of course, most valuable in determining conditions that existed in a relatively recent part of geologic history and is widely used in archeological research.

CLIMATIC CHANGE OVER GEOLOGIC TIME

Table 10.2 shows the geologic time divisions, naming the eras and periods, and identifies the times at which Earth was gripped by ice ages. Perhaps the most significant idea expressed in the table is the fact that average global climates have been warm throughout much of geologic time, and periodically the warmth has been interrupted by times of cooling ice ages.

Rocks of the Precambrian eon, extending back to the origin of Earth, provide few details of the climates that existed, and only the late Precambrian times can be reconstructed with any degree of confidence. It is known, however, that during the late Precambrian, much of the earth was glaciated, perhaps for the time span 850 million years before present (mbp) to 650 mbp. The Paleozoic (*old life*) era (570–225 mbp) represents a time when fossil and sedimentary evidence became more widespread and more details of the Paleozoic environments may be derived. Organisms and rocks suggest that the Cambrian and early Ordovician periods were largely warm, although there could have been a glaciation toward the end of the Ordovician. The withdrawal of ice and deglaciation at the end of this event made the climates of the Silurian and Devonian periods similar to those of today. The Pennsylvanian/Mississippian periods

TABLE 10.2 Geologic Time Divisions

Era	Period	Beginning (mbp)*	Ice Ages
Cenozoic	Quaternary	2–3	Pleistocene ice age
	Tertiary	65	
Mesozoic	Cretaceous	135	
	Jurassic	190	
	Triassic	225	
Paleozoic	Permian	280	Ice age at approximately 300 mbp
	Carboniferous	345	
	Devonian	400	
	Silurian	440	
	Ordovician	500	Ice age at approximately 450–430 mbp
	Cambrian	570	
Precambrian (eon)		> 570	Ice age at approximately 850–600 mbp

*mbp = millions of years before present.

were dominated by widespread humid climates with intermittent glaciation. Considerable evidence exists to indicate a major glaciation in the late Paleozoic.

The Mesozoic era (225–65 mbp) was essentially a time of widespread warmth and aridity. Climates of the Triassic period appear similar to those of the upper Paleozoic—cool and humid—but they gave way to a long period of warmth, especially marked during the Cretaceous period. This time of Earth history saw the world in its "greenhouse mode," when climate was predominantly warm, polar ice caps nonexistent, and sea level high. The change from this to an eventual "icehouse mode" may not have been smooth but rather episodic.

During the early Tertiary period (65–22.5 mbp), the warm temperatures of the Mesozoic began to decline. Long episodes of relatively warm climates were punctuated by abrupt drops in temperature. During this time, the first glaciers since the Paleozoic began to form in Antarctica. By the later Tertiary period (22.5–2 mbp), wide temperature swings occurred until the Pliocene, when the downward swings produced glaciations such as those associated with the Pleistocene.

The climates of the **Pleistocene** consisted of glacial and interglacial times, with polar ice advancing and retreating from numerous sources. The most recent full glacial period, known as the Wisconsin in North America, lasted from perhaps 30,000 to 12,000 years before the present. The coldest temperature during the period was 4°C to 6°C (7°F to 11°F) lower than present, which occurred about 18,000 years ago. Huge ice sheets extended as far south as 50° N in Scandinavia and 40° N in North America. Frigid polar water extended in the North Atlantic to 45° N, and sea levels were 120 m (400 ft) lower than today.

CLIMATE SINCE THE ICE RETREAT

The period from 18,000 to 5500 years ago corresponds to the deglaciation of the earth. By 12,000 years ago, only scattered areas of ice sheets remained in western North America, with the main ice sheet confined to eastern Canada. Approximately 10,200 years ago, a strange event occurred that affected Scandinavia and Scotland particularly: The margins of the remaining ice sheet expanded and some small ice sheets reappeared. This time, known as the Younger Dryas (named for a small flower found in cold climates), did not last long, and shortly afterward, climate conditions in the northern hemisphere resembled those of the present day. But the rapid temperature decline of the Younger Dryas illustrates that climatic change need not be a long, deliberate process.

After the cooling associated with the Younger Dryas, the climate began to warm. By 7000 years ago, conditions had improved such that only remnants of ice remained. The warm period peaked about 5500 years ago, and most ice disappeared leaving only the Greenland Ice Sheet and the Arctic Ice that we have today. This time has been described as the **Climatic Optimum**—a term originally applied to Scandinavia when temperatures were warm enough to favor more varied flora and fauna. During the summers of the optimum, it was generally warm in parts of the northern hemisphere as a result of astronomical climate forcing mechanisms.

EVIDENCE FROM THE HISTORICAL PERIOD

Many researchers have used historical records to establish climatic changes that have occurred during the brief existence of humans on Earth. Their findings have allowed fairly detailed reconstruction of climates over the past 6000 years. Medieval chronicles contain many references to prevailing weather conditions. Unfortunately, although not surprisingly, these pertain to exceptional weather events rather than day-to-day conditions. These events include such unusual phenomena as the freezing of the Tiber River in the 9th century and the formation of ice on the Nile River. Although these sources do not supply a continuous record, we can construct an overall view of the usual conditions by assessing the number of times given events are recorded. The freezing of the Thames River provides one such example. Between 800 and 1500, only one or two freezings per century were recorded. In the 16th century, the river froze at least four times, in the next century it froze eight times, and six freezing periods were recorded in the 18th century. One can suppose that progressive cooling increased the frequency. Guides to the conditions that existed are to be found in artwork and etchings of the periods.

A good example of the use of historic data for climate reconstruction comes from Iceland, where the following sources have been used:

1864 to present: actual meteorological instruments

1781–1845: a reconstruction of weather conditions as derived from the relative severity and frequency of drift ice in the vicinity of Iceland

1591–1780: historical records combined with incomplete drift ice data

900–1590: information from Icelandic sagas, indicating times of severe weather and related famines

Such a thorough record as this can be used as a base guide to the climate of the entire North Atlantic area. It is indicative of the methods that may be combined to complete a climatic record.

The Past 1000 Years

The time extending from about 950 to 1250 is known as the Little Climatic Optimum, or the Medieval Warm Period (Figure 10.5). Evidence of agriculture and other indicators has been used to reconstruct the climates that existed at the time Greenland was settled by the Vikings. Under the leadership of Eric the Red, the Vikings passed from Iceland, which they had settled in the 9th century, to Greenland. Although an icy land, it supported sufficient vegetation (dwarf willow, birch, bush berries, pasture land) for settlement. Two colonies were established, and farming was begun. The outposts thrived, and regular communications were established with Iceland.

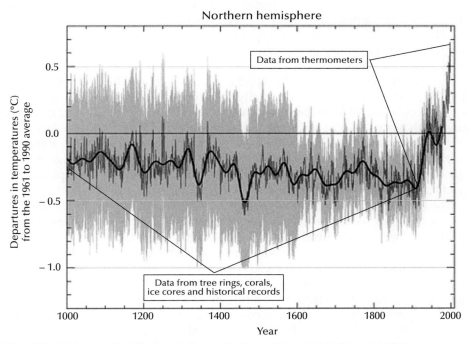

FIGURE 10.5 Variation of surface temperature in the northern hemisphere over the past 1000 years. (From IPCC, 2001.)

Between A.D. 1250 and 1450, climate deteriorated over wide areas. Iceland's population declined, and grains, which had been grown in the 10th century, were no longer produced there. Greenland was practically isolated from outside contact, with extensive drift ice preventing ships from reaching the settlements. In Europe, storminess resulted in the formation of the Zuider Zee, and the excessively wet, damp conditions led to a high incidence of the horrifying disease St. Anthony's Fire (ergotism).

Conditions continued to worsen, and from about 1450 to 1850, the time known as the **Little Ice Age** occurred. During this period, glaciers enlarged to the extent that some Alpine villages were overwhelmed by ice. At the same time, very cold winters led to the freezing of rivers and lakes that are seldom deeply frozen today. For example, in London during the 17th century, the ice was so thick on the River Thames that ice fairs were held on the frozen water.

The Little Ice Age marked the end of the Norse settlements in Greenland that had begun in the 10th century. In fact, in 1492, the Pope complained that none of his bishops had visited the Greenland outpost for 80 years because of ice in the northern seas. He was not aware that the settlements were gone.

By 1516, the settlements had practically been forgotten, and in 1540, a voyager reported seeing signs of the settlements but no signs of habitation. The settlers had perished. Whether their death was due to the deteriorating climate or invasion by other groups is not known, although a Danish archeological expedition to the sites in 1921 found evidence that deteriorating climate must have played a role in the population's demise. Graves were found in permafrost that had formed since the time of burial. Tree roots entangled in the coffins indicated that the graves were not originally in frozen ground and that the permafrost had moved progressively higher in the subsoil toward the surface. Examination of skeletons showed that food supplies had been insufficient. Most remains were deformed or dwarfed, and evidence of rickets and malnutrition was clear. All the evidence points to a climate that grew progressively cooler, leading eventually to the settlers' isolation and extinction.

Later in the cold spell, the colonies in the eastern United States were affected as well. The soldiers of the American Revolution suffered in the cold weather, although the unusual ice sometimes served as a useful tool. British troops, for example, were able to slide their cannons across the frozen river from Manhattan to Staten Island.

From North America comes a well-known account of life during the last years of the Little Ice Age—a description of the year 1816, known as "the year without a summer." The year began with excessively low temperatures across much of the eastern seaboard. As spring came, the weather seemed to be cool but not excessively so. In May, however, the temperatures plunged. Indiana had snow or sleet for 17 days, which killed off seedlings before they had a chance to grow. The cold weather continued in June, when snow again fell, totally devastating any remaining budding crops. No crops grew north of a line between the Ohio and Potomac Rivers, and returns were scanty south of this line. In the pioneer areas of Indiana and Illinois, the lack of crops meant that the settlers had to rely on fishing and hunting for their food. Reports suggest that raccoons, groundhogs, and the easily trapped passenger pigeons were a major source of food. The settlers also collected many edible wild plants, which proved hardier than cultivated crops.

The image of the period shown by artists of the time is very different from that of today. Paintings of scenery and activities in the Low Countries show winter scenes in which ice and snow are central to the theme. One famous painting by Pieter Brueghel the Elder, shown in Figure 10.6, provides an excellent example of this image.

Fortunately, by the end of the 19th century, the instrumental record shows that the climate was again improving. Although the Little Ice Age was devastating to regions of the northern hemisphere, it represented a cooling of just 1°C (1.8°F) and was not a global event. A reconstructed record of temperature is shown in Figure 10.7. If we consider 1961 to 1990 to be the baseline period, we see that the years prior to that time were cool, while those after were warmer. It is this latter trend that continues into the 21st century, indicating that the mean global temperature of Earth indeed is rising.

FIGURE 10.6 *The Hunters in the Snow* by Pieter Brueghel the Elder. (From the Kunsthistorisches Museum, Vienna.)

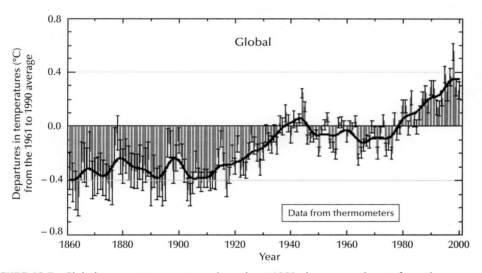

FIGURE 10.7 Global average temperatures since about 1860, shown as a change from the 1961–1990 average. (From IPCC, 2001.)

Summary

Reconstructing climates for the periods prior to instrumentation involves proxy data. Early evidence of past climates was derived from studying ice, but newer methods can more accurately describe a given climatic event. The interpretation of ice cores taken from ice sheets permits oxygen isotopic analysis to give insight into the temperatures at the time when ice formed. Surface evidence—including sediments, periglacial activity, and changes in sea level—also provides significant proxy data.

Both faunal and floral life of the past are used in climatic reconstruction. Invertebrates, especially those derived from deep-sea cores, are of particular significance. Dinosaurs are perhaps the most spectacular of the vertebrates studied, and climate change probably was involved in their extinction. Floral evidence relies on the physiology of plants, the study of pollen, and tree ring analysis. Floods, droughts, human migration, sailing records, ice cover, and a host of other sources of evidence provide additional clues to climate.

A review of climate over time shows that a number of ice ages have occurred throughout geologic history. The most recent of these, the Pleistocene, was responsible for widespread evidence of glaciation.

Deglaciation began about 18,000 years ago, and both warm and cool periods have occurred since that time. During the past 1000 years, the Little Ice Age was a dominant feature in parts of the northern hemisphere. The 20th century also had both warm and cool times, with the period from the 1980s to the present being an exceptionally warm period.

Key Terms

Climatic Optimum, *186*	Glaciation, *177*	Little Ice Age, *188*	Pleistocene, *186*
Dendrochronology, *184*	Ice core, *178*	Palynology, *184*	Proxy data, *175*
Evaporite, *181*	Isotope, *178*	Periglacial area, *181*	

Review Questions

1. What are proxy data? Provide some examples.
2. What types of evidence led Louis Agassiz to believe that parts of the earth had been glaciated?
3. What has happened to many European glaciers in the past 500 years?
4. What are isotopes, and why are they of value in climate reconstruction?
5. How does the area around the Great Salt Lake in Utah provide information about climate change?
6. What could cause sea level to fall?
7. How can the physiology of plants help us understand past climate?
8. What is palynology?
9. How can the study of tree rings explain past climates?
10. What are the major eras of geologic time?

Greenhouse Gases and Global Warming

It is evident that Earth's climate results from a wide variety of processes. Changes in the energy output of the sun, changes in the relative position of the sun and Earth, shifting locations of the continents, mountain building, volcanic eruptions, and changes in atmospheric composition all combine to cause climate to change. Most of the changes in climate of the past can be explained by a combination of all these processes. However, none of these natural changes, individually or collectively, explain the rapid change now taking place on earth (Figure 11.1).

There is something new and different taking place on the planet. Now these processes must be considered together with the impact of the human species. The species has grown to such an extent in numbers, and in per capita footprint, that the entire planet is being altered. That this is the case is well demonstrated by the extensive surface changes created by human activity. For instance, it has been known for decades that the human impact in cities is so great that a new set of climatic conditions is created. Now we know that the climate of the entire planet, from pole to pole, is being altered. Such extensive change has the potential to move our planet to a new

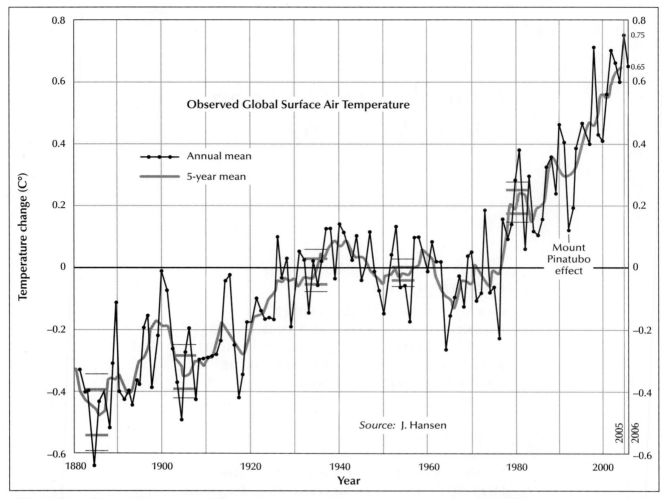

FIGURE 11.1 Global temperature trends from 1880 to 2006. The 0 baseline represents the 1951–1980 global average. Comparing annual temperatures and five-year mean temperatures gives a sense of overall trends. (From Christopherson R. W., *Geosystems: An Introduction to Physical Geography*, © 2009, Pearson Prentice Hall, Upper Saddle River, N.J.)

stage unknown in human history and to change the entire human economic and cultural systems. It is critical that we understand what causes climate change in general and specifically what is responsible for the current changes taking place.

CARBON IN THE ENVIRONMENT

Carbon is an element found throughout our environment. It is a primary element in the organic compounds that make up all plants and animals. It is the main ingredient in diamonds and is a part of a huge number of other common compounds.

By far the largest amount of carbon is stored in the earth as fossil fuels. Millions of years ago, before mammals were present on Earth, most of the planet had a tropical climate. In the warm and moist conditions, lush evergreen forests existed along with vast swamps in shallow inland waters. Vegetation thrived by taking **carbon dioxide** (CO_2) from the atmosphere and, with the aid of sunshine, incorporating it into organic carbon compounds. In the process, oxygen was released into the air. This process of carbon withdrawal and release of oxygen created the large store of oxygen in the atmosphere today. As the vegetation died and sank to the bottom, it rotted and was eventually covered with silt. With the passage of time, the organic debris was buried deeply by sediments deposited on top of it. With ages of time, and under heat and pressure, it was converted into coal, oil, and natural gas. This process took millions of years to complete. These huge deposits are the oil, coal, and gas fields of today. The precise amount is unknown.

Earth's vegetation is the second largest store of carbon, followed by the oceans and the atmosphere. Forests and grasslands store large amounts of carbon in organic compounds. Carbon is found dissolved in the oceans in the form of CO_2 and of course in marine organisms. The amount stored in the oceans, the second smallest store of carbon, is some 50 times greater than that of the pre-industrial atmosphere. Carbon is also found in the atmosphere in the form of CO_2. Relative to the other stores of CO_2, the atmosphere holds very little carbon.

Carbon is continually circulating between the biosphere, atmosphere, and ocean. Natural processes operate to move carbon from one reservoir to another. The ocean is a major absorber of atmospheric CO_2 and an important long-term sink for carbon from the atmosphere. CO_2 is soluble in sea water, and slowly but steadily it is removed from the atmosphere by solution at the sea surface. The amount of carbon the ocean takes up and releases depends on a variety of physical, chemical, and biological processes. For example, colder water absorbs more CO_2 than does warm water.

There is an exchange of CO_2 between the atmosphere and terrestrial vegetation that is about the same size as that between the atmosphere and the seas. The building blocks of photosynthesis are CO_2 and water. The process of manufacturing green plant material removes CO_2 from the air. Carbon is released when the vegetation is oxidized either by decomposition or burning.

CHANGING LEVELS OF CARBON DIOXIDE

CO_2 is one of the many variable gases found in the atmosphere. Atmospheric CO_2 is an active trace gas; it is found only in small quantities compared to the major constant gases of nitrogen and oxygen. It averages less than 0.04% of the atmosphere by volume. Under natural conditions, the amount of carbon dioxide in the atmosphere varies with time. There are regular daily and seasonal oscillations in the CO_2 content. During the day, photosynthesis withdraws CO_2 from the atmosphere, and at night respiration releases CO_2. Daily fluctuations during the growing season in midlatitudes are as high as 70 parts per million (ppm).

There is also a seasonal change in CO_2 in midlatitudes due to variation in the rate of photosynthesis. The carbon dioxide content rises to a peak in spring in each hemisphere and falls to a minimum in early fall. The seasonal change in atmospheric carbon dioxide reflects a very important factor affecting the atmosphere: the metabolism of all living matter. The seasonal change in carbon dioxide in the atmosphere results from the "pulse" of photosynthesis. The difference is only about 5 ppm in warm humid areas such as Mauna Loa, Hawaii. It is more than 15 ppm in central Long Island, New York. The difference is less near the equator where the vegetation is green most of the year. The difference is largest in midlatitudes where the vegetation is largely deciduous. The difference is also lower at higher elevations at all latitudes. These diurnal and seasonal changes in CO_2 have been a part of the natural atmosphere over geologic time. Irregular changes also take place over longer periods, due to changes in volcanic activity, the rate of chemical weathering, and the volume of living vegetation on Earth.

The CO_2 content of the atmosphere dropped steadily through geologic time until about 50 million years ago. Then, 3 to 4 million years ago, the decline slowed and stabilized. During the past 2 million years, carbon dioxide has varied over a relatively small range of between 200 and 280 ppm by volume (ppmv). The lowest concentrations were reached during the coldest part of the last glaciation, when carbon dioxide dropped to about 200 ppmv. The most likely reason for this drop in CO_2 is that cold water absorbs more carbon dioxide than does warm water. In the thousand years preceding 1850, the concentration of carbon dioxide stayed near 280 ppmv.

The coal, oil, and gas fields are the major sources of energy used in today's human culture to drive our industrialized world. The burning of fossil fuels adds carbon dioxide to the atmosphere. At the present time, the burning of organic compounds is adding far greater amounts of carbon dioxide to the atmosphere than photosynthesis or solution in the oceans can remove. As a result, the gas is steadily increasing in the atmosphere. What we are doing today is simply reversing the process of creation of these organic compounds, which took millions of years to form. In geologic terms, we are reversing this creation in an instant. When these fuels are burned, energy is released, as are the byproducts carbon dioxide and water vapor, into the atmosphere.

Carbon dioxide content of the atmosphere has been increasing since at least 1850. Just how much the total increase has been is uncertain due to the nature of the early measurements. The first attempt to directly measure changes in carbon dioxide on a continuous basis began in 1958. A monitoring station was set up on the Mauna Loa volcano on the island of Hawaii, under the direction of David Keeling. Within 10 years, it was not only clear that carbon dioxide content was increasing at a measurable rate but that it was also increasing rapidly. Soon studies from around the world verified the increase. The increase at Mauna Loa averaged 0.8 ppm per year, ranging from 0.5 to 1.5 ppm per year. A significant aspect of this change is that the rate of change is increasing. The CO_2 concentration increased from about 280 ppm in 1850 to about 387 ppm in 2008 (Table 11.1). The CO_2 concentration in 2008 was greater than at any other time in at least the past 650,000 years. In 2007, world emissions of CO_2 grew by 3%. China contributed more than half of the increase and passed the United States as the world's greatest contributor of greenhouse gases (Table 11.2). The concentration is now nearly 40% more than in the pre-industrial era from 1750 to 1800. It is increasing at a rate of about 1.8 ppm annually (Figure 11.2).

TABLE 11.1 Levels of Carbon Dioxide, 1744–2008

Year	PPM
1744	277
1791	280
1816	284
1843	287
1869	289
1878	290
1903	295
1915	301
1927	306
1943	308
1960	317
1970	326
1980	339
1990	354
2000	369
2004	377
2008	387

TABLE 11.2 Ten Leading Countries in Greenhouse Gas Emissions, 2007

1. China
2. United States
3. Russia
4. Japan
5. India
6. Germany
7. Canada
8. United Kingdom
9. South Korea
10. Italy

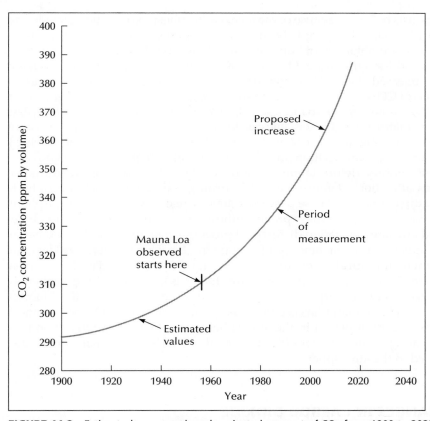

FIGURE 11.2 Estimated, measured, and projected amount of CO_2 from 1900 to 2025.

CARBON DIOXIDE AND GLOBAL WARMING

Carbon dioxide is partially responsible for the **greenhouse effect** of the atmosphere. The natural greenhouse effect of the atmosphere keeps Earth some 33°C (91°F) warmer than it would be otherwise. This increase in temperature results from the fact that the atmosphere is relatively transparent to solar radiation but quite opaque to infrared radiation. Earth's radiation is concentrated in the infrared band from 7 to

20 µm. Carbon dioxide is absorbent of radiant energy in the wavelengths in which the earth radiates heat away from the surface, and it is in the 15 to 20 µm range that most of the absorption takes place.

Perhaps the first direct scientific study of the relationship between carbon dioxide and global temperature was that of Svanti Arrenius, beginning in 1896. In the 1930s, a meteorologist named G.S. Callendar examined temperature data from around the world and detected increasing global temperatures. He believed the increase in temperatures was due to increasing carbon dioxide. Soon after Callendar's study, Gilbert Plass began working on the question of how carbon dioxide could increase atmospheric temperatures. He determined that infrared radiation from Earth was absorbed in the atmosphere by carbon dioxide. Because infrared radiation is a major means of removing energy from the earth's surface, it followed that increasing carbon dioxide would lead to increasing global temperatures. Plass presented his theory in 1956, and it is from his work that the term *greenhouse effect* came into use.

If carbon dioxide absorbs Earth radiation and if the amount of carbon dioxide in the atmosphere is increasing, then more Earth energy should be absorbed by the atmosphere. This absorption and reradiation back to the surface will shift the energy balance toward increased storage of energy, hence raising the temperature of Earth's surface and atmosphere.

Most scientists believe the cause of the increase in atmospheric carbon dioxide and ultimately for planetary warming is human activities. The main sources for the increased CO_2 are the combination of burning of fossil fuels and burning of natural vegetation to clear land for agriculture. Over the past century, fossil fuel use and cement manufacturing released about 200 billion tons of carbon into the atmosphere. Current global emissions of CO_2 from energy use are some 7 gigatons of carbon (GtC) each year. By the year 2025, this is expected to increase to between 8 and 15 GtC per year. For years beyond 2025, estimates are highly variable due to uncertainties about whether the countries with major emissions will take steps to reduce their emissions. Emission levels by 2100 could be about the same as those in 2000 or five times as high, around 36 GtC.

In the 20th century, deforestation might have released as much as an additional 115 billion tons of carbon. Deforestation and burning vegetation affect atmospheric CO_2 in at least three ways: (1) These processes directly release CO_2 to the atmosphere, (2) removing live vegetation reduces photosynthesis, which removes CO_2, and (3) deforestation and burning vegetation disturb soil processes that affect the CO_2 exchange with the atmosphere. Thus the amount of change in global air temperature depends on the rate at which natural mechanisms lower the CO_2 content. The faster these mechanisms operate, the lower the temperature change is likely to be. Approximately 40% of CO_2 placed in the atmosphere has been absorbed either by the earth's biomass or by the ocean. The remainder has accumulated in the atmosphere. Of the estimated 315 billion tons of carbon placed in the atmosphere since 1850, only 130 billion tons remain there. During the decade from 1981 to 1990, about half of the human-contributed emissions stayed in the atmosphere.

FUTURE CHANGES IN CARBON DIOXIDE

Uncertainties in the size of individual sources and sinks of CO_2 severely limit the accuracy of forecasts of future atmospheric concentrations. The task of predicting future abundance of atmospheric CO_2 requires scientific information from many scientific disciplines. We must understand how the CO_2 budget operates today. We also need to know how it responds to changes in climate and other environmental conditions. Present data suggest that CO_2 emissions will grow by about 1.8% annually until 2025 and then decline to a growth rate of around 1% per year. However, CO_2 emissions jumped to 3% from 2006 to 2007. It is both possible and probable that at some time the atmospheric level will reach 400 ppm. Forecasts made in the year 2000 placed the date of reaching the 400 ppm level at the year 2025. Once the concentration reaches 400 ppm, it will be at the highest level

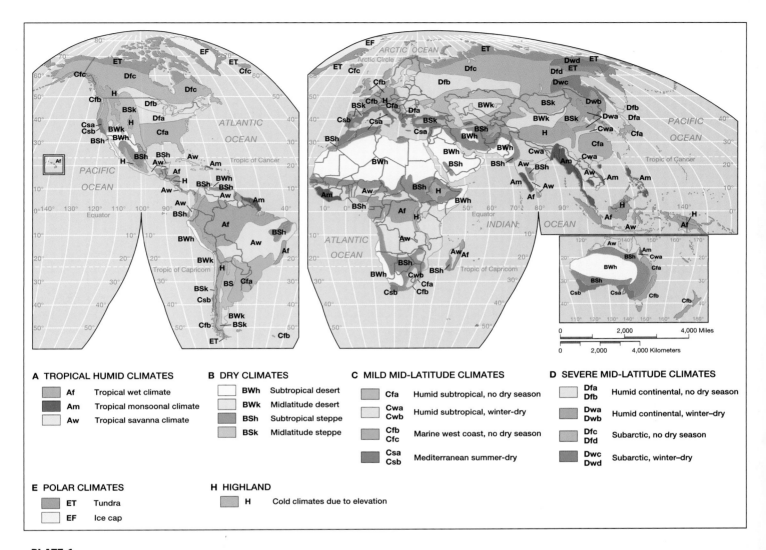

A TROPICAL HUMID CLIMATES

	Af	Tropical wet climate
	Am	Tropical monsoonal climate
	Aw	Tropical savanna climate

B DRY CLIMATES

	BWh	Subtropical desert
	BWk	Midlatitude desert
	BSh	Subtropical steppe
	BSk	Midlatitude steppe

C MILD MID-LATITUDE CLIMATES

	Cfa	Humid subtropical, no dry season
	Cwa Cwb	Humid subtropical, winter-dry
	Cfb Cfc	Marine west coast, no dry season
	Csa Csb	Mediterranean summer-dry

D SEVERE MID-LATITUDE CLIMATES

	Dfa Dfb	Humid continental, no dry season
	Dwa Dwb	Humid continental, winter–dry
	Dfc Dfd	Subarctic, no dry season
	Dwc Dwd	Subarctic, winter–dry

E POLAR CLIMATES

	ET	Tundra
	EF	Ice cap

H HIGHLAND

	H	Cold climates due to elevation

PLATE 1

World climates according to the Koppen classification system (from Aguado E. & Burt J. E., *Understanding Weather & Climate*, 4e., © 2007, Prentice Hall, Upper Saddle River, N.J.).

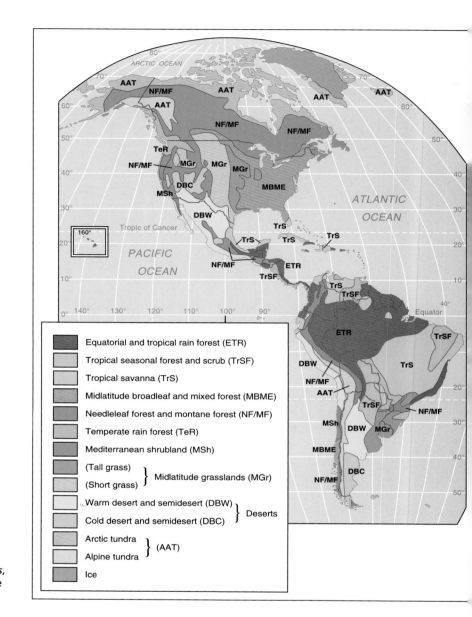

PLATE 2
The 10 major global terrestrial biomes (from Christopherson R. W., *Geosystems*, 7e., © 2009, Prentice Hall, Upper Saddle River, N.J.).

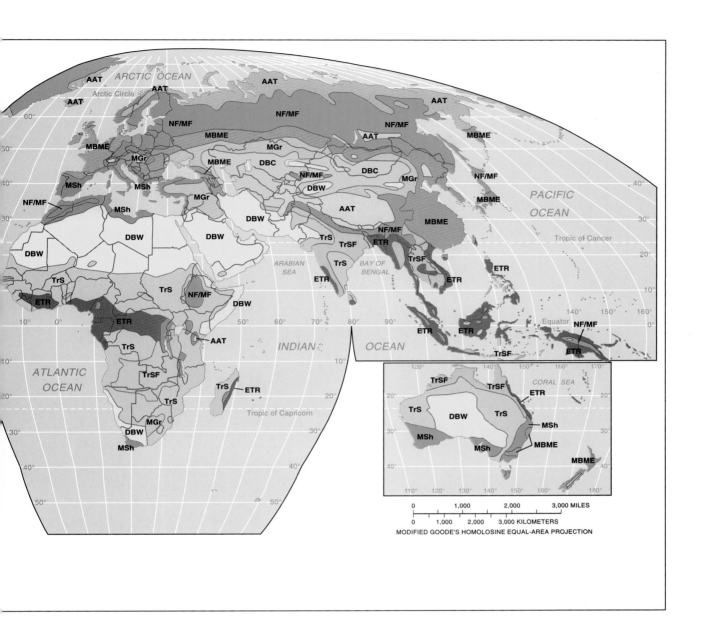

MODIFIED GOODE'S HOMOLOSINE EQUAL-AREA PROJECTION

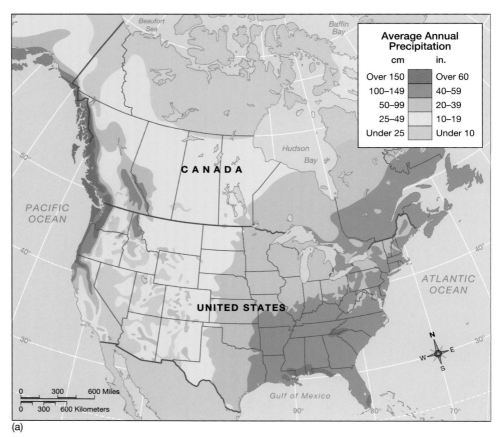

(a)

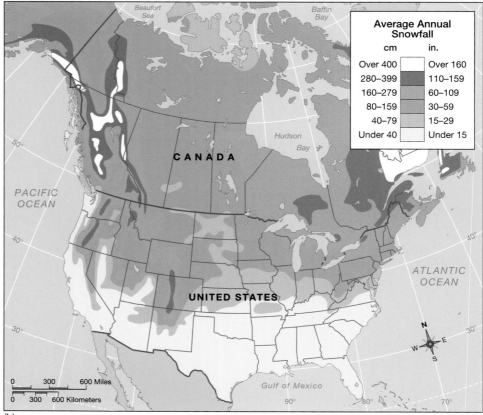

PLATE 3
(a) Average annual precipitation and (b) average annual snowfall in Canada and the United States (from Aguado and Burt, Understanding Weather and Climate, 5e., © 2010, Prentice Hall, Upper Saddle River, N.J.).

(b)

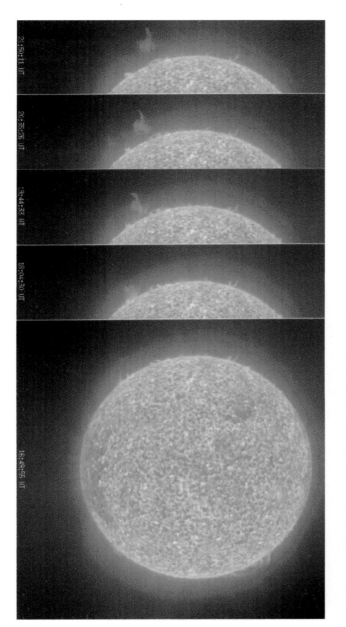

PLATE 4

Left: A sequence of images of the sun in ultraviolet light was taken by the Solar and Heliospheric Observatory (SOHO) spacecraft in 1996. An "eruptive prominence" of 60,000°C gas over 80,000 miles long is seen on the left of each image. Eruptions of this sort affect communications, navigation systems, and power grids on Earth (from JPL/NASA).

Right: A close up of the region around a sunspot. The granulation is the result the turbulent eruptions of energy at the surface (from JPL/NASA).

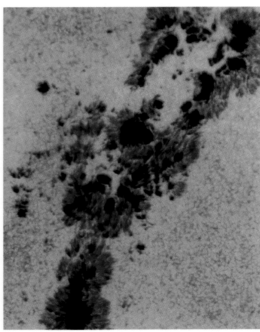

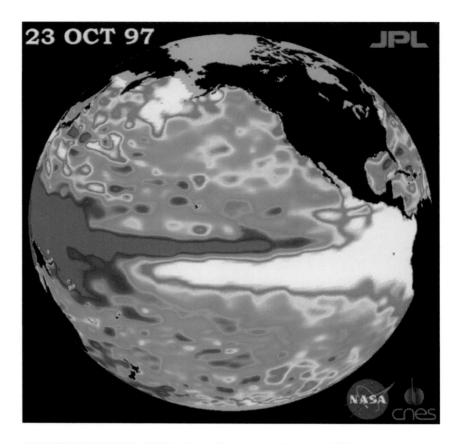

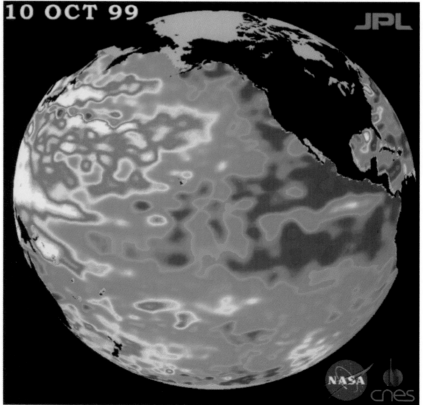

PLATE 5
Upper: A strong El Niño is shown in this October 1997 image taken by the US/French TOPEX/POSEIDON satellite. The image shows sea surface height relative to normal ocean conditions as the warm water associated with El Niño (in white) spreads both north and west (from JPL/NASA). Lower: La Niña conditions are seen in this TOPEX/POSEIDON image for October 1999. The height of the sea surface, an indicator of ocean temperature, shows the normal (green) temperatures and areas of cooler water (blue/purple) (from JPL/NASA).

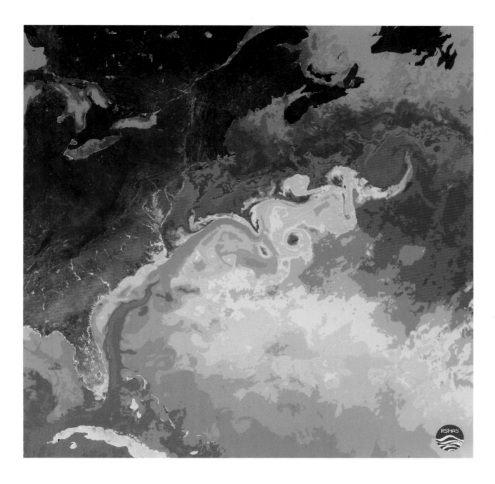

PLATE 6
An infrared satellite image of the Gulf Stream shows that it flows in a complex pattern with eddies of different size superimposed (from NASA/NOAA).

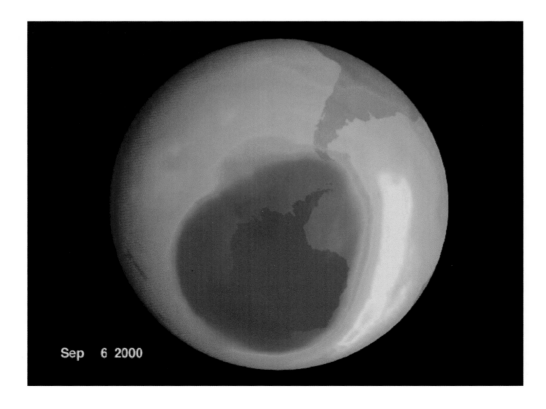

Sep 6 2000

PLATE 7
The largest single-day ozone hole, roughly three times the size of the U.S., was detected on September 6, 2000 by NASA's Total Ozone Mapping Spectrometer (from NASA: Goddard Space Flight Center).

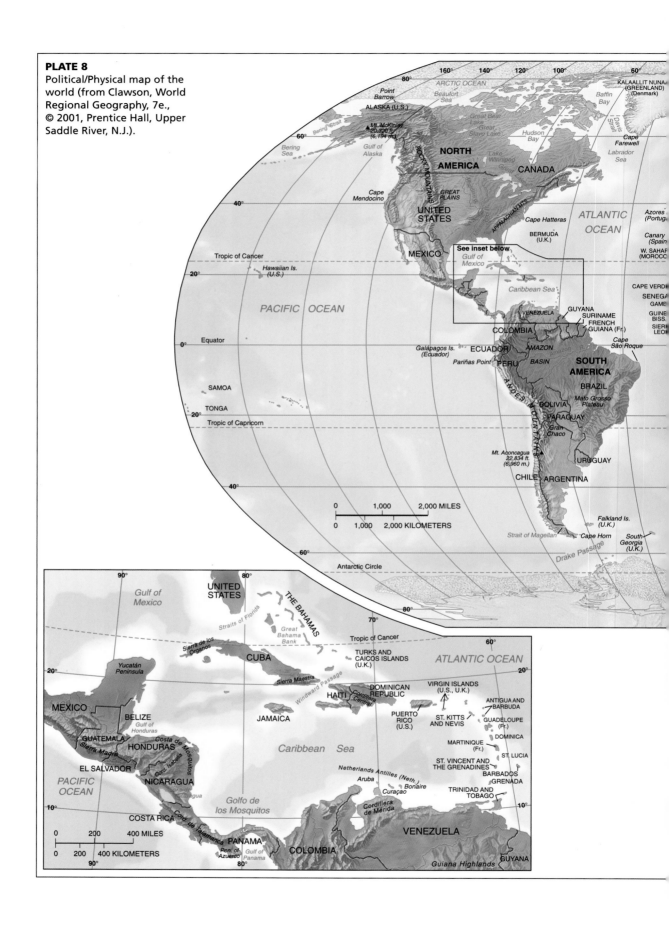

PLATE 8
Political/Physical map of the world (from Clawson, World Regional Geography, 7e., © 2001, Prentice Hall, Upper Saddle River, N.J.).

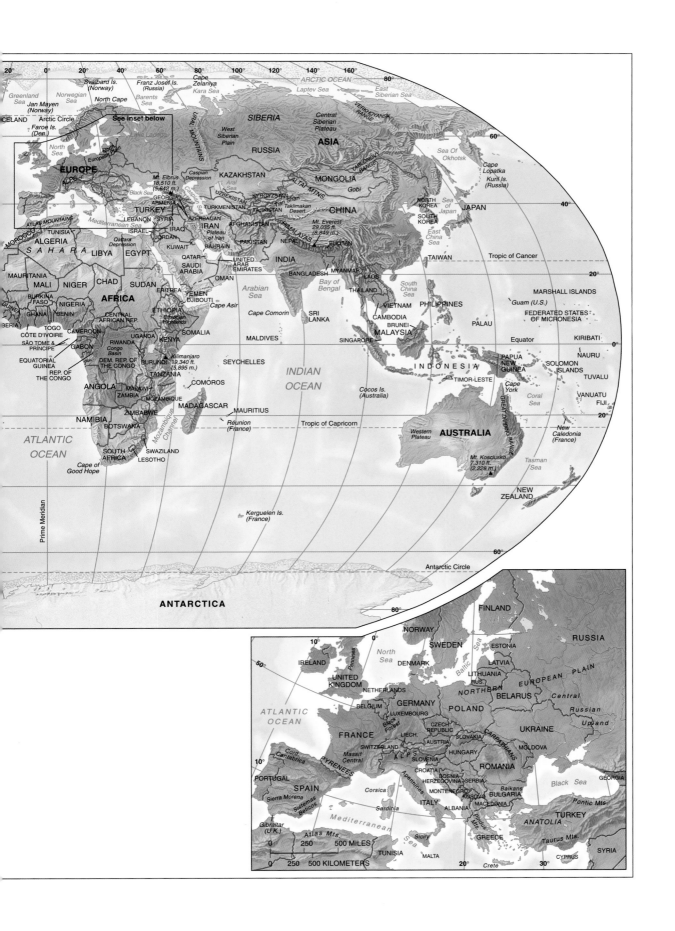

PLATE 9
A tropical rainforest biome typical of wet climates. This example is in Singapore (from istockphoto.com).

PLATE 10
A tropical grassland biome associated with wet-and-dry climates. This is Amboseli National Park, Kenya, with Mt. Kilimanjaro in the background (from enote/Shutterstock).

PLATE 11
A tropical desert biome in the Sahara Desert. In the photo are palm trees, which have their roots in water in an oasis (from Shutterstock © Pawel Kielpinski).

PLATE 12
A coastal desert in the tropics. Sand dunes from the Namib Desert meet the Atlantic Ocean (from istockphoto.com © Alexander Hafemann).

PLATE 13
A green, lush midlatitude forest in Oregon (from istockphoto.com © Jason Lugo).

PLATE 14
A midlatitude grassland in Waterton Lakes National Park, Alberta, Canada. There is enough rainfall here to support a tall-grass prairie (from Fallsview/Dreamstime LLC).

PLATE 15
A midlatitude grassland in a low rainfall area located in northeastern Colorado near the Wyoming border (from Marekuliasz/Dreamstime LLC).

PLATE 16
A midlatitude desert in the Arizona mountains. Most desert vegetation blooms in the spring (from Shutterstock © Paul B. Moore).

PLATE 17
A penguin group marches across the polar desert of Antarctica (from Shutterstock © Armin Rose).

PLATE 18
The foreground shows the tundra-heath vegetation with boulders in the Lac de Gras area of the Northwest Territories, Canada. (Reproduced with the permission of the Minister of Public Works and Government Services Canada, 2001 and Courtesy of the Geologic Survey of Canada).

PLATE 19
The boreal forest covers the Canadian Shield of northern Manitoba, Canada. (Reproduced with the permission of the Minister of Public Works and Government Services Canada, 2001 and Courtesy of the Geologic Survey of Canada).

PLATE 20
The barren, windswept Nightingale Island just off the coast of Tristan de Cuhna in the South Atlantic Ocean illustrates the wet polar climate (© John Ekwall).

TABLE 11.3 Historic Changes in Atmospheric Methane

Time	Parts per Billion by Volume
100,000 YBP	500
70,000 YBP	650
20,000 YBP	350
1750	700
19th century	800
1990s	1600
2005	1774

attained in the past million years; this level is significant even in geologic time. The volume of coal known to exist in reserves is far more than necessary to produce such an increase in CO_2. The amount may well double from the 1980 level in the years between 2050 and 2100. If all the known reserves of fossil fuels were burned, the CO_2 content in the atmosphere would triple. Once the level goes up, it will remain there for centuries because there is no rapid means of removing it from the atmosphere.

OTHER GREENHOUSE GASES

Methane

In addition to CO_2, other gases play a part in global warming. **Methane** is a greenhouse gas found in small amounts in the atmosphere under natural conditions. Each methane molecule is about 21 times as efficient at absorbing Earth radiation as is CO_2. It has begun to accumulate during the past two centuries. The level of methane has risen from about 700 parts per billion (ppb) in 1750 to around 1774 ppb in 2005. The level nearly doubled in the 20th century. About half of the increase has occurred since 1960. The current level is the highest in at least the past 650,000 years (Table 11.3). The additional methane comes from the raising of livestock, rice cultivation, industry, mining, and landfills (Table 11.4). Methane rises in the atmosphere, passing through the troposphere to the stratosphere. In the stratosphere, the sun breaks down methane, and the hydrogen oxidizes to form molecules of water vapor. Water vapor normally is not present in any significant amounts above the troposphere. Each methane molecule produces two water molecules. Computer models show that a doubling of methane in the atmosphere should increase water vapor in the stratosphere by approximately 30%. Enough water vapor exists to produce clouds made up of water and ice crystals at the cold temperatures found at heights of approximately 82 km (51 mi). These clouds reflect Earth radiation back to the surface, compounding the warming.

TABLE 11.4 Methane Emissions from Rice Cultivation by Country, 2005

Country	Millions of Metric Tons 2005
China	224
India	97
Indonesia	51
Thailand	46
USA	8

TABLE 11.5 Global Greenhouse Emissions Since the Beginning of the Industrial Era

Gas	Percentage of Total Emissions
Carbon dioxide	69.6
Methane	23.0
Nitrous oxides	7.0
Other	0.4

In latitudes from about 50° to 60° in the northern hemisphere, much of the land surface is covered by permafrost. Permafrost is ground that is frozen in winter to considerable depth. In summer the surface melts, but it does not melt down far enough to thaw all the soil. The permafrost provides a barrier for gases beneath and water above. A large amount of methane is trapped beneath this frozen soil. Because global warming is most rapid in subarctic regions, rapid melting of the permafrost is taking place in some areas and releasing methane into the atmosphere. The release of methane in subpolar regions is self-perpetuating. If more methane is released into the atmosphere, it will lead to still more warming and more methane release. On a global basis, methane is now second to CO_2 in volume of emissions (Table 11.5). It accounts for about 23% of greenhouse gases. The CO_2 and methane combined account for 93% of greenhouse gas emissions. In the United States, the proportion of greenhouse gases is a little different, as shown in Table 11.6. In the United States, CO_2 is by far the most abundant greenhouse gas placed into the atmosphere.

Nitrogen Oxides

In addition to carbon dioxide and methane, much of the remainder of greenhouse gases is in the form of **nitrous oxide** (N_2O). The main source of N_2O is from the natural soil processes of adding and removing nitrogen. One-third or more of N_2O is now from agriculture. The rest is from non-agricultural land. Concentration of the gas has increased by about 20% since pre-industrial time. Other gases may also have the potential to add to global warming. However, carbon dioxide, methane, and nitrous oxide are now believed to be the main contributors.

Chlorofluorocarbons

The **chlorofluorocarbons (CFCs)** are a group of gases that play a part in the greenhouse gas process. These chemicals, first formulated in the first half of the 20th century, have been widely used in air conditioning, as solvents, and in manufacturing foams.

TABLE 11.6 Greenhouse Gas Emissions in the United States, 2003

Gas	Percentage of Total Emissions
Carbon dioxide	84.6
Methane	7.9
Nitrous oxide	5.5
Other	2.0

These gases are responsible for the ozone hole over Antarctica. Although present in small amounts compared to other gases, their impact on the energy balance is large. Each molecule is 10,000 times more effective at absorbing Earth radiation than is a molecule of carbon dioxide.

A quadrupling of CFCs could produce an increase from 1 to 2°F. Major international efforts to control the increase in CFCs are in place, but in the fall of 2006, the ozone hole was as large as or larger than it has ever been.

Thermal Pollution

Thermal pollution resulting from increasing production and use of energy and the inevitable discharge of waste heat into the atmosphere or ocean is another contributor to global warming. Although it is not yet significant on a global scale, this heat source may become appreciable by the middle of the 22nd century. If future energy production is concentrated in large nuclear power parks, the natural heat balance might be upset even sooner.

CHANGES IN THE TEMPERATURE RECORD

By the year 2000, many scientists and scientific organizations had concurred that Earth was warming at an unusual rate. Evidence of the warming is present in many different forms. The most obvious and mathematically supportable evidence is in the temperature records. Analysis of temperature data gathered over the planet indicates rapidly rising temperatures over the past 140 years. Table 11.7 provides temperature data for the 20th century that illustrates the rise, particularly toward the end of the record. For example, the 20th century was the warmest century in the past millennium, the 1990s were the warmest decade of the century, and 1998 was the warmest year of the century. Temperatures in the world ocean are also increasing to some depth, not just at the surface.

The year 1880 marks the beginning of the historical record. Urbanization has increased the temperature at about one-third of the weather stations that have long records. In addition, most of the earth is covered with an ocean for which little data are available. Global circulation models predict the most warming for the region around the Arctic Basin. Climatic data show that average surface temperatures increased over the past century by 0.4 to 0.8°C (0.7 to 1.4°F). Most of the increases took place in the 1990s.

Other temperature-related data also indicate planetary warming. There has been a reduction in the change in temperature from day to night. Most of this reduced diurnal range is attributed to nights staying warmer—a function of the night air trapping more of Earth's radiation. In parts of the northeastern United States, the warmer

---

TABLE 11.7 Global Temperature Data

2005 was the warmest year on record

2007 tied with 1998 as the second warmest year on record

The 10 warmest years on record have occurred since 1983

Seven of the 10 warmest years on record have occurred since 1990

The 1990s were the warmest decade on record

The 1980s were the second warmest decade on record

The 20th century was the warmest century of the millennium

The mean temperature of Earth increased about 0.6°C (1°F) in the 20th century

---

TABLE 11.8 Warmest Years in the United States, Based on the Annual Average Temperature

Year	Average Temperature (°F)
2006	55.01
1998	54.94
1934	54.91
1999	54.53
1921	54.49
1931	54.34
1990	54.24
2000	54.23
1953	54.18
1954	54.13

nights have increased the growing season by 11 days since 1970. The length of snow cover has shortened considerably in some parts of the northern hemisphere, as spring snow melt is taking place earlier in the year. This in turn affects the annual distribution of rivers and streams in areas that usually have some snow cover in the winter. It is also fairly evident that the warming may be changing the distribution of El Niño and La Niña events. The frequency of El Niño events has increased in the past 25 years. Global temperatures are now warmer than they have been any time in the past 650,000 years. Both Earth-based and space-based data support the most recent global changes in temperature.

The year 2005 was the warmest year on a global basis since records began being kept on a wide scale. It was warmer than 1998, which held the record prior to 2005. The average global temperature in 2005 was about 58.3°F. Earth has been warming at a rate of 0.36°F per decade for the past 30 years. In this time, Earth has warmed slightly more than 0.6°C (1°F).

The year 2006 was the warmest year on record in the United States (Table 11.8). The previous warmest year on record for the United States was 1998. Of the 25 warmest years on record, 15 have occurred since 1981. The warmth appeared most in higher nighttime lows. More than 3000 times, the highest overnight lows were tied or broken in cities around the country. More than 100 locations reported the highest overnight lows for any day of the year. Other events related to temperatures also took place. New York City broke a 129-year record for the latest first snow. The first snow, which was a light one, did not occur until January in 2007. The year 2006 saw the worst wildfire season ever recorded in the United States; approximately 9.8 million acres burned, mostly in the West.

FUTURE CHANGES IN GLOBAL TEMPERATURE

The mean temperature of Earth results from a wide variety of processes. Natural changes in temperature due to unknown interannual variations could be as large as changes predicted due to increases in CO_2. It is also possible that global warming due to CO_2 forcing is much larger than indicated by mean Earth temperatures. Other processes may be cooling the atmosphere.

The climate is always changing. For almost all of the 4.6 billion years the planet has been in existence, the climate has changed due to natural processes. Now, however, climate change is being driven by human actions. Climate change in the near future will primarily result from human activities. So great has become the global population and the impacts from its economic processes that it now dominates changes in our atmosphere on a global basis. If the CO_2 content increases, so will global temperatures.

Estimates vary significantly as to how much the temperature will increase. The estimates vary because they come from different mathematical models of atmospheric processes. If CO_2 concentrations double from the pre-industrial level of 280 ppm, mean world temperatures will likely increase about 1.5 to 4.5°C (3 to 8°F) by 2100. New data issued in the fall of 2008 indicated that with the 3% increase in CO_2 in 2007, global temperatures could reach 2.4 to 6.3°C (4 to 11°F) by the year 2100.

The problem of global warming due to CO_2 is a complex one. Many factors affect atmospheric CO_2. Part of the problem is that other factors cause the temperature to oscillate, independently of the CO_2 in the atmosphere. One atmospheric component that affects global temperature is pollutants in the form of particulates. One of these pollutants is sulfur dioxide. Sulfur dioxide particles serve as very small nuclei for cloud particles. Clouds made of very fine particles reflect more solar radiation than clouds made of large particles. Because sulfur dioxide and carbon dioxide are both products of burning fossil fuels, they work in opposite directions on the energy balance. One leads to warming and the other to cooling. It is possible that increased reflectivity of the atmosphere may be offsetting as much as half of the potential greenhouse warming.

As mean global temperatures increase, the warming will not be the same everyplace on the globe. The subpolar latitudes will warm two or three times as much as equatorial regions. The northern hemisphere has a zone where the southern limit of snow cover is shifting northward from year to year. In this zone, mean temperatures will increase 4°C (7°F) if CO_2 content doubles. If other variables do not play a part, the exponential rise in the atmospheric CO_2 content will become significant. Based on the projected increase in CO_2, the mean temperature at the surface of the earth could increase an additional 2.8°C (5°F).

One set of projections forecasts an increase in global temperature under a "business-as-usual" scenario. Assuming that no controls are placed on greenhouse gas emissions, temperatures will be about 1°C (1.8°F) above present levels (2°C or 3.6°F above pre-industrial levels) by 2025. Before the end of the 21st century, global mean temperature will be 3°C (5.4°F) above the current level. There are several uncertainties in the projected temperature increase. First, we still have an incomplete understanding of the nature of sinks for carbon dioxide. In addition, as may be anticipated, future additions of CO_2 to the atmosphere will depend largely on the actions and policies of governments in relation to fossil fuel use. Finally, the roles of clouds and aerosols in a warmer climate are not yet fully understood. Whether clouds will be more efficient reflectors of short-wave radiation or absorbers of long-wave radiation is not yet clear. Also, the role of the oceans, which influence both the pattern and timing of climatic change, has yet to be fully determined. One thing is certain: Even if major reductions in carbon emissions into the atmosphere are undertaken now, atmospheric warming will continue for several centuries.

To many, it might seem that an increase of but a fraction of a degree of temperature is of little consequence in the scheme of a climate where day-to-day changes are in terms of tens of degrees. Of major significance is that the change would lead to a modification of temperature gradients over the earth's surface, and this in turn would lead to a modification of the general circulation pattern. Changes in precipitation and storm tracks are some of the related consequences. Model calculations indicate that evaporation from water surfaces will increase, and this will lead to greater cloud cover and precipitation for the planet as a whole. Much of this additional precipitation will be over the oceans, with some land areas becoming much drier. In the United States, precipitation from 1970 to 2000 was about 5% greater than in the prior years of the 20th century. However, the frequency of extreme rainfall events (more than 2 in. per day) over the United States appears to have increased substantially in the 1990s.

POLITICS OF CLIMATE CHANGE

Almost all of the debate concerning global warming focuses on the addition of greenhouse gases to the atmosphere. There has been controversy, with some scientists taking a view that the case for such warming and associated change is overstated. Nonetheless, there is a general consensus that global warming is real, with the major areas of disagreement being the interpretation of the rate at which it is occurring and the extent of the impacts of that increase.

Perhaps the best view of the controversy is provided by the final report of the working group of the Intergovernmental Panel on Climate Change (IPCC), which is sponsored jointly by the World Meteorological Organization and the United Nations Environment Programme. The report is based on the scientific assessment of climatic change by several hundred scientists in 25 countries to provide perhaps the most comprehensive view of the subject. The scientists participating in the study are confident that carbon dioxide has been responsible for over half of the enhanced greenhouse effect in the past and is likely to remain so in the future. The atmospheric concentrations of long-lived gases including carbon dioxide, nitrous oxide, and the CFCs adjust only slowly to changes in emissions. Continued emission of these gases at present rates will commit the atmosphere to increased concentrations for centuries to come. To stabilize concentrations at their present levels, emissions of long-lived gases need be decreased by at least 60%.

Since what we call civilization began on this planet some 6000 years ago, the mean temperature has not varied more than 1°C from the average. The forecast change in temperature of from 1.5 to 4°C (2.7 to 7°F) by 2100 has no equal in the recent history of the planet. Based on data related to the temporary storage of heat in the ocean, some scientists are forecasting a change of 3°C (5.4°F) or more. Changes in the global mean temperature of 2°C (3.6°F) or more would have a tremendous impact on global society. The stress of adjustment would be unparalleled: It would alter virtually all aspects of life.

The phenomenon of global warming has become a political issue. Thousands of scientists worldwide agree that warming is occurring. A small but very vocal number of scientists argue that warming is not occurring. In the United States, the large oil, coal, and automobile companies have maintained a continuous campaign to denounce the reality of global warming and advocate that no action should be taken to reduce carbon emissions. This is in the face of the fact that the United States contributes more carbon to the atmosphere than 150 of the smaller countries of the world combined. Acting on their own, some nations are taking steps to reduce their emission of carbon. Denmark currently supplies 25% of its electricity from wind power. It is part of a plan to cut greenhouse emissions by 50% by the year 2030 and meet 40% of power needs from wind power.

Summary

A natural storage process operating in the atmosphere raises the temperature at the Earth's surface from below freezing to about 15°C (59°F). This storage process is the greenhouse effect. It results from some atmospheric gases being fairly transparent to solar radiation but more opaque to Earth radiation. This results in energy flowing back and forth between the surface and the lower atmosphere.

Carbon is a primary element on the planet. It is found in many compounds, including all organic matter. Carbon moves from place to place and from compound to compound. The major reservoirs are in fossil fuels, vegetation, the atmosphere, and the oceans. The amount of CO_2 in the atmosphere is increasing rapidly due to the burning of fossil fuels. Global temperatures have increased since the middle of the 19th century. It is probable that global warming due to anthropogenic CO_2 forcing is responsible for the warming and is much larger than indicated by mean Earth temperatures.

Key Terms

Carbon, *192*
Carbon dioxide, *192*

Chlorofluorocarbons
(CFCs), *198*

Greenhouse effect, *195*
Methane, *197*

Nitrous oxide, *198*

Review Questions

1. What is the largest store of carbon in the earth environment?
2. What is the major sink, or means of removal, of carbon dioxide from the atmosphere?
3. Explain the mechanisms responsible for the daily and annual periodicities in CO_2.
4. How does atmospheric carbon dioxide act differently on solar radiation than on Earth radiation?
5. What are the three ways in which deforestation contributes to the accumulation of carbon dioxide in the atmosphere?
6. What two greenhouse gases account for over 90% of the total emissions?
7. How does a molecule of methane compare to a molecule of carbon dioxide in ability to absorb earth radiation?
8. The daily range in temperature is decreasing. What is the main process responsible for this reduction?
9. Natural and human activities both contribute to climate change. At present, which of these groups of processes is driving climate change?
10. In the past century, which decade recorded the highest average temperatures?

Global Warming and the Physical Environment

Earth is the only planet that we are sure has water in all three physical states—solid, liquid, and gaseous. The balance between the three is delicate. Any global change in temperature will alter the balance between the amount of ice, liquid, and gas. When warming occurs, there is evidence of the net melting of the ice and snow and an increase in the relative amount of water that is present in the liquid and gaseous forms. Environmental data indicate that over much of Earth's history, the planet was warmer than it is now. In the past, when global temperatures warmed, the snow and ice melted, and sea level rose. Water is now present in solid form as continental **glaciers**, **sea ice**, and **snow pack**. Across the earth at the present time, there is a rapid and massive conversion of snow and ice to water. Winter snow pack is declining; mountain glaciers and continental **ice sheets** are melting. Sea ice is declining in areas and getting thinner.

Melting of snow and ice from the ice caps and mountain glaciers has been particularly rapid since 1950. The rate of melting far exceeds the rate for thousands of years before the present. So rapid is the melting that in many areas, the mountain glaciers will be gone before the end of the century. This includes those in Glacier National Park and the famed snows of Kilimanjaro.

The polar regions are where global warming was expected to appear first and show the most rapid change. This is because these two areas typically have the greatest difference between summer and winter and because an equal amount of additional heat will produce a greater change in a cold environment than it would in an already warm one, such as in the tropics.

Continental glaciers contain a wealth of information about the recent history of the planet. Each winter season adds a new layer of snow on top of the ice. In some years, the deposit of new snow is thick, and in other years, it is thin. This provides a record of annual precipitation on the ice. Embedded in these layers are thin layers of dust and volcanic ash. The layers of dust represent times of drought, when the winds were picking up debris on the continents and carrying it over the ice sheets. The volcanic ash indicates when there were eruptions and how violent they were.

There are also chemical and biological differences in each layer of ice. Different isotopes of oxygen form in the ice, depending on how warm or cold the weather was at the time. Bubbles of air trapped in the ice indicate the chemistry of the atmosphere at the time the bubbles were trapped. Different spores and seeds also indicate the type of vegetation that was growing upwind of the ice.

ANTARCTIC WARMING

The Antarctic region and the Arctic region are very different from each other. The Antarctic is a continent surrounded by sea. The largest continental ice sheets are on the Antarctic continent. These glaciers show signs of decay and rapid movement. The **West Antarctic Peninsula**, which extends toward South America, warmed by more than 2.5°C (4.5°F) over the last half of the 20th century. Midwinter temperatures increased between 4 and 5°C (7 and 9°F) on the western Antarctic Peninsula during the same time period. One of the large ice sheets on Antarctica is the West Antarctic Ice Sheet. Much of the huge mass of ice sits on ground that is below sea level. The ice sheet contains some 29,400 km^3 (11,500 mi^3) of ice. Were all of the West Antarctic Ice Sheet to melt, it would raise sea level 4.5 to 6 m (15 to 20 ft). This would become a global problem of major proportions. The biggest ice sheet is on Antarctica. This ice sheet is not likely to disappear altogether, but it might be substantially reduced.

The huge ice shelves on the edges of the continent are the key to Antarctic glacial retreat. Ice shelves are floating masses of ice frozen to the land mass. These ice shelves have always shed icebergs. Large pieces of ice occasionally break off the edges of these floating ice masses and drift away from the land mass. The largest one ever noted broke from the shelf in 1956. What caused this piece to break off is unknown, but it was probably due to natural attrition. In March 2000, the second largest iceberg ever measured, named B-15, broke free from the Ross Ice Shelf and floated into the Ross Sea. The huge iceberg was approximately 11,000 km^2 (4247 mi^2) and was about 300 km (180 mi) long and 37 km (18 mi) wide. Within a week of this iceberg's formation, three more large pieces broke free from the ice shelf.

Ice shelves along the West Antarctic Peninsula are either rapidly melting or have broken up altogether. Three of the ice shelves disintegrated in the last decade of the 20th century. Five of the ice shelves located on either side of the peninsula have been melting at an ever-increasing rate. These ice shelves are the Prince Gustave Channel, Larsen A, Larsen Inlet, Muller Ice Shelf, and Wordie Ice Shelf. The Larsen B and Wilkes shelves had an area of about 21,000 km^2 (8100 mi^2) in 1997. These two shelves lost 3000 km^2

TABLE 12.1 Recent Collapse of Ice Shelves

Ice Shelf	Location	Year	Size
Ayles	Arctic	2005	106 km^2 (41 mi^2)
Larsen A	Antarctic Peninsula	1995	4200 km^2 (1620 mi^2)
Larsen B	Antarctic Peninsula	2002	3100 km^2 (1200 mi^2)
Markham	Arctic	2008	49 km^2 (19 mi^2)
Ross	Antarctic	1987	4,750 km^2 (1834 mi^2)
Ross	Antarctic	2000	11,000 km^2 (4,250 mi^2)
Serson	Arctic	2008	lost 50% of area
Ward Hunt	Arctic	2003	384 km^2 (150 mi^2)
Wilkes	Antarctic	2008	416 km^2 (160 mi^2)

(1180 mi^2) of ice in just two years from 1998 to 2000. In 2002, a section broke from the Larsen Ice Shelf that was 3100 km^2 (1200 mi^2) in size. On February 28, 2008 a piece of ice approximately 416 km^2 (160 mi^2) broke off the edge of the Wilkes Ice Shelf. In contrast to the huge pieces breaking off from the Ross Ice Shelf, these shelves are disintegrating in small pieces, which is usually the case with warming temperatures (Table 12.1).

The breakup of these ice shelves will not have a major effect on sea level because they are already floating in water. The density of the ice is about 9% less than that of fresh water. The volume of water in the iceberg is almost the same as the volume of water the iceberg displaces. However, in many areas, these ice shelves hold back glaciers in the ice sheet from advancing seaward. The Ross Ice Shelf is an example. It is one of the largest shelves on the West Antarctic Ice Sheet. The tides lift and lower the shelf twice a day. When the shelf is lifted during high tide, the glacier behind the shelf moves forward. If the Ross Ice shelf melts, it will allow the glaciers behind to move forward, adding ice to the sea and raising sea level. What is of most concern is how the breakup of these ice shelves will affect the glaciers that are driving the ice shelves seaward. It is possible that the removal of these ice shelves will result in more rapid advance and faster melting of the glaciers. The Marr Ice Piedmont, Anvers Island, has retreated about 0.5 km (0.3 mi) in the past 50 years. Winter temperatures in the area have increased some 5°C (9°F) in the same time.

ARCTIC WARMING

Greenland

Greenland is Earth's largest island. At the present time, the island is some 85% covered with ice in the form of glaciers. The ice on Greenland is about 3.2 km (2 mi) thick in some places. However, the glaciers are melting rapidly. On the west coast, approximately 296 km (185 mi) north of the Arctic Circle, is the town of Ilulissat (which means "place of the icebergs"). In this part of Greenland, the mean winter temperatures have gone up 5°C (9°F) in the past 15 years. Generally, the effects of global warming are greatest in higher latitudes, but this rise is phenomenal. It is highly unlikely to occur, but if Greenland's ice cap were to completely melt, it would result in a rise of 7 to 9 m (23 to 30 ft) in the global ocean. The Ilulissat Glacier is a World Heritage site. In the past, it moved toward the sea in the summer, calving huge icebergs. It ceased moving in the winter. Now it slides into the sea year-round. In 1992, the glacier was moving seaward at a rate of 5.6 km (3.5 mi) per year. In 2007, it was moving nearly twice as fast.

Melting of the Greenland Ice Sheet increased 30% between 1979 and 2007. Melting in 2007 was 10% greater than in any past year. The loss of ice is now nearing 200 km^3 (48 mi^3) per year. The Greenland ice sheet is becoming increasingly porous, and there is increasing calving of ice around the edges of the ice sheet. This ice sheet is not likely to disappear altogether, but melting may reduce it to nearly one-third its present size. Because this ice is all above sea level, when it melts, the meltwater adds to the volume of water in the ocean, causing sea level to rise. The projected melting could raise sea level 0.15 to 0.6 m (0.5 to 2 ft). If there is substantial slipping at the base, it will rise more. If the entire ice sheet melted, it would raise sea level between 7 and 9 m (23 and 30 ft.).

There is now a proposed project to drill ice cores all the way to the bottom of the ice. This is expected to produce a continuous record of climatic data going back perhaps as far as 130,000 years. Between 115,000 and 130,000 years ago, the climate of Greenland was some 5°C (9°F) warmer than it is now.

Greenland is a self-governing territory of Denmark. The human population is approximately 56,000. Eastern Greenland was settled in the year 982 by Norse farmers from Iceland. The Inuit settled western Greenland at about the same time. Some 85% of the current permanent residents are Inuit. Many, if not most, have made their living from fishing. However, tourism is a rapidly growing source of income. Airlines fly passengers from Denmark and New York to the island in the summer. The winter season has shortened a great deal. In the past, winter travel was mostly by dogsled. Traditionally, sledding began in October, but now it is as late as February before there is enough snow and ice for sleds. The harbors are now ice free for longer periods, which extends the fishing season. The changing climate is changing a whole way of life for the permanent residents of the town. A plus is the regularly scheduled commercial passenger service that operates in the summer and the growing tourist trade.

Sea Ice

In contrast to the Antarctic, the Arctic region is essentially a sea surrounded by land. In the winter months, the sea is covered with a broken sheet of floating ice up to perhaps 4 m (13 ft) thick. Even in winter, there is always some open water. There are open cracks in the ice as the rotation of the earth causes stress in the ice. These cracks continually open and refreeze.

In summer, the ice melts away from the shores of the continents and islands for some distance. The ice is thinner on the edges and so it thaws more readily. Over much of the sea, away from shore, there is sea ice that does not melt in the summer. This is thicker than ice near land masses. The sea ice has declined throughout the Arctic Basin, but it has receded more in some places than others. Similar melting has taken place among the Canadian islands. The ice has been thawing further from shore, and the remaining perennial ice pack has been getting thinner. In some areas, it is only half as thick as it was a few decades ago. The present area of sea ice is about 125,000 km^2 (50,000 mi^2). This ice pack has been melting faster in summer at least since the beginning of the 20th century. The summer ice pack has lost 40% of its area since 1958. Over the past three decades, more than 2.5 million km^2 (1 million mi^2) of ice has been lost in summer. The rate of loss is now about 10% per decade. The melting of sea ice has been taking place at an ever-increasing rate, and it is now taking place much faster than predicted even a decade ago.

Satellite images of the Arctic ice have been available since the 1970s. On August 17, 2007 these images show the least sea ice ever recorded—76,000 km^2 (30,000 mi^2) less ice than had previously been detected (Figure 12.1). The months of June and July were particularly cloud free. This is the time of year when sunshine is most intense in the Arctic. Around the summer solstice in late June, the sun is highest in the sky, and the days are

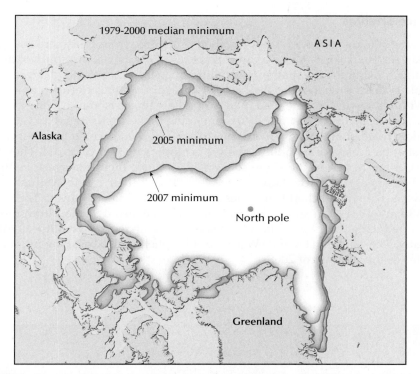

FIGURE 12.1   A comparison between the extent of sea ice at the end of the summer melting period in 2007 and the average for 1979 to 2000. The extent of the sea ice in September 2007 was 39% below the long-term average for 1979–2000. (From Tarbuck E. J., Lutgens F. K., & Tasa D., *Earth Science*, 12th ed., © 2009, Pearson Prentice Hall, Upper Saddle River, N.J.)

the longest in the northern hemisphere. The long days and intense sunlight are a primary factor in the melting at this time of year. The effect of the sun on snow and ice is very different from that of sun on open water. When the sea is covered with snow and ice, the bright surface reflects approximately 80%, or more, of the solar radiation. This reflected sunlight does not alter the frozen surface—it simply bounces off. However, when the surface is open water, the water absorbs up to 90% of the radiation. Much of this absorbed solar radiation is converted to heat energy and warms the water.

The maximum extent of the sea ice in the winter season has also been declining. The six winters ending with the 2008-2009 winter all had below the average sea ice cover dating back some 30 years. These last six years were also the six years with the lowest maximum sea cover.

Early forecasting of Arctic warming suggested a possible ice-free Arctic Sea in the summer months perhaps between 2050 and 2100. Based on recent rates of melting, forecasts suggested a possible ice-free sea in the summer as soon as 2030. However, to the astonishment of nearly everyone, at the end of the summer melt period in 2007, there was an open passage of water circling the Arctic Sea. It was hailed as the opening of the Northwest Passage from the Atlantic to the Pacific. In August 2008, sea ice had shrunk to a level surpassed only by that of 2007. In September 2007, Arctic sea ice was only 4.27 million km^2 (1.65 million mi^2). In August 2008, the area covered by ice was measured at 5.26 million km^2 (2.03 million mi^2). Based on the data from these two summers, the Arctic could be ice free in summer by 2020. The evidence of change in sea ice parallels other evidence that global warming is taking place much faster than climate models forecast. Whether the passage remains open in subsequent years remains to be seen. At its current rate of melting, the Arctic Ocean could be totally ice free within this century.

Arctic Ice Shelves

In 2005, the 106 km² (41 mi²) Ayles Ice Shelf broke free from Ellesmere Island in the Canadian Arctic. The **ice shelf** was approximately 800 km (500 mi) south of the North Pole. In fact, using a combination of satellite imagery and seismic data, researchers were able to determine that the shelf actually broke away the afternoon of August 13, 2005. Within a few hours, the mass had formed a new ice island, trailing bits of broken ice. The ice shelf was just one of six remaining in the Arctic region of Canada. The ice island drifted some 49 km (30 mi) out to sea and then became frozen into the sea ice. Most of the ice making up these shelves is more than 3000 years old. They float on the sea but are frozen to the shore (Figure 12.2). Since their discovery in 1906, just over 100 years ago, the ice shelves have lost 90% of their ice. In August 2008, the Markham Ice Shelf detached from Ellesmere Island and floated away. The shelf was about 48 km² (19 mi²) in size. Two large pieces of ice also detached from the Serson Ice Shelf, reducing the overall size of the shelf to less than half what it was the previous winter. The 440 km² (170 mi²) Ward Hunt Ice Shelf is also breaking up. It lost 10% of its area in the summer of 2008.

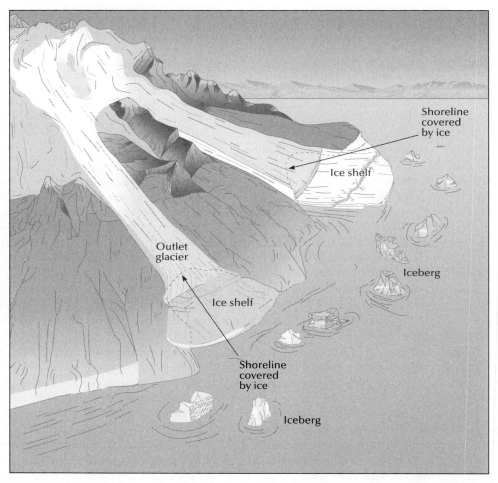

FIGURE 12.2 When either an ice sheet or an outlet glacier reaches the ocean, some of the ice may float out over water as an ice shelf. When the ice shelf breaks up, the pieces float away as icebergs. (From McKnight T. L. & Hess D., *Physical Geography: A Landscape Appreciation*, © 2005, Pearson Prentice Hall, Upper Saddle River, N.J.)

Melting Permafrost

In Alaska, the **permafrost** is melting. The mean annual temperature of the layer above the frozen subsoil rose 3.5°C (6.3°F) between the late 1980s and 1998 in the area north of the Yukon River. The permafrost layer warmed 0.6 to 1.5°C (1 to 2.7°F) south of the Yukon River. In this region, not all the land is underlain by permafrost. Along the boundary between the areas with and without permafrost, the temperatures within the soil are close to the freeze/thaw temperature. Just a slight increase in temperature can melt permafrost over large areas. Along this boundary, the permafrost is melting at a rate of about 1 m (3.3 ft) every 10 years. Around the base of Mt. McKinley, large areas of Arctic meadow have turned into a mix of ponds containing water-loving plants and dry meadows. Much of the melting of the permafrost took place between 1980 and 2000. Permafrost is not restricted to North America. Somewhere between 20 and 25% of Earth's land area contains permafrost. In China, the permafrost is estimated to be melting northward at a rate of about 1.6 km (1 mi) per year (Figure 12.3).

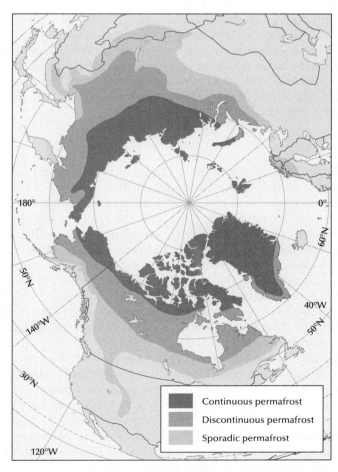

FIGURE 12.3 Extent of permafrost in the northern hemisphere. (From McKnight T. L. & Hess D., *Physical Geography: A Landscape Appreciation*, © 2005, Pearson Prentice Hall, Upper Saddle River, N.J.)

MELTING OF EARTH'S MOUNTAIN GLACIERS

Asia

The Himalaya Mountains are home to approximately 18,000 glaciers. These glaciers cover an area of about 33,000 km² (13,000 mi²). In the adjacent Tibetan Plateau, another 46,000 glaciers cover an additional 155,000 km² (60,000 mi²). Some, if not most, of these glaciers are melting at an astonishing rate. In the past 50 years, the areas covered by glaciers in the Tibetan Plateau shrank by more than 5%, and their volume was reduced by more than 7%. The meltwater has contributed some 360 km³ (90 mi³) of water to the sea.

The changes in runoff are creating severe physical and economic problems. In many parts of the world, the summer melting of mountain glaciers provides household water and water for irrigation schemes on the lower slopes of the mountains or on surrounding lowlands. Rivers that are fed by the mountain glaciers are experiencing greater variation in the seasonal flow. There is more water flowing downstream in the peak runoff season, causing flooding. Early summer melting and flooding is countered by low winter flow and drought. The warmer summers also result in a shorter runoff season. This in turn reduces the length of the irrigation season. Four of the great rivers of India—the Brahma Putra, Ganges, Indus, and Sutlej—are among those affected by the changing glacier melting.

The meltwater from these glaciers is forming mountain lakes faster than they can be identified and catalogued. These lakes represent a major hazard to areas downstream from them. The reason is most are dammed by rocks, gravel, and other debris pushed forward by the glaciers as they advanced down the valleys during colder times. The material making up these dams is loose and subject to rapid disintegration when flooded. When these earthen dams break, they can unleash millions of cubic meters of water downstream in the space of a few hours. The downstream flooding can be catastrophic in terms of lives and property. One of these lakes, Imja Tsho, is now about 1.6 km (1 mi) long and in places 90 m (295 ft) deep. This lake did not even exist 50 years ago. There are many others just like it.

Africa

The melting of snow and ice in mountain glaciers is taking place near the equator. In Kenya there are two mountains near the equator that sustained glaciers through historic times. On both mountains, the glaciers are melting rapidly. On Mount Kenya, which stretches to a height of 5153 m (16,902 ft), the glaciers have shrunk some 80%. They are now barely visible from the surrounding countryside. The rivers that drained these glaciers used to run most of the year, and the water was used for irrigating farm crops on the lower slopes of the mountain. The meltwater has been decreasing in recent years, and the rivers are declining to very low levels. One river, the Naro Moru, now vanishes into the sands. Livestock is dying of thirst, and nomadic farmers are raiding upstream projects to destroy diversion schemes for irrigated agriculture. There simply is no longer enough water to go around, and survival depends on water.

Mt. Kilimanjaro is about 320 km (200 mi) south of Mt. Kenya and just south of the equator. The Snows of Kilimanjaro were first described by unbelieving Europeans in 1848. The snows became famous through the media, and partly because the mountain is visible from the capital city of Nairobi. As on Mt. Kenya, the glaciers that have topped Kilimanjaro for so long have all by disappeared over the past 100 years. Just as on Mt. Kenya, there have been irrigation schemes on the lower slopes of the mountain using the meltwater, and a very productive agricultural industry has prospered. Water is now in short supply here, and fighting over water rights is commonplace.

The same retreat of glaciers has been taking place in central Africa, in the Rwenzori Mountains of Uganda. The final melting of glaciers in these mountains is forecast to take place in the next 20 to 40 years. Estimates are that the three highest

TABLE 12.2 Melting of Earth's Ice

- The edge of the West Antarctic Ice Sheet is shrinking at the rate of 120 m (400 ft) each year.

- The Larsen Ice Shelf on Antarctica disintegrated in January 1995.

- Much of Antarctica's Larsen B and Wilkes ice shelves disintegrated in 1998–1999.

- The average elevation of glaciers in the Southern Alps of New Zealand moved upward 91 m (300 ft) in the 20th century.

- In the Tien Shan Mountains of China, glacial ice has decreased by nearly 25% over the past 40 years.

- In the Caucasus Mountains of Russia, half of all glacial ice has melted away in the past 40 years.

- In the Garhwal Himalayas of India, glaciers are rapidly retreating.

- In the Andes Mountains of Peru, glacial retreat increased sevenfold from 1978 to 1995.

- The area of sea ice in the Bering Sea has decreased 5% since 1975.

- The area of sea ice in the Arctic Ocean has shrunk by 35,000 km^2 (14,000 mi^2) since 1978.

- The largest glacier on Mt. Kenya almost completely melted away in the 20th century.

- Bering Glacier in Alaska is retreating.

- Glaciers in Glacier National Park, Montana, are melting rapidly.

- Glaciers in the Alps Mountains of Europe shrank by about 50% in the 20th century. Gruben Glacier and Aletch Glacier in the Alps are among those rapidly melting.

mountains have lost 80% of their ice since 1850, with most of the loss occurring in the past 40 years.

Europe and the Americas

The glaciers in the Alps of Europe might be gone within 30 to 40 years. In the province of Tyrol in Austria, glaciers have been shrinking about 3% per year. Rapid melting and flooding from the glaciers caused a great deal of damage in Switzerland in 2005. Measurements in the Alps indicated that the ice decreased in thickness an average of 70 cm (27.5 in.) in 2004 and 60 cm (23.5 in.) in 2005. In the great European heat wave of 2003, an estimated average of 2.1 m (7 ft) of ice melted from the glaciers. The Pasterze Glacier in Austria has been closely monitored since 1889. It has been slowly receding ever since. In recent years, the ice has been retreating an average of 15 m (50 ft) each year. However, in 2003, the glacier retreated 30 m (100 ft) in length and 6.3 m (21 ft) in thickness.

In South America, most mountain glaciers are retreating rapidly. The Cordillera Blanca, which has the most glaciers in the tropical regions, is shrinking rapidly.

OCEAN PROPERTIES

The global ocean is being altered by atmospheric warming in a number of ways: the water is warming, salinity is changing, acidity is changing, circulation patterns of ocean currents are changing, frequency and intensity of marine storms and El Niño and La Niña events are changing, and marine ecosystems are changing.

Ocean Temperatures

Earth's ocean warmed rapidly over the last four decades of the 20th century. An extensive study was conducted of ocean temperatures over the last half of the 1900s, using data from all over the world. Millions of individual temperature recordings were gathered and combined into a single large group. The analysis of the data shows a clear pattern. In the period from 1955 to 1995, the world ocean warmed by an average of 0.6°C (1°F) from the surface down to about 3000 m (9900 ft). The Indian Ocean warmed 0.3°C (0.5°F) down to a depth of 800 m (2600 ft) in a period of 20 years. The depth to which warming can be detected is also increasing. Ocean warming represents the additional storage of tremendous amounts of heat. Warming of the ocean is temporarily slowing the rate of atmospheric warming. As this additional heat stored in the surface waters of the ocean is released to the atmosphere, it will cause global air temperatures to rise at least another 0.5°C (0.9°F) by 2100. This is separate from the warming caused by the greenhouse effect of the atmosphere itself.

Rising Sea Level

The processes that determine sea level are many and complex. The level of the ocean can change if the volume of the ocean basins changes. The accumulation of sediments and addition of magma to the ocean floor are examples of changes in the volume of ocean basins. At any specific coastal location, it is very difficult to measure actual sea level change because both tectonic and climatic changes determine the relative height of sea level. Most changes in the ocean basins require long periods of time. Submergence and emergence of coastal areas and marine terraces point to the amount of water previously stored as ice. Sea level can rise also as a result of the subsidence of coastal land areas.

Changes in the ocean level can occur if the volume of water in the ocean changes. Rapid changes result mostly from ice alternately forming and melting. Several warm periods in the past have provided information about environmental conditions that would accompany global warming:

- The Climatic Optimum that occurred 100 million years ago
- The Pliocene, from 2 to 5 million years ago
- The last interglacial, 125,000 years ago
- The Holocene Climatic Optimum, 6000 years ago
- The Medieval Warm Period in the 12th century

Analysis of various kinds of data shows that the last interglacial occurred some 125,000 years ago. On Greenland, the ice sheet melted away enough to expose the land beneath in some areas. The ice sheet was smaller in both area and depth. Coral fossils from the last interglacial are about 6 m (20 ft) above present sea level. This shows that sea level was at least 6 m (20 ft) higher during the last interglacial than it is now. If the remaining ice on Greenland at present were to melt, sea level would rise the same amount. The Pleistocene Period provides much information about temperature change. During the Pleistocene, there were several episodes of rapid growth and melting of ice. The height of the ocean is the net record of melting of the ice sheets. As ice on the continents alternately advanced and retreated, sea level fell and then rose in response. During periods of warmer climate, sea level rose. Glaciers melted, adding water to the ocean, and the water expanded from the additional heat. The fall and rise of sea level during the Pleistocene is an important guide to glacial and interglacial periods. With appropriate dating methods, such evidence provides a guide to glacial advance and retreat and hence global temperatures.

The end of the last glacial advance was about 21,000 years ago. Some 5% of all planetary water was in the form of ice. This amounted to 58 million km^3 (14,000 mi^3) of water withdrawn from the ocean. There was a staggering 50 million km^3 (12 million mi^3) of water stored in the northern hemisphere ice sheets alone. This lowered sea level by a large amount. Evidence from the Gulf of Mexico suggests that the ocean was between 125 and 140 m (413 and 462 ft) below its present level.

At the height of the last glacial advance, continental margins were very different from those of today. Once the ice sheets began to melt 11,500 years ago, sea level rose rapidly, if not steadily, 50 to 100 m (165 to 330 ft). The level stabilized for a time, reflecting the cooler conditions around the North Atlantic Ocean. When the climate continued to warm, sea level rose again. It has been rising since then. Over the past 6000 years, sea level has been within about 1 m (33 ft) of its present level.

Sea level rose 10 to 25 cm (4 to 10 in.) in the 20th century. The rise was due mostly to expansion of the sea water due to warming. Sea water has its greatest density at about the freezing point. As sea water warms above this point, the water slowly expands. Melting of glaciers also contributed to the rise.

Future Sea Level

The causes of sea level rise, the absolute amount of sea level rise, and the rate and timing of the rise are uncertain. As the earth warms, sea level will rise due to expansion of the water and the addition of melt water. The effect of global warming will be to warm the water in the ocean. Warmer temperatures cause sea level to rise due to the thermal expansion of sea water. A temperature increase between 0.6 and 1.0°C (1.2 and 1.8°F) might cause mean sea level to rise 40 to 80 mm (1.6 to 3.2 in.) by the year 2025. The melting of glacial ice is most likely to be the reason for sea level rise in the coming centuries. If there is a rapid melting of sea ice, sea level rise could increase 1 or 2 m (3 to 6.5 ft) by 2100. A rise of even half this much would lead to flooding of 12,500 km^2 (5,000 mi^2) of dry land and another 10,000 km^2 (4000 mi^2) of wetlands. There is a real hazard of rapid sea level rise if mean global temperatures rise 2°C (3.6°F) or more.

Two major geographic areas may contribute to rising sea level: the ice on the southern part of Greenland and the West Antarctic Ice Sheet. The West Antarctic Ice Sheet has its base on the sea floor and hence would respond rapidly to changes in water and atmospheric temperatures. At present, the summer temperature over the West Antarctic Ice Sheet averages −5°C (23°F). If summer temperatures rise above freezing, very rapid melting would occur. This would result in a rise of sea level of 5 to 6 m (16 to 20 ft). Also, if any sudden surge of part of a continental ice sheet took place, sea level would rise quite rapidly.

If the main Antarctic Ice Sheet were to melt, there would be enough additional water to cause a sea level rise of at least 60 m (200 ft). The removal of ice from such a large land area would result in an upward movement of the land where the ice melted away. Complete melting of both the Antarctic and Greenland Ice Sheets would raise sea level about 70 m (230 ft). However, it would take many of years for the ice sheets to respond to a global warming of several degrees. This may be beyond the range of the possible effects of the CO_2 changes. It is not even certain if warming of the seas would cause the ice caps to grow or to shrink. The water might warm slightly, but not enough to raise the temperature of the atmosphere over the ice above freezing. In this case, snowfall could increase, and glaciers might expand causing sea level to drop. Table 12.3 provides data on potential sea level rise due to thermal expansion and melting of ice sheets.

TABLE 12.3 Potential Sea Level Rise

Source	Amount
Thermal expansion	0.09 to 0.88 m (4 to 35 in.)
Greenland Ice Sheet	7 to 9 m (23 to 30 ft)
West Antarctic Ice Sheet	4.5 to 6 m (15 to 20 ft)
Entire Antarctic Ice Sheet	70 m (230 ft)

CHANGING WEATHER SYSTEMS

One of the widespread aspects of climate change is the documented warming. Another widespread phenomenon associated with the warming is the increase in extreme weather events. Most areas of the world are experiencing changes. The following are just a few examples.

European countries are experiencing rapid change in weather. The speed and amount of change are higher than in much of the rest of the world. In the summer of 2003, an extreme heat wave spread over much of Europe. More than 35,000 people perished, and there was a huge loss of crops due to the heat. A large contributing factor in the high death toll was warmer nighttime temperatures. As a result of the warmer-than-normal nighttime temperatures, people without air conditioning could not cool down during the night, and the heat stress accumulated over time.

The United States has experienced record storms. There is growing evidence that the strongest hurricanes have become stronger since 1980. There are more Category 4 and 5 hurricanes. From 1981 to 2006, the highest wind speeds increased from 225 to 250 kph (140 to 156 mph). Warmer air over warmer water holds more water vapor. As this added vapor condenses, it adds more heat and more water to the atmosphere. This creates higher wind velocities and more precipitation. Hurricanes in recent years could have produced as much as 8% more precipitation due to global warming.

There is also some evidence that the frequency of hurricanes may be increasing. Prior to 2005, the previous record for the number of hurricanes was 14 in a season. There were 26 named storms in 2005, and 14 of them became hurricanes. Among these were 5 storms that reached Category 5. Four of the named storms reached the United States, which set another record. Among these was Katrina. Whether there has indeed been an increase in frequency in tropical storms is not yet certain.

In the United States, the number of weather-related events resulting in at least $1 billion of insured damage increased from 3 in 1980 to 25 in the 1990s. In May 2003, there were more than 400 tornadoes in just 10 days.

Data indicate that El Niños are increasing in frequency and intensity. Formerly they averaged every 7 years. Starting in the 1970s, the frequency began increasing. The pool of warm water in the western Pacific is getting larger and warmer. The El Niño of 1997 and 1998 caused tremendous disruptions. There was extreme flooding on the Korean Peninsula in the summer of 1998. Flooding on the Yangtze River caused 3000 fatalities. Hurricane Mitch in the Caribbean Sea drove 230 million people from their homes and resulted in approximately 11,000 deaths in Central America.

The frequency and severity of droughts, floods, and severe storms has been increasing on all the continents.

Summary

Warming is a global phenomenon. Air temperatures are increasing worldwide, most rapidly in upper midlatitudes. Winter temperatures are warming faster than summer temperatures. In the Antarctic, sea ice is retreating and icesheets are melting due to the warming. The same is true in the Arctic basin. The Greenland ice sheet shows much evidence of retreating in some areas and thinning in others. An open water route opened across the Arctic from the Atlantic to the Pacific in 2007.

Mountain glaciers and snow fields are exhibiting signs of rapid melting, even near the equator. The glaciers are thinning and retreating. The melt water is producing new lakes and also spring flooding downstream. The result in many areas is reduced runoff in summer with major rivers drying up at times.

The oceans are warming. This leads to the expansion of the sea water and a rise in sea level. Additional sea level rise is resulting from the melting of glaciers. The change in the energy in the atmosphere, and the distribution of that energy is resulting in more varied weather, more extreme storms, and distinct changes in seasonal weather in some areas.

Key Terms

Glacier, *205*	Ice shelf, *210*	Sea ice, *205*	West Antarctic
Ice sheet, *205*	Permafrost, *211*	Snow pack, *205*	Peninsula, *206*

Review Questions

1. List four ways in which global warming affects the world ocean.
2. What is the difference between a glacier and an ice shelf?
3. What evidence is there that global warming is affecting fresh-water lakes?
4. Why are the polar regions experiencing greater change due to warming than are tropical regions?
5. Why is the West Antarctic Peninsula warming faster than the rest of Antarctica?
6. Why is current melting of the ice sheet on Antarctica and Greenland so important in regard to the hydrologic cycle?
7. What are the two processes that will contribute to future rise in sea level?
8. How much is sea level forecast to rise by the year 2100?
9. In which two countries is most of the earth's permafrost located?
10. How does the collapse of ice shelves affect sea level?

CHAPTER

13 Climate Change and the Living World

TERRESTRIAL ECOSYSTEMS

The probable effects of the increase in carbon dioxide on the biosphere are considerable. The current increase in CO_2 is not a direct threat to humans. Carbon dioxide is not normally toxic, as humans can tolerate concentrations of up to 15 times present levels before lethal results occur. Carbon dioxide is, however, a growth stimulant to green plants. Plants grow bigger and faster, contain more vitamin C and sugar, and are more disease resistant when grown in atmospheres with high concentrations of CO_2. Experiments show that some vegetables may grow to a weight as much as 800% above normal in carbon dioxide–rich environments.

Climate change is another matter. Historic climate changes, such as the ice ages, have led to extinction of many species. As the planet warms and climate changes, plant and animal species must adapt or perish. Evidence of migration and adaptation are widespread. Forests, other plants, and animals may have difficulty migrating as fast as climate zones migrate. The presence of urban areas, agricultural lands, and roads restrict habitats and block migration of some species. These obstacles also make it harder for plants and wildlife to survive climate changes. Thus the population size and geographic range of many species are changing as temperature and rainfall change. Many plant, animal, and insect species are moving poleward. In mountain areas, animals are moving to higher elevations as warmer temperatures move upward. Recently, human activities have accelerated the rate of species extinction. Climate warming will most likely lead to an even greater loss of species. Natural ecosystems may suffer in negative ways and in ways we cannot predict.

Growing Season Length

Some 35% of growing global vegetation is in the latitudinal zone from 45°N to 70°N latitude in the month of August. Thus more than one-third of growing vegetation is in the midlatitudes, which have a seasonal temperature regime. Many plant species time their growth season to the length of day. Many others time their leafing and blooming on their own internal temperature. In North America and elsewhere in the northern hemisphere, spring is coming earlier. It is primarily plants that time their leafing and blooming to temperature that are starting their spring growth cycle early. In the northern United States, the spring green-up has been taking place an average of eight hours a year earlier since 1982, and it is occurring more rapidly with time. The greater growth and earlier green-up are the result of warmer temperatures.

While the earlier spring may benefit some plants, it can cause the virtual extinction of others. The problem is the symbiotic relationship between some plants and animals. For example, many species of birds are changing their time of spring migration northward from warm environments. At the Seney National Wildlife Refuge in Michigan, migratory birds are arriving an average of 21 days earlier than they did in 1965. If plant species green-up and bloom before animal species arrive in their migration pattern, or before they come out of their winter shelter, the animals may die of starvation if they cannot find another source of food. In the fall, the time of the first frost is also moving later in the year, but at a slower rate than the spring green-up.

In 2006, in response to warmer temperatures in parts of North America, the National Arbor Day Foundation issued a new hardiness zone map for the United States. The temperature zones are based on average annual low temperatures.

North American Forests

Forests occupy one-third of the land area of the United States. Temperature and precipitation are of major importance in the distribution of forests. Climate change could move the southern boundary of hemlock and sugar maple northward by 600 to 700 km (370–435 mi). Poleward migration of the forests lags behind shifts in climate zones. If there is a migration rate of 100 km (60 mi) per century, or double the known historic rate, the geographic area of forests will drop because the southern boundary may advance more quickly than the northern boundary.

Climate change may significantly alter forest composition and reduce the area of healthy forests. Seedlings of trees now growing in the southeastern areas may not grow because of higher temperatures and dry soil conditions. In central Michigan, grasslands may replace forests now dominated by sugar maple and oak. In northern Minnesota, the mixed boreal and northern hardwood forests could become entirely northern hardwoods. The process of change in species composition would most likely continue for centuries.

Other factors will also influence forests. The health of forests is not determined by climate change alone. The drier soils expected to accompany climate change could lead to more fires. Warmer climates may cause changes in forest pests and pathogens. On the Kenai Peninsula in Alaska, warmer temperatures are allowing spruce pine beetles to reproduce at twice their historic rate. More than 4 million acres of spruce forest has been killed. This is the largest amount of forest killed by an insect anywhere in the United States. The aspen forests of the Rocky Mountains are now being affected by the same kind of insect problem, just a different insect. The survival of these forests also stand in doubt.

Joshua Tree National Park is being adversely affected by warmer temperatures. Joshua trees depend on winter freezes to set flowering and seeding. There are now less frequent freezes in Joshua Tree National Park, and the trees no longer regularly flower and seed and so are dying off. Changes in air pollution levels could reduce the resilience of forests. Continued depletion of stratospheric ozone would also further stress forests.

Climate Change and Agriculture

At present, mountain glaciers are losing ice at a rapid rate. This often produces floods in the summer months. The loss of glacial ice in mountainous regions around the world will have, and is already having, a negative effect on the inhabitants living downstream at lower elevations. Glaciers alternately store and release water with the change of seasons. Meltwater is utilized downstream for irrigation and in some cases for power production. Glacial melt tends to even out the flow of water through the year.

In Asia, four of the major rivers have their source in the Himalayas, and most of the glaciers providing the water are melting fast. Declining summer runoff is already affecting villages in China, India, Pakistan, Bhutan, and Nepal. The agricultural economies of the Indus, Brahma Putra, Ganges, and Sutleg Rivers are all affected. One of the Himalaya glaciers, Gangotri, supplies 70% of the water in the Ganges River during the dry season. In Asia, the ice melt in spring and fall at either end of the monsoon lengthens the seasonal flow of water for irrigation. The Ganges, along with other glacier-fed streams, could dry up at times if the glaciers melt substantially. Certainly, with a shorter runoff season, the system of agriculture utilizing double-cropping will break down, greatly reducing agricultural production. The result may well be drought and famine.

In Africa there are already feuds over declining irrigation water around the base of Mt. Kenya and Mt. Kilimanjaro. Some oases in China's western desert regions may dry up as glacial meltwater ceases to flow to feed the groundwater.

There will undoubtedly be major shifts in the general circulation if the atmosphere warms. There will also be changes in the length of the growing season in midlatitudes, and there will be changes in precipitation patterns. We cannot predict what these changes will be or what net effect they will have on world agriculture. Some changes will be offsetting. An increased length of growing season may be offset by lower rainfall. Crop yields could be reduced, although the combined effects of climate and CO_2 would depend on the severity of climate change. In most regions of the United States, climate change alone could reduce dry-land yields of corn, wheat, and soybeans. Site-to-site losses would range from negligible amounts to 80%. These decreases would be primarily the result of higher temperatures, which would cause considerable heat and water stress in the plants. In southern areas where heat stress is already a problem, yields will drop. In areas where rainfall decreases, crop yields could decline. In response to the shift in relative yields, grain crop acreage in Appalachia, the Southeast, and the southern Great Plains may decrease.

The combined effects of climate change and rising CO_2 may increase yields in some areas. This would be the case in northern areas or in areas where rainfall is abundant. There would be warmer temperatures and a longer growing season. Acreage in the northern Great Lakes states, the northern Great Plains, and the Pacific northwest might increase.

A change in agriculture would affect not only the livelihood of farmers but also agricultural infrastructure and other support services. Even under scenarios of extreme climate change, the production capacity of U.S. agriculture is estimated to be adequate to meet domestic needs. A decline in crop production would reduce exports, which could have serious implications for food-importing nations. Climate change in southeast Australia has already decreased that country's rice production by 98%. Australia was one of the five major grain exporting countries, and the decline in production has contributed to a world-wide shortage of grain.

Improvements in crop yields could offset some negative effects of climatic change. In many regions, the demand for irrigation is likely to increase as a result of higher prices for agricultural commodities. Farmers might also switch to more heat- and drought-resistant crop varieties, plant two crops during a growing season, and plant and harvest earlier. Whether these adjustments would counter climate change depends on the severity of the climate change.

AQUATIC ECOSYSTEMS

Coral Reefs

Coral reefs are some of the richest ecosystems on the planet, and they are dying at an unprecedented rate. More than 30 nations have reported losses to offshore reefs. The dying of the reefs is attributed to a process caused by bleaching that occurs when sea temperatures become abnormally high. The bleaching is actually the result of the death of microscopic algae that both color and feed the coral. When sea water gets too warm for prolonged periods of time, corals become stressed, causing them to expel symbiotic microalgae that live in the coral tissue and provide the coral with food. The algae supply it with sugars and oxygen in return for shelter and carbon dioxide. This expelling of microorganisms leaves the coral appearing bleached or whitened. It takes only a week or so of bleaching to kill coral (Figure 13.1).

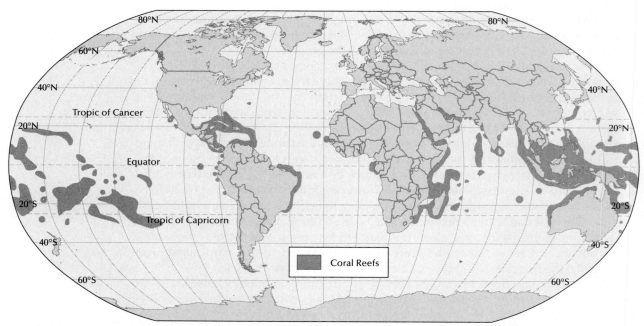

FIGURE 13.1 Distribution of coral reefs and other coralline structures. (From McKnight T. L. & Hess D., *Physical Geography: A Landscape Appreciation*, © 2005, Pearson Prentice Hall, Upper Saddle River, N.J.)

The United Nations Environment Programme reports that one-third of the world's coral is dead or dying. It predicts that 60% of all reefs will be lost by 2030. Another increase of 1°C (2°F) in global temperatures will increase bleaching substantially. Coral reefs are major fish nurseries. Coral bleaching in the tropical oceans by 2030 may alter the global ocean ecology. Large numbers of species of fish and other organisms will cease to exist. There is considerable doubt that the protecting coral reefs around the Maldives can survive if water temperatures rise much higher.

Records show that the water temperature in 2005 was the warmest in the past century. In the fall of 2005, there was massive bleaching in the Gulf of Mexico and the Caribbean, attributed to sea temperatures being 1°C (2°F) above normal monthly temperatures. At Culebra Island, Puerto Rico, up to 97% of the coral colonies surveyed were bleached when water temperatures rose to 32°C (89°F).

Warming brings many changes in marine organisms other than coral reefs. The population of Adelie penguins in Antarctica dropped 40% in the last quarter of the 20th century. The hypothesis is that the warmer temperatures make it harder for the penguins to find food and to breed. Off the coast of California, the sea surface has warmed as much as 1.5°C (2.7°F) since 1951. This has led to an 80% decline in the zooplankton, which is a basic component of the marine food chain. The full impacts on marine species are not known. The loss of coastal wetlands could certainly reduce fish populations, especially shellfish. Increased salinity in estuaries could reduce the abundance of fresh-water species but could increase the presence of marine species.

Fresh-Water Systems

Global precipitation is likely to increase. However, it is not known how regional rainfall patterns will change. Some regions may have more rainfall, while others may have less. Furthermore, higher temperatures would most likely increase evaporation. These changes would create new stresses for many water management systems.

Results of hydrology studies show that it is possible in some regions to identify the direction of change in water supplies and quality. For example, in California, higher temperatures would reduce the snow pack and cause earlier melting. Earlier runoff from mountains would increase winter flooding and reduce deliveries of water in the dry season. In the Great Lakes, reduced snow pack combined with potentially higher evaporation could lower lake levels.

Changes in water supply could significantly affect water quality. Wherever river flow and lake levels decline, there would be less water to dilute pollutants. Where there is more water available, water quality may improve. Higher temperatures may enhance thermal stratification in some lakes and increase algal production, degrading water quality. Changes in the amount of runoff, leaching from farms, and increases in irrigation would affect water quality in many areas.

Fresh-water fish populations may grow in some areas and decline in others. Fish in large water bodies such as the Great Lakes might grow faster and migrate to new habitats. Increased amounts of plankton could provide more forage for fish.

Fresh-water lakes are also being affected. Temperatures have warmed suddenly in Lake Superior. In 1998, temperatures at the surface were 11°C (20°F) above normal—the warmest ever recorded. The surface temperature was over 20°C (68°F). The warming was not just restricted to the very surface: The water was warmer by as much as 8°C (15°F) down as far as 30 m (100 ft). The impact of this warming on life in the lake is phenomenal. Warmer temperatures affect the level of the lake itself. The warmer the water, the more evaporation there is from the surface. Evaporation is the primary means by which water is removed from the lake. Lake levels in 1999 were the lowest in 30 years. Lakes Huron and Michigan dropped 0.6 m (2 ft) between 1998 and 1999. The lower lake levels affect shipping and power generation, among other economic activities.

Changing temperatures will have a detrimental effect on living organisms in Lake Superior. The warmer water kills the plankton on which higher forms of life depend. Already the plankton has descended to lower depths, where the water is colder. At greater depths, they grow more slowly due to the lack of sunlight. Less plankton will mean less fish, and it also encourages many alien species that are prevalent in the other Great Lakes. The cold temperatures in Lake Superior have kept out some undesirable organisms, such as zebra mussels, which are a serious problem in the other lakes. In addition, non-native species will move into Lake Superior if the temperature structure changes, and some of them are competitive enough to destroy the existing plants and animals.

Because of increases in habitat size or reduction in population of competitors, some species might benefit from climate change.

Higher temperatures may lead to more aquatic growth, such as algal blooms. Decreased vertical mixing in lakes would deplete oxygen levels in shallow areas of the Great Lakes. This would make some lakes, such as Lake Erie, less habitable for fish. Fish in small lakes and streams may be unable to adapt to changing temperatures. In other cases, their habitats may simply disappear.

THE IMPACT OF RISING SEA LEVEL

The one result of a warming planet that will be the same everywhere over the planet is the rise in sea level. One of the great threats to sea level is the possible melting of the Greenland and Antarctic Ice Sheets. About 80% of all the world's glacial ice is located in these two places. If either the Greenland or the West Antarctic Ice Sheets melts in its entirety, sea level will rise an additional 6 m (20 ft). The impact of this rise on coastal cities would be phenomenal.

Over half of the human population lives within 100 km (60 mi) of the sea. Most of this population lives in urban areas that serve as seaports. A measurable rise in sea level will have a severe economic impact on low-lying coastal areas and islands. A rapid rise in sea level of several meters would flood large sections of the earth's coastal plains, including many of the world's great cities. The North American coastline would change drastically. A considerable part of Florida and New York City would be submerged. The small rise in sea level now taking place is already causing major problems in some areas. London, England, Venice, Italy, and The Netherlands are taking remedial measures to reduce the risk of flooding from the sea.

Water levels have been rising for years in Venice, as the city gradually sinks into the mud and the sea level rises. St. Marks Square has been flooded many times in recent decades. The highest water level recorded since recordkeeping began was in 1966. On December 1, 2008 water rose rapidly due to a high tide and a wind-blown surge of water north on the Adriatic Sea. It reached the fourth highest level yet recorded. Water was deep enough in the square that a motor boat crossed the square. Temporary elevated walkways that pedestrians normally use to move about the square would have been useless in this flood as they would have floated. The city's water buses could not operate, as the loading platforms were under water. The flood receded with the next low tide and a shift in wind direction. The city debated for many years over a plan to build movable floodgates among the offshore islands to reduce flooding in the city. The project, known as the Moses Project, is under construction and due to be completed in 2011.

The melting of glacial ice produces drastic changes on coastlines and island nations. Any rise in sea level will have a significant impact on low-lying coastal land. Some areas that would feel the impact are islands such as the Maldives and low coastal areas such as Bangladesh, the Mississippi Delta, and the state of Florida.

The Maldives is a small group of islands in the Indian Ocean that is already suffering the effects of rising sea level and changes in storm tracks. The islands are largely remnants of coral reefs, and most of the land area is within 1 m (3.3 ft) of mean sea level. As sea level rises, these islands will be affected in a variety of ways.

In April 1987, a large storm in the Indian Ocean generated large swells that traveled through the Maldives. The islands suffered huge losses in terms of the economy, public utilities, and the land. Coral reefs surrounding the island act as natural break-waters. Recent storms have damaged the reefs, and waves have been able to attack the shoreline more frequently, resulting in extensive beach damage. Beach erosion is becoming a serious problem. There is a net loss of beach, and hence land area, on over half the islands. A United Nations study projects that under a high-sea-level-rise scenario, the island of Mali could be completely flooded at extreme high tide by the year 2100.

It is worth noting that the Maldives was the first country to sign the Kyoto Protocol on March 16, 1998 and the first nation to ratify the protocol on December 30, 1998. In February 2008, the United Nations held hearings on climate change. Tuvalu is a small island nation consisting of some 11,000 people living on nine small atolls in the South Pacific. A representative of that nation spoke at a recent UN conference, asking for help in adapting to rising sea level.

The horizontal effect of rising sea level can be greater than the actual rise would indicate. As sea level rises, erosion increases along coastlines, particularly along flat coastlines. On emergent coastlines, a rise of 10 mm (0.4 in.) can increase beach erosion inland by 1 m (3.3 ft). A rise in sea level would push salt water inland. A 10 mm (0.4 in.) rise in sea level would push the boundary between salt water and fresh water 1 km (0.6 mi) upstream along coastal rivers. Salt water would also displace fresh groundwater for a substantial distance inland. Oceanic islands would be severely affected. Costs of keeping out the sea may be prohibitive, and movement inland may be impossible. In these cases, the only solution may be to move people and their institutions to some other location.

CLIMATE CHANGE AND HUMAN HEALTH

Global warming is having a significant effect on human health. One reason is that a warmer environment is expected to be more variable and conducive to greater extremes. Both a warmer environment and more extremes of weather associated with warming affect vitality and distribution of many diseases. Extremes of heat and cold and hot and dry conditions have been frequent in recent years. Down through history, such extremes have often been followed by outbreaks of disease. Such has also been the case in recent decades. Mortality rates, particularly for the elderly and seriously ill, are affected by frequency and severity of extreme temperatures. Respiratory infections, diarrhea, and other diseases also increase when people are forced into crowded shelters. Extremes of weather favor the insects, bacteria, protozoa, and viruses that spread disease. Changes in temperature and rainfall affect the life cycles of disease-carrying insects such as mosquitoes and ticks. Changes in weather affect contagious diseases such as influenza and pneumonia, as well as allergic diseases such as asthma. An obvious effect of a warmer environment is more severe heat waves. In the European heat of 2003, there were an estimated 35,000 more deaths than would be expected in the first two weeks of August. Many of the deaths resulted from cardiovascular problems among the elderly. If heat waves become more prevalent in the future, so will cardiovascular problems.

Extreme heat waves can also devastate agriculture. In Europe during the heat wave of 2003, temperatures averaged 5.5°C (10°F) above normal. In Italy, corn yields dropped 36% below average. In France, fruit yields fell 25% and wine production was down 10%. Heat also affects the rate of plant pollination. A 2°C (3.6°F) rise in rice-growing areas would cut rice production in half.

Warmer summers will also increase other less obvious problems. Kidney stones are most prevalent in warm climates, such as the southern United States. Sweating removes water from the body and increases salt concentration in urine. Too much salt in the urine often results in the growth of kidney stones. Passing stones can be extremely painful. With hotter weather in summers, especially in the Midwest, there will be more frequent cases of kidney stones.

Finally, increased air pollution can heighten the incidence and severity of such respiratory diseases as emphysema and asthma. Severe storms and flooding also contribute to increased outbreaks of disease. Respiratory infections, diarrhea, and other diseases increase when people are forced into crowded shelters and water supplies become contaminated.

Changes in climate as well as in habitat may alter the regional prevalence of vector-borne diseases. For example, some forests may become grasslands, thereby modifying the incidence of vector-borne diseases. Changes in summer rainfall could alter the amount of ragweed growing on cultivated land, and changes in humidity may affect the incidence and severity of skin infections and infestations such as ringworm, candidiasis, and scabies. Increases in the persistence and level of air pollution episodes associated with climate change would have other harmful health effects.

Warmer temperatures will also result in increased ailments due to undesirable vegetal products. In U.S. cities, the carbon dioxide content of the atmosphere is higher than in rural areas, and so are temperatures. The higher temperatures are the result of several factors. On a particularly bad day, the carbon dioxide content of the air in New York City can reach 700 parts per million, nearly twice the global average. Ragweed grows much bigger in cities. Not only is it bigger, but it produces nearly 10 times as much pollen as rural ragweed and for a longer period during the year. This has a major effect on hay fever sufferers. Higher temperatures are increasing the pollen season by about one day per year. Higher levels of carbon dioxide also cause poison ivy to grow better. In addition, higher levels of carbon dioxide increase the toxicity of the urushiol oil in poison ivy, which causes the rash. So poison ivy grows better and becomes more poisonous with greater amounts of carbon dioxide.

Higher summer temperatures increase the ozone concentration, which affects asthma sufferers and causes lung and heart damage similar to that which occurs in a person living with a cigarette smoker.

SPREAD OF TROPICAL DISEASES

The planet's warming is expected to increase the range of tropical diseases. Tropical diseases have become more prevalent in their traditional zones and have broken out into areas where they were previously unknown. Diseases that have increased in frequency or severity include malaria, cholera, yellow fever, the plague, dengue fever, meningococcal meningitis, influenza, diarrhea disease, hantavirus, West Nile encephalitis virus, and Chikungunya. Rising temperatures may lead to increased frequency of some of these diseases in the United States.

Malaria

Malaria is perhaps the most commonly known of the tropical diseases. Malaria is currently the most widespread of all mosquito-borne diseases affecting humans. The symptoms are fever and chills, anemia, enlarged spleen, severe stomach pain and headaches, general weakness, and increased susceptibility to other disease. The disease is caused by one of four species of protozoa of the genus *Plasmodium*. Female Anopheles mosquitoes carry this one-cell parasite, or plasmodia. A single plasmodia can cause the death of a human. Of the four species of these parasites, the most common and most dangerous is *Plasmodium falciparum*.

Most people contracting malaria survive. For some, the exposure to the parasite provides some immunity or resistance to additional attacks. The parasite can remain in the system in a dormant form and then activate in weeks, months or even years. A person can also be reinfected over and over again by subsequent bites. The disease can also be transmitted by blood transfusion and by contaminated needles that are reused.

More people suffer from malaria today than at any time in history. It exists in some 106 countries and is a serious health problem in more than 90 countries. These countries contain some 40% of Earth's population—more than 2.4 billion people. The World Health Organization estimated that in the year 2000, between 300 and 500 million people suffered from malaria. The number suffering the debilitating effects of malaria is in the tens of millions. The total number of deaths each year ranges as high as 2.5 million. It is also estimated that the disease kills more than 1.6 million children under the age of 14 each year, perhaps as many as 3000 each day. Most of the deaths are in children under age five.

The parasite needs temperatures above 18°C (64°F) to develop. Malaria is generally found only in areas where winter temperatures average above 16°C (61°F). The mosquitoes need stagnant, sunlit water to breed. Where crops are flooded for periods of time, as in rice-growing areas, malaria is a constant problem. As irrigation and water storage have greatly increased in the past few decades, so has the incidence of malaria.

Extreme weather favors mosquito reproduction. Flooding leaves behind stagnant ponds that warm during periods of drought. Ponds are ideal places for mosquitoes to breed. Warmer weather makes mosquitoes bite more frequently because it gets their reproductive juices flowing. Also, the microorganisms that cause malaria mature more quickly. At 20°C (68°F), it takes the parasite 26 days to mature. At 25°C (77°F), it takes just 13 days to mature. This translates into more frequent transmission of the parasite and more malaria victims. In malaria-infected areas, more frequent drought will further increase malaria because the drought will reduce the numbers of mosquito predators such as damselflies and dragonflies.

Malaria is increasing in numbers of people infected and spreading rapidly. It is expanding toward the poles and to higher altitudes in mountainous regions. Between 1993 and 1998, the incidence of the disease increased 400%. Many African countries have experienced epidemics in the past decade. Europe has had outbreaks in recent years, and in the United States 1000 to 1200 cases have been reported each year.

The people most affected by malaria are inhabitants of Africa. More than 90% of all cases and 75% of fatalities are reported in Sub-Saharan Africa. Some researchers and health care workers believe that nearly every child over the age of one year in Sub-Saharan Africa is infected with malaria.

The highlands of Africa in the past were free of malaria, but not anymore. Higher temperatures, deforestation, and population growth have all contributed to the increase in malaria. On Mt. Kenya, 80 km (50 mi) north of Nairobi, the snow and glaciers have been melting rapidly in the past decade. At the same time, the frequency of malaria has been increasing. There have always been cases that were carried onto the mountain by travelers. One nurse working in the hospital in the town of Karatina, on the slopes of Mt. Kenya, estimated the number of malaria cases to have multiplied five times in just the past several years. In another district in western Kenya, in the period from 1986 to 1998, the number of malaria cases increased from 16 per 1000 population to 120 per 1000. Because of the warmer weather in recent years, malaria has been particularly severe. Increased global temperatures result in an increased range of the Anopheles mosquito. A rise in global temperature of 2°C (3.6°F), which is well within the range of possibility in the next 50 years, may expand the geographic area in which malaria is found from the present 42% to 60% of the planet.

Not only does the rising temperature increase the range of the Anopheles mosquito, it increases their biting rate. A 2°C (3.6°F) rise in temperature would more than double the mosquitoes' rate of metabolism, leading them to feed more frequently. It would result in an increase in fatalities of more than 1 million per year by 2050. Mosquitoes in many areas are becoming resistant to the most common pesticides, and the increased temperatures seem certain to increase the perils of malaria.

West Nile Encephalitis Virus

The effect of the **West Nile encephalitis virus** on humans is generally minimal. It can cause extreme weakness of muscles and general confusion. In the worst case, it can cause permanent brain damage or death. The effects are severe in a very small percentage of cases. Most infected individuals suffer no symptoms at all or have mild flu-like symptoms, such as headache, sore muscles, swollen glands, or rashes.

West Nile is not a new virus; it is significant only because it has now appeared in the United States and carries over from one summer to the next. It first appeared in the United States in 1999, when it killed seven people in New York. It was found in birds and mosquitoes in summer 2000 and again in 2001. The virus is carried by birds that have been infected by mosquitoes carrying it. The apparent reason for the continuity of the disease is that some mosquitoes that carry the virus survive the winter. They probably inhabit buildings in which the temperatures stay warm enough for them to survive. The disease first appeared in birds in the Carolinas and Georgia. By the summer of 2001, the disease was detected in birds in some 20 states east of the Mississippi River. The infection rate among mosquitoes is only about 1%.

Preventive measures consist of wearing long-sleeved tops and long pants and applying insect repellant. However, repellants containing DEET should not be applied to children under five years of age. These measures should be taken in areas with substantial mosquito infestations, especially around dusk, which is a primary feeding time for mosquitoes. Elderly people are the age group most likely to be affected by the disease. People should check around the outside of their dwellings for items containing standing water, such as flower pots, pails, and bird baths. The water in bird baths should be changed regularly.

Hantavirus Pulmonary Syndrome

A relatively recent disease to appear on the scene is **hantavirus**, a flu-like virus, whose symptoms are fever, muscle aches, shortness of breath, and coughing. The disease is carried by rodents, especially deer mice. It is not contagious in the United States. Ordinarily, hospitalization is required. A long drought followed by heavy rainfall in the U.S. Southwest led to an outbreak of a very serious form of pulmonary hantavirus that proved fatal in about half of the identified cases. When warm wet winters occur with El Niño, the population of deer mice increases.

Dengue Fever

Dengue fever is a mosquito-borne disease that appeared in South America in the last decade of the 20th century. Dengue fever is a viral infection carried by day-biting mosquitoes. There is now a worldwide pandemic of dengue fever, and it is most common in India, Vietnam, and the Caribbean. The symptoms are fever, excruciating pain in the bones, muscle aches, and rash. Most cases are not serious and last only a few weeks. A more severe form, known as dengue hemorrhagic fever, has similar symptoms but also causes internal bleeding and bleeding through the nose and gums. This form of the disease can lead to shock and death.

Most cases of dengue fever are found in urban areas. It has existed on the Pacific coast of Costa Rica for some time. There were epidemics there in 1993 and 1994. In 1995, unusually warm weather allowed the mosquitoes to cross the mountains into the rest of the country. Outbreaks of the disease occurred in other areas of South America. Thousands died of the epidemic, and nearly 150,000 were infected. The disease was not limited to Central and South America. There was an outbreak in Australia in 1992 following a particularly rainy season that favored the breeding of mosquitoes.

In the fall of 2007, dengue fever spread across the Caribbean in one of the worst outbreaks in decades. Health agencies declared that the disease was at near record levels in 2007. It affected hundreds of thousands of individuals and killed more than 200. Part of the problem is that mosquitoes that carry the fever are increasing in numbers in urban areas where stagnant pools of water form in unused containers and puddles along roads and in residential areas.

Every year, cases of dengue fever are diagnosed in the United States in persons who have traveled to regions where the disease is endemic. The Centers for Disease Control and Prevention (CDC) reported that 1167 cases were reported between 1996 and 2005. When blood tests were conducted by the agency on residents of Brownsville, Texas, 38% showed signs of past dengue infection; 11% of these had not been outside the country and so had to have been infected in the local area, and 2.5% of those tested showed signs of recent infection.

Chikungunya (Chicken Guinea)

Chikungunya is a relatively new mosquito-borne viral fever. The symptoms, which may last for a week or more, are sudden high fever reaching 39°C (102°F), a variety of skin rashes and other skin irregularities, arthritis of the joints, and headache. Severe joint pain may last for weeks or months. In the recent epidemic in India, high fever and joint pain were the most common symptoms. Fever normally lasts about 48 hours before breaking.

Both dengue fever and Chikungunya are spread by the *Aedes aegypti* and *Aedes albopictus* mosquito, both of which are found in the United States, especially in the South. The CDC has reported that more than 30 cases have developed in the United States in persons who have traveled to the areas in Asia where outbreaks were occurring.

The first known outbreak of Chikungunya was along the border of Tanganyika and Mozambique in 1952, on the Makende Plateau. It appeared in Malaysia in 1999. In the fall of 2004, thousands of cases were reported in southern India. Since March 2005, it has infected more than 1 million people in India and another 300,000 on the islands of the Indian Ocean. In September and October 2006, outbreaks of mosquito-borne diseases swamped hospitals and clinics in India. In northern India, most of the cases were dengue fever, and in the south they were Chikungunya. Both outbreaks are believed to be related to the monsoon, which left widespread wet land and stagnant water in which mosquitoes breed.

There are no specific means of prevention or treatment yet. Chloroquin has been used effectively in some cases. The most effective means of preventing or lessening the risk of further attacks is to use bed nets treated with a specific pesticide. Using chemical repellants also reduces the risk of infection.

Cholera, Plague, and Yellow Fever

Cholera is another disease of warm climates. It, too, seems to be spreading. A cholera outbreak occurred in 1991 along the coast of Peru. There was already serious pollution of the water, and cholera was introduced by wastewater pumped from the hull of a freight-carrying ship from Asia. The cholera organism thrived in the unusually warm water and algae bloom.

An ancient nemesis, the **plague**, made its appearance in Asia once again at the end of the 20th century. In 1994, northern India had a long rainy monsoon season. When the rains subsided, there were three months of weather with temperatures around 55°C (100°F). Rats, which are the primary hosts of the flea that carry the plague, converged on the cities. The result was an epidemic of pneumonic plague, a form that is transmitted from person to person through the air.

In February 2008, the World Health Organization confirmed cases of yellow fever in Latin America. This mosquito-borne disease can spread extremely rapidly in urban areas.

Global warming is compounding the disease problem. The same elements associated with warming that increase the vitality of diseases reduce human immunity to them by weakening our natural defenses. Rising temperatures may lead to increased frequency of some of these diseases in the United States.

Summary

If global temperatures rise by 1°C or more, the environment will see significant changes. There will be changes in sea level, changes in the distribution of biomes, and changes in biodiversity. A change in temperature of 1.5 to 4.5°C (2.6 to 8°F) in a century or two has no equal in the recent history of the earth. Such a change in global mean temperature would have a significant impact on world society and would affect nearly all aspects of life. The stress of adjustment would be phenomenal. These changes would in turn result in changes in global economic systems. Changes in economic systems are accompanied by changes in social systems and widespread human stress. As the human population grows, the number of people affected increases.

Key Terms

Chikungunya, *229*

Cholera, *229*

Dengue
  fever, *228*

Hantavirus, *228*

Plague, *229*

West Nile encephalitis
  virus, *227*

Review Questions

1. Why will warmer temperatures aid in the spread of tropical diseases?
2. How will warmer temperatures accelerate the extinction of species?
3. How does increased CO_2 affect plant growth?
4. Name two tropical diseases that have appeared in the United States since 1980.
5. Is there any evidence that fresh-water lakes are increasing in temperature?
6. How do the rates of migration of plants and animals compare?
7. Is there any evidence that plant and animal populations are shifting their range up or down in elevation in mountainous regions?
8. In what way is the length of growing season for plants in midlatitudes changing?
9. Why is the melting of mountain glaciers in Asia a threat to the human population?
10. Why is climate change a threat to coral reefs around the world?

CHAPTER
14

Global Changes in Atmospheric Chemistry: Acid Precipitation and Ozone Depletion

Although oxygen and nitrogen are by far the major constituents of the atmosphere, many trace elements play an important part in the way in which energy is transferred through the atmosphere. Gases such as sulfur dioxide, hydrogen dioxide, and carbon monoxide are continually released into the air as by-products of natural occurrences such as volcanic activity, decay of vegetation, and forest fires. It makes no sense in today's world to consider the atmosphere in its natural form because human activity is rapidly altering atmospheric chemistry. Technological processes inject into the air a wide variety of solids, liquids, and gases collectively called *pollutants*. Air pollution is neither new nor necessarily associated with industrial processes. Even before factory chimneys and automobiles added their output to the atmosphere, pollution existed. Juan Rodriguez Cabrillo, sailing into San Pedro Bay in 1542, named the Los Angeles Basin the "Bay of Smokes." Prevailing weather conditions caused the smoke from scattered human fires to hang as a pall of pollution over the area. As we now know, this was a harbinger of things to come.

Some **effluents** introduced into the atmosphere are in fact natural constituents. These materials become pollutants only when they are placed into the atmosphere in abnormally large amounts. The volume of effluents placed in the atmosphere has reached the extent where levels of some constituents are increasing beyond natural limits. Such is clearly the case with CO_2. The Environmental Protection Agency (EPA) released a report in 1989, indicating the magnitude of the industrial injection of chemicals into the atmosphere. In the United States, there are at least 1600 industrial facilities in 46 states that release into the air significant amounts of chemicals suspected of being carcinogenic. Some 125 of these plants release more than 400,000 pounds of chemicals each year. There are some 30 industrial facilities that emit more than 1 million pounds a year each. For instance, 2.7 billion pounds of pollutants were placed into the atmosphere in 1987. Of this amount, 360 million pounds are suspected of being **carcinogenic**.

A phenomenal amount of pollutants is pumped into the atmosphere every year. The potential for ill health resulting from air pollution is well known, and the EPA is responsible for monitoring and enforcing laws pertaining to the environment. Air quality standards have been used to produce a Pollution Standards Index (PSI), which is computed for a defined location and uses the value of the pollutant that occurs in the highest concentration in relation to the national air quality standards. If one of the pollutants is at the primary standard, a PSI value of 100 is given. Below that value, the air is considered moderate or good; above the 100 level, the categories pass from unhealthy to hazardous.

ATMOSPHERIC POLLUTANTS

The most common air pollutants are carbon monoxide, sulfur dioxide, nitrogen dioxide, and surface ozone. All are harmful to human health, built structures, and the environment. Particulates and surface ozone are the most harmful to human health.

Particulates

The primary sources of particulates are diesel trucks and buses, factory and electric utility smokestacks, car exhausts, burning wood, mining, and construction. In 1994, nearly 23 million Americans, or 9% of the total population, lived in areas where **particulates** such as soot and acid aerosols exceed EPA standards. In these areas, the air often is thick and hazy, especially in the summer. Particulates can impair the functioning of the lungs and seriously threaten health. They mainly affect people with chronic respiratory illnesses such as asthma, bronchitis, and emphysema. Lung disease is the nation's third leading cause of death. California has the strictest laws governing particulate air pollution. The EPA standard for particulates is 150 $\mu g/m^3$ of air. California has a maximum of 51 $\mu g/m^3$. Dust-sized particulates of less than 10 μm in diameter have been linked to bronchitis, asthma, pneumonia, and pleurisy in children.

Carbon Monoxide

By volume, the greatest emission from human activity is **carbon monoxide (CO)**. This colorless, odorless, and tasteless gas is formed primarily by incomplete combustion of organic fuels. A wide variety of sources produce CO. The largest single source is automobiles; automobiles running in attached garages or left running in traffic are major sources. Other sources are unvented gas furnaces, stoves, fireplaces, water heaters, space heaters, and generators. Unvented kerosene heaters and fireplaces are also sources of the gas. The gas begins to affect the human body at concentrations of about 100 parts per million (ppm). At this concentration, people develop headaches and may become dizzy. They can also become fatigued or nauseated, develop shortness of breath, and become fatigued. Higher levels and prolonged exposure result in more severe effects. Initially the symptoms are general confusion, loss of muscular coordination, and vomiting. Eventually, death may result. Most fatalities occur from being in a confined space with a fuel-burning device. Fatalities are not high, averaging fewer that 200 per year. Levels of 100 ppm have been observed outdoors in some urban areas, and concentrations of 370 ppm have been recorded inside vehicles trapped in traffic jams. Carbon monoxide has a residence time of several days. Eventually the carbon monoxide combines with oxygen to form carbon dioxide.

Sulfur Compounds

Sulfur oxides are the second most abundant pollutant. Sulfur dioxide (SO_2) is one of the major oxides of sulfur. This heavy, pungent, colorless gas forms from the combination of sulfur from emissions of coal-burning industries and atmospheric oxygen. Sulfur dioxide is highly reactive and hence is not cumulative. The maximum residence time is probably 10 days. In 2002, SO_2 emissions exceeded 14 million tons in the United States. The vast majority (95%) of the emissions came from three sectors of business: electric power generation, burning of fossil fuels in internal combustion engines, and industrial processes. Electricity generation was by far the greatest source. Of the total emissions, 72% came from burning low-grade coal and petroleum in electric power plants. Sulfur dioxide emissions from electric power plants have declined by about 30% since 1975 due to the use of higher-grade coal and cleaner-burning plants. Most of the severe offenders are located in the Ohio River Valley.

Because SO_2 reacts with other substances in the air, it causes a variety of health problems. Groups most affected are those with respiratory problems, children, and elderly people. SO_2 is a major irritant to the eyes and respiratory system, and it is lethal at a few parts per million. High levels of SO_2 in the air can cause breathing problems for people with asthma, especially those who spend much time outdoors.

SO_2 combines with atmospheric water to form sulfuric acid. Atmospheric sulfuric acid causes the leaves of plants to turn yellow and, with long exposure, can change the makeup of plant communities. It dissolves limestone and marble, and it is highly corrosive of iron and steel. It also reduces atmospheric visibility and blocks out sunlight. Sulfate particles are the major cause of the haze that reduces visibility across the United States.

Hydrogen sulfide (H_2S) is another sulfur compound that forms in the atmosphere. It forms from organic decay when there is not enough oxygen present to oxidize the organic material. The main sources of hydrogen sulfide are swamps. It has a very bad smell, like rotten eggs, but fortunately has a short residence time. In the atmosphere, it darkens lead in oil-based house paints. It is also responsible for tarnishing copper and silver.

Nitrogen Oxides

Nitrogen and oxygen do not normally interact at standard environmental temperatures. Substantial quantities of nitrogen oxides (NO_x) result from combustion at high temperatures. Nitrogen dioxide (NO_2) is the only widespread pollutant that has a color. It is yellow-brown in color and has a pungent sweet odor. The average residence time is about three days. Because the gas has a residence time of several days, the gas and the products created from it can travel far from the source and spread over large areas. The end product of nitric oxides is nitric acid (HNO_3). Nitrogen oxides (NO_x) are also major contributors to acid rain and surface ozone.

The three main sources of NO_x are automobiles and trucks, generation of electricity, and off-road internal combustion engines. These three sources accounted for 82% of emissions in 2002 in the United States. Motor vehicles added the most, contributing 39% of the total. Nitrogen oxides are distributed as vehicle traffic is distributed. It is emitted over a much broader area and more evenly than is sulfur dioxide.

Nitric oxides have a number of detrimental effects, including the following:

- They contribute to the formation of surface ozone.
- They react with other substances to form acid rain.
- NO_2 is a greenhouse gas.
- They contribute to the regional haze that reduces visibility.
- Combined in small particles, they contribute to emphysema and bronchitis.

Surface Ozone

Ozone (O_3) is a form of oxygen that contains three atoms of oxygen instead of the usual two. The gas is colorless and odorless except at very high concentrations. A major ingredient in smog, ozone is formed near the ground when pollutants such as unburned petroleum hydrocarbons and nitrogen oxides from automobile exhausts and fossil fuel power plants react in sunlight. The chemical reactions are faster on hot sunny days. It takes several hours after the sun rises for the chemical reactions to reach the level where ozone accumulates. It usually begins to form about 10 A.M. solar time. The ozone that forms in sunlight usually breaks down at night.

Ground levels of ozone reached their highest levels on record in 1988. About half of the U.S. population lives where ozone exceeds the EPA standard at least part of the time. Ozone is highest in the states east of the Mississippi River and in California and Texas.

Breathing ozone may cause respiratory problems for those who exercise outdoors. Ozone is an irritant to the lungs and air passages. It is especially irritating to those who engage in vigorous exercise. Some individuals are affected almost immediately if they exercise in air with elevated ozone levels. They may cough or experience chest pain and shortness of breath. In general, the more and harder you exercise, the greater the intake of air and therefore of ozone. To reduce the chance of respiratory irritation, it is best to take precautions while exercising in warm weather when there is a risk of ozone accumulation. It is best to exercise before 10 A.M. and to avoid jogging along major traffic thoroughfares.

Atmospheric Brown Clouds

In the early 1940s, the United States was heavily involved in World War II, and large numbers of military aircraft were being flown across the Atlantic Ocean for delivery to Great Britain. Pilots ferrying these planes across the ocean reported sighting a brown haze in the atmosphere at altitudes of a few thousand feet. The haze was the product of all the particulates and chemicals being placed in the atmosphere by the human population. This haze gradually expanded over the years until the haze was present over much of the earth. It has become readily visible from commercial aircraft as they climb out of airports.

This brown haze has become particularly thick over some areas, where it is referred to as atmospheric brown clouds. In some places, and at some times, such as over the high mountain ranges of Asia, the layer is now more that 2 km (1.2 mi) thick, and it reaches almost 3 km (1.9 mi) thick at times. The layer is often visible from the ground over Asia, tropical Africa and South America, and the Middle East. The severity of the problem was brought to public attention in the summer of 2008 as the atmospheric brown clouds were a threat to the 2008 Summer Olympics held in Beijing.

The tremendous growth of the human population is a major culprit in the growth of these clouds. The primary sources are the burning of fossil fuels and the burning of vegetation in the process of deforestation. Another significant source is the burning of wood for heat and cooking in developing countries. The layer consists of a mix of soot and other particulates, ozone, and chemical pollutants. The ozone involved in brown clouds is surface ozone, not to be confused with high altitude ozone. The clouds tend to stay over some areas, but individual clouds move with the general circulation. In 2001, such a cloud developed over China, moved eastward over Japan and Korea, and crossed the Pacific Ocean.

The atmospheric brown clouds are a threat to the physical environment and to human health. In conjunction with the greenhouse gases, this layer is changing the earth's weather and climate. Fallout of particulate matter on the glaciers of Asian mountains increases the rate of melting. The glaciers in these mountains are absorbing more sunlight and melting faster. More rapid melting increases the problems outlined in chapters 12 and 13. The plumes absorb a great deal of solar radiation in the lower atmosphere. This adds heat to the warming that is already taking place. Counter to this, the haze reduces the amount of solar radiation reaching the ground which reduces heating of the surface. In parts of Asia, the dimming of sunlight has been observed since the 1950s. The clouds are particularly thick over some of the world's large cities (Table 14.1). In some Asian cities, sunlight is reduced by as much as 20%.

TABLE 14.1 Large Cities Experiencing Reduced Sunlight Due to Atmospheric Brown Clouds

Bangkok, Thailand

Beijing, China

Cairo, Egypt

Dhaka, Bangladesh

Karachi, Pakistan

Kolkata, India

Lagos, Nigeria

Mumbai, India

New Delhi, India

Seoul, South Korea

Shanghai, China

Shenzhen, China

Teheran, Iran

Atmospheric brown clouds contribute to the impact of climate change on agriculture, the hydrologic cycle, and human health. The clouds may also be affecting the monsoon circulation over Asia. Monsoon rainfall has been declining in recent decades, and spring flooding is more frequent. Rice production in Asia may have declined 5% since the 1960s. Health problems related to the mix are increased respiratory and cardiovascular diseases. The number of premature deaths attributable to atmospheric brown clouds may be up to 350,000 per year in Asia.

It should be noted that efforts to curb greenhouse gases will also reduce the atmospheric brown clouds.

ACID PRECIPITATION

Acid rain is a phrase that applies to a process resulting in the deposition of acid on the surface of the earth. All precipitation is slightly acidic in nature. One index of measuring acidity is the concentration of hydrogen ions (pH). A neutral solution has a pH of 7.0. The lower the pH, the more acidic the water. For each unit the pH drops, the acidity increases by a factor of 10. Thus, a pH of 6 represents an acidic element 10 times that of a pH of 7. A pH of 5 represents 100 times the acidity of water with a pH of 7.

Natural precipitation has a pH near 5.6. The term *acid rain* was first used by a British chemist, Angus Smith, in 1858. It refers to precipitation with a pH less than 5.6. When precipitation has a pH less than 5.6, it is usually due to the injection of sulphur compounds or nitrogen oxides into the atmosphere. Coal-burning electric power plants, industrial furnaces, and motor vehicles inject large amounts of these chemicals into the atmosphere. In the atmosphere, the chemicals combine with water to form sulfuric acid and nitric acid. These droplets may be transported great distances by wind before they precipitate to the ground.

The acidity of precipitation has increased over North America, and the pH is less than 4.6 over most of the continent east of a line from Houston, Texas, to the southern tip of Hudson Bay. In 1980, the pH of precipitation dropped to an average of 4.1 over part of the Ohio River Valley and the Adirondack Mountains. In the Great Smoky Mountains, precipitation with a pH of 3.3 was measured. At Wheeling, West Virginia, in 1980, rainfall with a pH of 1.4 was recorded. Ordinary battery acid has a pH of 1.1.

Much of the attention on the pollution causing acid rain in the northeastern United States and in Canada has focused on the Ohio River Valley and other areas of the Midwest, where there are large concentrations of coal-fired power plants. Nine states—Georgia, Illinois, Indiana, Kentucky, Missouri, Ohio, Pennsylvania, Tennessee, and West Virginia—produce 52% of the sulfur dioxide emissions in the United States.

Global Distribution of Acid Rain

Acid rain has become a worldwide problem. In Europe, Norway, Sweden, Denmark, The Netherlands, and West Germany all have problems with acid rain. In Sweden alone, an estimated 18,000 lakes are more acidic than natural rainfall. The continental nations accuse Great Britain of being the main source of the pollutants. Great Britain has admitted to being a source of sulfur dioxides and nitrogen oxides.

Great Britain has also had its share of **acid precipitation** problems. The worst case of acid mist occurred in 1989. The mist came in over the east coast of Great Britain on September 9, affecting a 2600 km^2 (1000 mi^2) area. The area was mainly in Norfolk and Lincolnshire. The mist, estimated to have a pH of 2.0, was so acidic that it corroded aluminum instruments. It damaged thousands of trees, killing the leaves and turning them brown overnight. The source of the sulfuric and nitric acid particles is believed to be automobile traffic on the continent. This incident was worse than the incident at Pitlochry, Scotland, in 1974. The Pitlochry acid mist had a pH level of 2.4—stronger than vinegar.

Precipitation in other parts of the world is also acidic. In the city of Guiyang, China, concentrations of the sulfate ion are about 6 times greater than in New York City and 20 to 100 times greater than those over Katherine, Australia (an area little affected by industrial pollution). The higher concentration of sulfates in China is due to the heavy use of coal as a primary fuel for home cooking, heating, and generation of electricity. There are virtually no controls on the use of coal as a fuel in China. Although concentrations of sulfates are higher in China than in the United States, there is a lower concentration of nitrates there, primarily due to the low number of automobiles. The nitrate concentration is highest in Beijing, where there are the most automobiles.

The Impact of Acid Precipitation on Aquatic Environments

The impact of acid rain on aquatic environments, particularly fresh-water lakes, has been clearly established. Aquatic systems are very susceptible to acidification. Fish are very susceptible to acidification. Fish become endangered when the pH drops to about 5.5. Most species of fish stop reproducing at pH levels between 5.3 and 5.6. Fish are hurt by acidification in a number of ways. As acidity increases, more trace metals are dissolved in the water.

Aluminum is one such metal. Young fish are particularly susceptible to increased aluminum concentrations. The aluminum collects in their gills. In trying to get rid of it, the young fish strangle in their own mucus. Above-normal acidity also prevents fish from absorbing calcium and sodium. The lack of calcium weakens their bone structure, and their skeletons become deformed and are easily damaged. Lack of sodium causes convulsions that kill the fish. Fish are also susceptible to acid shock, the sudden introduction of large amounts of acid, which is commonly associated with spring snowmelt in mountain regions. The acid is deposited in the snow crystals and remains on the ground for periods of up to several months. With spring melting, large amounts of acid enter the streams and lakes, resulting in a sudden, if temporary, increase in acidity. In the northern United States, some winters, such as the winter of 1993–1994, have a lot of snow accumulation, and there may be widespread acid shock in the spring. Other years, it may be minimal. In parts of Canada, snow accumulates in most winters, and there are annual episodes of acid shock.

By the time the pH of a lake drops to 5.0, between 30 and 50% of the natural biota cease to exist. The most susceptible are the smaller organisms such as mollusks and minnows. Many lakes contain water with a pH of less than 4.5. At 4.5, all fish are gone, and the water supports completely different organisms from normal lake water. High acidity favors the growth of sphagnum mosses and filamentous algae.

The sensitivity of lakes to acidification depends on their natural ability to neutralize the acidic runoff into the lake. Lakes located in areas where the parent rock is igneous and metamorphic containing lots of silicates are most sensitive to acid deposition. The dissolved minerals from these rocks result in acidic runoff. In regions where the parent rock is high in the mineral salts such as calcium, magnesium, and phosphorous, lakes can better tolerate the acid runoff. The reason is that the soil solution tends more toward alkaline and the salts neutralize the acid.

On a global scale, there may be more than 1000 lakes that have become too acidic to support life. There may be several thousand that receive episodes of acid shock; the greatest share of these are in eastern Canada and Scandinavia. Acidification is a particularly severe problem in the Adirondack Mountains of New York and in New England. The parent rock in these areas is high in the silicates. In these areas, soils are thin and acidic under natural conditions, and so the runoff from the acid rain remains highly acidic. Acidity in lakes in these areas has increased sharply since 1950.

The Impact of Acid Precipitation on Terrestrial Systems

The impact of acid precipitation on terrestrial ecosystems is less clear than for aquatic systems. A major area of controversy is whether acid clouds and acid precipitation are damaging world forests. In September 1990, the National Acid Precipitation Assessment Program (NAPAP) released the results of a 10-year study. One of the study's conclusions was that there is now widespread forest damage in North America that can be directly linked to acid rain. Many scientists are convinced that acid rain is the leading cause or at least the catalyst in widespread forest damage in midlatitudes. Evidence of damage to vegetation in North America is beginning to accumulate.

One area where rapid dieback of the forests is occurring is around Mount Mitchell in North Carolina. **Dieback** is the gradual dying of a tree or trees, either from the crown downward or from the tips of the branches inward toward the trunk. Acid rain and acid fog are factors suggested for the problem. Fog over the mountains has frequently been measured with a pH of between 2.5 and 3.5, and rain with a pH of 2.2 was measured in 1986. The trees affected are a variety of spruce, the remnant of a once widespread forest that was logged off more than a century ago. Dieback of this forest has been observed only since 1983. It began at the summit and has progressed down the mountain to lower elevations. Sections of the dead timber can now be seen from the Blue Ridge Parkway that skirts the mountain.

A second area where tree damage has been documented is in New York and New England. Stands of spruce on Whiteface Mountain in New York are dying, and Dr. Hubert Vogelman, a botanist at the University of Vermont, found a 19% decline in the number of sugar maple trees over a 20-year period. In 1965, researchers counted 345,493 maple seedlings in a 2-acre area on Camels Hump Mountain near Duxbury, Vermont. By 1983, the number had dropped to 53,400. Dr. Vogelman also reported that wood samples show increasingly high concentrations of residues of industrial chemicals and hydrocarbons.

Forests in other parts of the world have been affected by acid precipitation. In 1983, it was estimated that one-third of the forests in West Germany were damaged by acid rain. The extent of forests suffering from acid rain in Germany is increasing at a geometric rate. Five percent of the forests were damaged enough to be essentially dead.

It may be that the damage to forests is done through acidification of the soil. When soils become more acidic, there are more dissolved metals in the soil water taken in by the tree roots, and there is less decomposition of organic matter in the soil.

Acid Rain and Human Health

Dr. David V. Bates, a University of British Columbia physician, found that several years of hospital records indicate increased admissions as atmospheric sulfate levels rose in one urban area of southern Ontario containing some 6 million people. The correlation between admissions for ailments including pneumonia and asthma were significantly related to sulfate levels. Bates' study indicated that 13% of the variations in admissions could be explained by changes in sulfate levels. Other researchers suggest that it contributes to emphysema and other respiratory diseases as well, particularly in children.

There may be a health hazard in eating fish taken from streams and lakes with increased acidity. Fish from these waters often have high levels of aluminum, copper, lead, mercury, and zinc. Intake of these metals can affect health. Aluminum may be linked to the onset of Alzheimer's disease. People who die of Alzheimer's disease often have high enough concentrations of aluminum to reduce neural function.

FIGURE 14.1   Statues of the maidens on the porch of the Erechtheum. The Erechtheum is one of the structures on the Acropolis in Athens, Greece. These statues are made of fiberglass. The originals were moved to a museum because they were being destroyed by acid precipitation. (From Hidore J. J., *Global Environmental Change*, © 1996, Prentice Hall, Upper Saddle River, N.J.)

Impact of Acid Rain on Structures

Limestone and marble are soluble in acids. Many major buildings and sculptures are made of limestone and marble, and they can be damaged by acid rain. In Great Britain, acid rain is dissolving the exteriors of major historical buildings. Forty-four of the flying buttresses on Westminster Abbey need to be replaced. These buttresses were rebuilt less than 100 years ago, but the limestone is badly eroded due to acid rain. Another famous structure suffering from solution by acidic precipitation is the Taj Mahal in India. The world-renowned structure, which is built of marble, is rapidly being destroyed. Replacement of damaged panels cannot keep up with the rate of damage. The Erechtheum is a small structure that stands near the Parthenon on the Acropolis in Athens. Its porch roof is supported by six marble statues of maidens carved by the Greek sculptor Phidias in the 5th century B.C. The statues had to be removed because of rapid deterioration due to acid rain. They were replaced in 1977 with fiberglass copies (Figure 14.1).

STRATOSPHERIC OZONE AND ULTRAVIOLET RADIATION

Ozone is a form of oxygen in which three atoms of oxygen combine to form a single molecule of ozone. Ozone normally is not abundant in the lower atmosphere under natural conditions. It does, however, form in smog through the action of sunlight on oxides of nitrogen and organic compounds. This ozone does not stay in the air for very long. It reacts with other gases in the atmosphere and changes to normal oxygen molecules.

Stratospheric Ozone

Ozone exists in the stratosphere, though the total amount is small. It is concentrated in a layer or layers between altitudes of 12 and 50 km (7 and 30 mi). The ozone is continually formed and then removed. The process that forms ozone is the absorption of ultraviolet radiation in the range from 0.1 to 0.3 μm. This absorption of radiant energy breaks oxygen molecules apart into single oxygen atoms. Some of the single atoms combine with an oxygen molecule to form ozone. Absorption of additional radiation breaks up the ozone molecules. Most ozone forms over the tropical latitudes. It is here that most solar radiation, and the most intense solar radiation, reaches Earth. The upper atmospheric circulation carries the ozone toward the poles.

The atmosphere absorbs ultraviolet radiation at all altitudes. Single atoms of oxygen absorb the shortest wavelengths (less than 1 μm) at altitudes above 160 km (99 mi). From 110 to 160 km (68 to 99 mi), oxygen molecules absorb radiation in the range from 0.1 to 0.2 μm in length. Below 110 km (68 mi), ozone absorbs the longer-wavelength ultraviolet radiation. The most ultraviolet radiation is absorbed at heights from 20 to 50 km (12 to 31 mi) (Figure 14.2). Most of the ozone is at these altitudes. Ozone has a broad absorption band that peaks at 0.255 μm. The energy absorbed adds to the increase

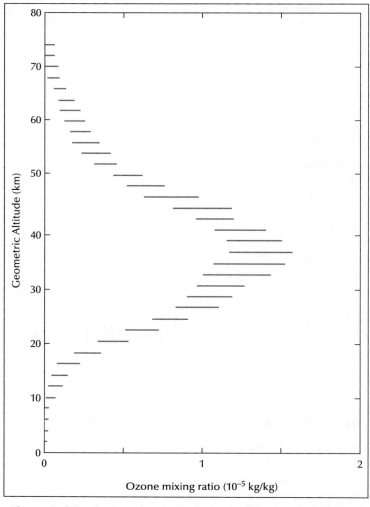

FIGURE 14.2 The vertical distribution of ozone in the lower 80 km of the atmosphere.

in temperature with height in the stratosphere. The atmosphere, mainly as a result of ozone, absorbs about 98% of the ultraviolet radiation reaching Earth. Although the upper atmosphere absorbs most ultraviolet radiation, some reaches the surface. The 2% that reaches the surface of the planet is critical to life on Earth.

Chlorofluorocarbons

In 1974, scientists warned there was evidence to suggest that chlorofluorocarbons (CFCs) have a depleting effect on stratospheric ozone layers. First synthesized in 1928, these compounds promised to have many uses. They are odorless, nonflammable, nontoxic, and chemically inert. The primary CFCs are listed in Table 14.2, along with some of their characteristics. They first came into use in refrigerators during the 1930s. Since World War II, they have been used as propellants in deodorants and hair sprays, in producing plastic foams, and in cleaning electronic parts. The United States, Japan, and Europe produce and consume most of the chemicals. At one time in the United States, per capita use reached 1.1 kg (2.4 lb) per person per year.

CFCs are not natural compounds. They do not react with most products dispersed in spray cans. They are transparent to sunlight in the visible range. They are insoluble in water and are inert to chemical reaction in the lower atmosphere. It is for these reasons that they are valuable compounds. It is for these same reasons that the chemicals are a problem in the stratosphere. The average lifetime of a CFC-11 molecule is between 40 and 80 years. A CFC-12 molecule may last from 80 to 150 years.

CFCs rise into the upper atmosphere, where they break apart under ultraviolet radiation. The breakdown takes place when the compounds are exposed to radiation wavelengths of less than 230 nm. Ultraviolet radiation of this wavelength or shorter does not reach the troposphere because it is absorbed at altitudes of 20 to 40 km (12 to 24 mi). This breakdown releases chlorine, which interacts with oxygen atoms to reduce the ozone concentration. The process ends with the chlorine atom once again free in the atmosphere. Each atom of chlorine may persist for years, acting as a catalyst that may remove 100,000 molecules of ozone. The maximum rate of ozone destruction takes place at an altitude of 40 km (25 mi). The final means of disposing of the chlorine is a slow drift downward into the troposphere, where it combines with water molecules and falls out as hydrochloric acid.

TABLE 14.2 Common Chlorofluorocarbons

Compound (chemical formula)	Ozone Depletion Potential*	Atmospheric Lifetime (years)	Major Uses
CFC-11 ($CFCl_3$)	1.0	64	Rigid and flexible foams, refrigeration
CFC-12 (CF_2Cl_2)	1.0	108	Air conditioning, refrigeration, rigid foam
CFC-113 ($C_2F_3Cl_3$)	0.8	88	Solvent

* Ozone depletion potentials represent the destructiveness of each compound. They are measured relative to CFC-11, which is given a value of 1.0.

Source: U.S. Environmental Protection Agency.

The Antarctic Ozone Hole

The most disturbing change in atmospheric ozone is the **ozone hole** (Figure 14.3) over Antarctica. The ozone hole is defined as the area of the atmosphere where the level of ozone drops below 220 Dobson units. The Dobson unit is a measure of the total column of ozone above the earth. The value 220 was selected as the average amount of ozone above the earth. Any measurement of ozone below 220 Dobson units is attributed to depletion by chemical emissions. The ozone hole is a loss of stratospheric ozone over Antarctica, which has occurred in September and October since the late 1970s. The hole appears in September, when sunlight first reaches the region, and ends in October, when the general circulation brings final summer warming over Antarctica. During the Antarctic spring, there is a decrease in ozone north from the pole to nearly 45° S latitude.

In August and September 1987, the amount of ozone over the Antarctic reached the lowest level recorded to date. On September 17, 1987, Nimbus-7 recorded a large area in which the ozone concentration was only about half that of the surrounding region. That fall, the ozone hole, the area of maximum depletion, covered nearly half of the Antarctic continent.

In the winter over Antarctica, a very large mass of extremely cold dry air keeps out warmer air surrounding the continent. This cold air gets even colder during the months when there is no sunlight. Temperatures drop as low as –84°C (–119°F). In the extreme cold, moisture condenses into ice crystals, and nitric acid crystals form. These crystals form very high thin clouds called **polar stratospheric clouds (PSCs)**.

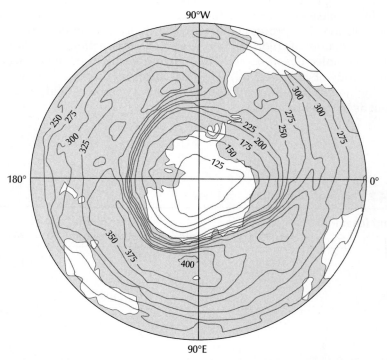

FIGURE 14.3 The Antarctic ozone hole over the southern hemisphere on September 28, 1992, as measured by the Total Ozone Mapping Spectrometer (TOMS) aboard the Nimbus-7 satellite. Notice that the area of lowest ozone concentration is larger than Antarctica and that the ozone hole is nearly centered over the South Pole. (From Hidore J. J., *Global Environmental Change*, © 1996, Prentice Hall, Upper Saddle River, N.J.)

The cloud crystals play a very important role in the chemistry of the CFCs and ozone depletion. The nitrogen oxide crystals drop out of the stratosphere, leaving behind the chlorine and bromine compounds and the ice crystals. Each ice crystal provides a place for accelerated chemical reactions. The chemical processes are more rapid where there is a surface on which the reaction can take place. Ice particles are around 10 times as efficient as the surface of water droplets. This partially explains the speed with which the process takes place in the Antarctic spring. It also explains why the process is less effective in low latitudes.

The chemical process begins when sunlight appears in the spring. The warming increases the rate of chemical reactions, and chlorine destroys ozone at a rapid rate. The depletion actually first begins near the Antarctic Circle, where sunlight begins to penetrate the stratosphere. It may begin here by mid-August. Spring over the south pole occurs in September and October. During this time, the ozone level drops until there is no more ozone left or the clouds evaporate. There may be a total loss of up to 60% of the ozone in the center of the Antarctic hole. At some altitudes, it is 90%. Eventually, air from surrounding regions flows into the area, and ozone levels recover. Polar stratospheric clouds disappear with the spring warm-up. The same process takes place elsewhere in the atmosphere, but at higher altitudes and at slower rates. The depletion usually takes place at altitudes of 12 to 22 km (7 to 14 mi), which is the range where most of the polar stratospheric clouds form.

The Global Decline in Ozone

Ozone depletion is not as great outside the Antarctic region because the stratospheric aerosols are less abundant and consist of liquid sulfuric acid droplets rather than ice. This difference is significant. There is no Arctic ozone hole like that of the Antarctic. Temperatures are warmer and there is more variable weather in the Arctic, which provides less favorable conditions for the necessary chemical and circulation processes. Ozone levels in the high latitudes of the northern hemisphere have dropped 5% since 1971. In 1988, researchers in Thule, Greenland, measured increased concentrations of reactive chlorine compounds, the compounds known to be present over the Antarctic while depletion takes place. Experiments in 1989 showed the presence of the ozone hole over Antarctica and provided detailed measurement of the amount and extent of the depletion.

In the low- and midlatitude stratosphere, there is greater solar radiation during the winter months, and there is a general absence of polar stratospheric clouds. In this part of the atmosphere, the destruction of ozone is due to a combination of chemical processes. Models of the atmosphere show that nitrogen oxides play a leading role in the ozone destruction. Particular nitrogen oxides react on the surface of sulfuric acid solutions, which are similar to stratospheric aerosol particles. Sulfate particles exist throughout the lower stratosphere. They form from biological and volcanic activity. One of the processes is not dependent on extremely cold conditions, so it operates much of the time to deplete ozone.

Ultraviolet Radiation and Living Organisms

The United Nations Environment Programme called a conference in Montreal, Canada, in September 1987 that drafted a treaty restricting the production of CFCs. The agreement is officially termed the *Montreal Protocol*. Concern over the possible connection between CFCs and ozone loss led to a ban on the use of these compounds in the United States, effective in 1978. There are two areas of concern about the possible reduction in the ozone layer and an increase in ultraviolet radiation reaching ground level. The first problem concerns public health. The second is the role of CFCs in global warming.

Stratospheric ozone filters out ultraviolet radiation in the 280 to 320 nm range. This is the high-energy portion of the ultraviolet radiation spectrum known as ultraviolet-B (UVB). UVB radiation is very harmful to living organisms. Although the atmosphere blocks most UVB radiation, it does not block all of it. Plants did not flourish on Earth until there was enough atmosphere and ozone to block much of the UVB radiation.

All plants and animals now existing and living in sunlight on Earth have adapted to ultraviolet radiation. Plants vary widely in their tolerance of UVB. Plants that developed in climates with high-intensity sunlight show a variety of defense mechanisms for UVB. Some produce clear or nearly clear pigments that absorb UVB radiation. For example, marijuana plants produce protective chemicals called *cannabinoids* (which are also the main hallucinogenic ingredients in marijuana). In arid climates, plants develop thick, shiny leaves. Cacti and olive trees are examples.

Although sunlight is essential to most life, there is a limit to how much sunlight is good. One of the effects of ozone depletion is to let more ultraviolet radiation through the atmosphere to the surface. Ultraviolet-B can damage DNA, the genetic code in every living cell. Most living organisms are subject to damage by UVB radiation. Because plants cannot adjust their behavior to changing solar radiation, some are damaged by UVB radiation. The soybean is one such plant. Excessive amounts of UVB slow growth and reduce yields. Soybean yields may drop 1% for each 1% drop in ozone.

Animals and humans also have adapted to UVB radiation. Nearly 90% of marine species living in the surface water surrounding the Antarctic continent produce some form of chemical sunscreen. Humans manufacture melanin in the skin. This is a pigment that blocks ultraviolet radiation. A summer tan results from increased production of melanin. Persons with very fair skin do not readily manufacture melanin and sunburn very easily.

Ultraviolet-B and Human Health

Exposure to ultraviolet radiation results in tan skin, sunburn, aged skin, skin cancer, and a weakened immune system. The risk of skin cancer is much greater from overexposure to UVB, as in a sunburn, than from steady low doses. A single blistering sunburn in a person 20 to 30 years of age triples the risk of skin cancer.

MELANOMA   One form of skin cancer is **melanoma**. It may start in or near a mole. This involves the cells that give the skin its color and often are a mixture of black or brown, sometimes with red or blue areas. These moles continue to grow and have irregular borders. Melanoma is the least common but most lethal form of skin cancer. The fatality rate from melanoma is about 25%. It is almost always fatal if it spreads to other parts of the body. Early treatment results in a survival rate greater than 80%.

The highest incidence of melanoma occurs in individuals who do not tan easily. The incidence of melanoma in the United States and other countries where sunbathing and tanning salons are in vogue is increasing at about 4% each year. Younger and younger persons are being diagnosed as having melanoma. When first regularly reported, it was in persons ages 40 or over. By 1990, it was frequent in the age group from 20 to 40 years. The incidence of melanoma remained level from 2000 to 2004 for American men and women combined. It increased from 1981 to 2004 among women. Fatality rates from 1998 to 2004 remained level for men and women combined.

BASAL AND SQUAMOUS CELL CARCINOMA   Basal cell carcinoma is the most common form of skin cancer. It is a slow-growing cancer that usually begins with a small, shiny, pearly bump or nodule on the head, neck, or hands. It can bleed, crust over, and then open again. It is not life threatening. Squamous cell carcinoma may start as nodules or red patches with well-defined outlines. It typically develops on the lips, face, or tips of the ears. It can spread to other parts of the body and enlarge.

These skin cancers can be removed by simple surgery and are rarely fatal. It is likely that nearly everyone in the United States over age 30 has some skin damage from solar radiation. At current rates, at least one in seven Americans will develop some form of skin cancer.

REDUCING THE RISK OF SKIN CANCER.   The risk of getting skin cancer can be reduced with reasonable care. The first rule is to avoid exposure to the midday sun. The most dangerous hours are between 10 A.M. and 2 P.M. local time. There is an old saying: "Only mad dogs and Englishmen go out in the noonday sun." If exposure to the sun is necessary, use a sunscreen with a rating of 15 based on ultraviolet-B radiation. Ultraviolet-A is also harmful to health, but not nearly as much as ultraviolet-B. Lotions with a rating of 15 provide protection from both UVA and UVB radiation. Avoid tanning parlors because the radiation is as bad as or worse than that of natural sunlight.

Future Depletion of Ozone

Despite the international efforts to reduce the emission of ozone-reducing gases into the atmosphere, the depletion of ozone over Antarctica continues. In the spring of 2006, the size of the Antarctic ozone hole reached a record average size of 27 million km^2. In the Antarctic spring of 2008, the area dropped to 25 million km^2, and total ozone levels dropped to about half of the standard of 220 Dobson units. One of the problems associated with ozone depletion is that maximum depletion may not occur until between 2010 and 2020. The Antarctic ozone hole may not fill until as late as 2075. Each CFC molecule has a lifetime of up to 30 years. It takes these molecules 6 to 8 years to rise to the stratosphere. Even if present production of CFCs and related compounds stops, there is a huge quantity of CFCs in old refrigerators, air conditioners, and foam packaging. Production has not stopped; under present global arrangements, nearly half as much CFC can be produced in the future as the total produced since the chemicals were first introduced.

Compliance with existing agreements will slow but not stop the accumulation of the chemicals in the stratosphere. Concentrations may grow to as much as 30 times the 1986 levels. Computer models show that at least an 85% reduction in CFC use is needed to stabilize the level of the chemicals in the atmosphere. The chlorine already released will continue to remove ozone for at least a century. If the release of CFCs stops, there will be a lag of several decades before maximum ozone depletion takes place. Only after that time can the rate of removal begin to decline. The chlorine content may continue to rise until it reaches a level as much as six times the 1986 levels. If the release of the compounds continues unabated, a reduction in ozone of about 10% will take place, with a possible range of 2 to 20%.

Summary

Earth's atmosphere is no longer a natural one. That is, it has been greatly altered by human activity. The major gases of the atmosphere are still nitrogen and oxygen, but the variety and amounts of the variable gases have changed. Large amounts of dust, soot, metals, and organic matter are injected into the atmosphere each day. Large volumes of gases such as carbon monoxide, carbon dioxide, and nitrogen dioxide are also added each day. Many of the gases or their by-products are harmful to plants and animals.

Sulfur compounds and nitrogen dioxide combine with water droplets in clouds to become acid precipitation. Acid precipitation is harmful to most ecosystems. It is harmful to human health, commerce, and structures. Control of acid precipitation is difficult because the atmosphere is so mobile. The precipitation may fall hundreds or thousands of kilometers from the place where the pollutants are injected into the atmosphere.

Ultraviolet radiation from the sun reaching Earth's atmosphere creates a layer of triatomic oxygen called *ozone*. The ozone absorbs about 98% of the high-energy radiation. One of the many classes of chemical compounds developed by humans are chlorofluorocarbons.

These chemicals escape into the atmosphere and rise into the stratosphere. In the stratosphere, radiation breaks them down in a fashion that releases chlorine. The chlorine removes oxygen ions and reduces the amount of ozone. With a reduction of ozone, there is increased ultraviolet radiation reaching the surface, which is detrimental to the health of plants and animals.

Key Terms

Acid precipitation, 236
Carbon monoxide (CO), 233

Carcinogenic, 232
Dieback, 238
Effluent, 232

Melanoma, 244
Ozone hole, 242
Particulate, 232

Polar stratospheric cloud (PSC), 242

Review Questions

1. What is the primary source of carbon monoxide in the atmosphere?
2. A neutral solution has a pH of 7. What is the pH of normal precipitation?
3. Below what level of acidity in fresh water do fish not survive?
4. What locations in the United States have forests suffering from dieback due to atmospheric pollution?
5. How does acid rain affect structures made from marble and limestone?

6. What process creates the stratospheric ozone layer?
7. How do the CFCs that are relatively stable in the troposphere affect the ozone layer in the stratosphere?
8. Why does ozone depletion take place more readily over the polar regions than in the midlatitudes and tropics?
9. Why is ultraviolet-B radiation damaging to living organisms?
10. What are the three most common forms of skin cancer?

PART III

Regional Climatology

Regional Climates: Scales of Study

Climates can be identified and analyzed over a broad range of areal units from the small vegetable garden up to the continental size regions. The climate over a plowed field is different from that over a field of clover or pasture. The climate of a city differs from that of a rural area, and the climate of a desert differs from that of a rain forest. Significant differences occur in the climate at all of these different scales. Climatologists must deal with these differences at all levels. Often the analytic approach used varies with the relative size of the study area. Before considering major world climates, it is desirable to briefly examine some other scales of study.

DEFINITIONS

There has been much confusion in identifying and naming the range of scale in climatic studies largely as a result of the difficulty of separating the atmospheric continuum into discrete units. The identification problem has been compounded by the historic evolution of climatic studies. Researchers in different countries have used names for areas that, in other countries, are described by different terms. Thus, we find such terms as *ecoclimate* and *topoclimate* referring to areas of somewhat similar size. At the same time, the dimensions suggested as suitable boundaries for identified scales frequently differ from one source to another.

In a discussion of scale in climatology, the Japanese climatologist M. M. Yoshino derived a general consensus of definitions, and the description given here follows his grouping. Figure 15.1 provides examples of the scales that he suggests. The major subdivisions include the following:

- *Microclimate:* This category is characterized by the climate that might occur in an individual field or around a single building. This scale describes the climates of an area that might extend horizontally from less than 1 m to 100 m (3.28 ft to 328 ft). Vertically the area can extend from the surface up to 100 m (328 ft).
- *Local climate:* This category comprises a number of microclimatic areas that make up a distinctive group. The climate in and above a forest or that of a city may be classed in this division. Horizontal dimensions can extend from 100 m to 10,000 m (328 ft to 32,000 ft), and the vertical extent is up to 1000 m (3280 ft).
- *Mesoclimate:* Such climates might range horizontally from 100 m to 20,000 m (328 ft to 65,616 ft) and vertically from the surface to 6000 m (19,685 ft). As Figure 15.1 illustrates, a great variety of individual landscapes are considered in this category.
- *Macroclimate:* The largest of the areas studied in climatology, macroclimates extend horizontally for distances more than 20,000 m (65,616 ft) and vertically to heights in excess of 6000 m (19,685 ft). Such areas can be continental in extent.

The dimensions provided in this list are basic guides, with many areas studied overlapping the identified groups. Perhaps the best way to illustrate the role of scale in regional studies is to examine specific examples.

MICROCLIMATES

Studies of **microclimates**, small areal units, inevitably begin with fieldwork. The preliminary step is the accumulation of data. Such a procedure is needed because published climatological data are almost entirely comprised of readings taken at the standard height of the instrument shelter. The microclimatologist is often concerned with the state of the atmosphere below that level. Furthermore, most of the stations that are used to report climatological data are widely spread, with perhaps one station being the representative of many square kilometers. To obtain data such that differences over small distances can be derived, it is necessary to place a fairly dense network of instruments within a small area.

Many studies of microclimatic environments have provided a generalized picture of climatic conditions near the ground. These findings indicate the types of conditions that will occur over a bare surface. Through analysis of surfaces covered by vegetation or synthetic materials, it is possible to generate a more complete understanding of the variation of climatological processes that occurs at the microscale.

General Characteristics of Microclimates

A large diurnal temperature range occurs at the near-Earth levels. During the day, interface temperatures have been found to be as much as 10°C (18°F) warmer than the air only 1 m (3.28 ft) above the surface. At night, the situation is reversed. An inverted lapse rate occurs, with temperatures at the surface cooler than those immediately above. Such a response is to be anticipated. During daylight hours, incoming

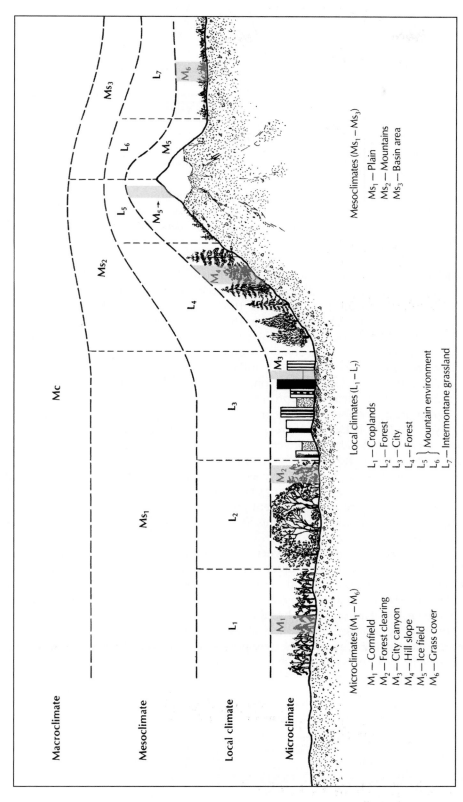

Macroclimate

Mesoclimate

Local climate

Microclimate

Mc

Ms₁

Ms₂

Ms₃

L₁ L₂ L₃ L₄ L₅ L₆ L₇

M₁ M₂ M₃ M₄ M₅ M₆

Microclimates (M₁–M₆)

M₁ — Cornfield
M₂ — Forest clearing
M₃ — City canyon
M₄ — Hill slope
M₅ — Ice field
M₆ — Grass cover

Local climates (L₁–L₇)

L₁ — Croplands
L₂ — Forest
L₃ — City
L₄ — Forest
L₅ — } Mountain environment
L₆ —
L₇ — Intermontane grassland

Mesoclimates (Ms₁–Ms₃)

Ms₁ — Plain
Ms₂ — Mountains
Ms₃ — Basin area

FIGURE 15.1   Area scales of climatic investigation.

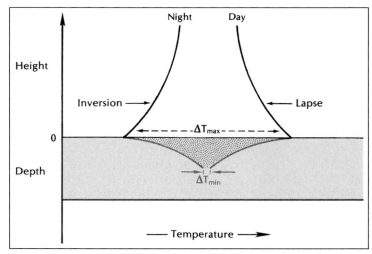

FIGURE 15.2   Idealized profiles of temperatures near the boundary layer in clear weather. (After Oke, 1978.)

solar radiation warms the surface and heat diffuses into the atmosphere, raising the temperature of the air by smaller amounts at increasing altitude above the ground. After sunset, under clear-sky conditions, the surface cools rapidly by radiation and the atmosphere loses heat by diffusion to the cold surface.

Temperature changes in soils decrease with depth. As Figure 15.2 illustrates, the amount of diurnal temperature change is greater near the surface and decreases with depth until equilibrium is attained. A similar pattern is obtained for the annual cycle, although, of course, it follows the seasonal rather than the diurnal cycle.

Figure 15.3 is an idealized representation of wind near the surface. At the interface between the atmosphere and the ground, a thin layer of air adheres to the surface. In this layer, the flow is **laminar**, streamlines are parallel to the surface, and they lack the cross-stream component of convective currents. The depth of this layer depends on the surface roughness and wind speed, but it is seldom more than 1 mm (0.04 in.) in thickness. Its significance lies in its role as an insulating barrier in which all nonradiative transfer is by molecular diffusion rather than the turbulent transfer typical of most of the lower atmosphere. The turbulent surface layer comprises a complex flow of swirling eddies extending upward some 50 m (165 ft). In this zone, a general increase of wind speed occurs with height.

The exchange of moisture between the surface and air above is reflected in humidity measurements at various levels. Figure 15.4 provides a generalized profile of water vapor concentration day and night. During the day, the concentration decreases away from the surface in a similar way to the temperature profile. At night, if the dew point of the air is not attained, the humidity profile is somewhat similar to that in daytime. This occurs because evaporation will continue during the night hours but at a lower rate than during the day. If the dew point is reached, an inverted moisture profile

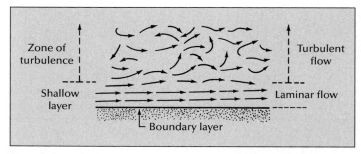

FIGURE 15.3   Flow of air at the boundary layer from laminar to turbulent flow.

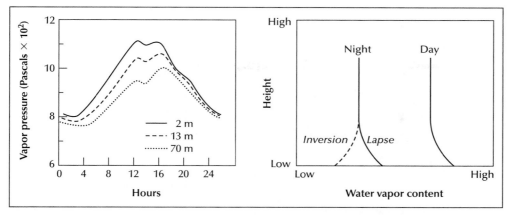

FIGURE 15.4   Generalized profile of water vapor concentration near the boundary layer. (After Oke, 1978.)

will exist. The deposition of dew causes a lowering of near-surface moisture content. If the moisture is replaced by downward movement of air, dew formation will continue. Downward movement will occur through slight turbulence. If this turbulence does not occur, the near-surface air is not replenished with moisture, and dew fall ceases.

Although it is possible to generalize about the nature of the microclimate above a bare soil surface, it is important to note that the actual characteristics depend, at least in part, on the type of soil surface exposed and the amount of water it contains. For example, sandy soils have a lower heat capacity than clays. This means that a sandy soil will heat up rapidly in its top layers during the day, but at night will be cooler than less sandy soils because it undergoes rapid radiational cooling. Organic matter in the soil has low specific heat and the dark color increases the absorption of solar radiation. At night, such soils are relatively warm.

The presence of water in the soil or at the surface greatly modifies the exchange of energy that occurs. When water is present, incoming energy is used for evaporation, making it less available for sensible heat. A comparison of the energy budgets for an irrigated field and one that is dry shows that the temperatures over a moist soil are usually lower than those over dry ground.

These general principles apply to all microclimates. However, modification of the bare surface through vegetative growth or human interference alters the intensity and rates at which ongoing processes occur. Such changes are described in the following section.

The Role of Surface Cover

The role of surface covering in creating microclimates is a response to the way in which incoming energy is disposed at the surface and the way in which the surface modifies airflow. The differences that exist can best be shown through illustrative data. Figure 15.5 shows, in schematic form, some modifications that occur when plants cover the surface. Of particular significance is the creation of a **canopy layer**, in which the microenvironment reflects the nature and extent of the canopy. Clearly, the relative continuity of the canopy plays a major role so that plants with large leaves horizontal to the surface form a more effective canopy than those whose leaves are small and aligned vertically. The canopy becomes most effective when a plant stand has grown to the extent that the ground is shaded. This growth causes the highest temperature zone to move away from the surface to the canopy area. Hence, the top of the canopy, rather than the soil surface, becomes the energy exchange layer.

For example, a building or fence has a marked effect on the microclimate. A building creates a **climatic sheath** (Figure 15.5c) in which temperature variations occur as a result of shading, humidity anomalies are found, and even a variation in

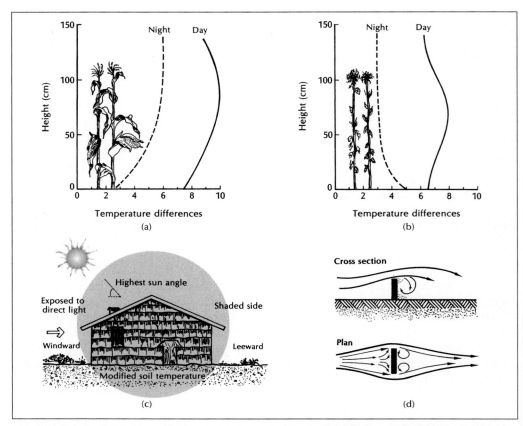

FIGURE 15.5 Schematic representation of microclimatic modifications caused by surface cover. (a) Temperature profiles in a crop whose growth shades the ground. (b) Temperature profiles for crops with essentially vertical growth. (c) Climatic sheath around a building. In the sheath, many microclimatic variations occur. Examples show creation of windward–leeward sides, sunlit and shaded sides, and soil modification. (d) Airflow across a barrier at right angles to wind direction. Cross-section shows how such an obstruction is used as a snow fence.

precipitation might occur. When the changes in the microenvironment of a single building are assessed, it is not surprising that cities consisting of numerous buildings can create their own climate.

The modification of wind in the microenvironment is well demonstrated by effects of a vertical structure on air flow. As the diagram in Figure 15.5d shows, the patterns change with the direction from which the wind is blowing. Such modification has been put to good use in the construction of snow fences to keep roads and other areas clear of drifting snow. The same procedure is used to prevent sand from drifting over roads in arid regions.

The Role of Topography

It was noted earlier that topographic aspect is of importance in determining the distribution of temperature at a given location. The height of the snow line on equator-facing slopes compared to those facing poleward is located at different elevations. Most people have seen the effect during winter when snow on one side of a small hill remains in place long after that on the opposite slope has melted.

Topographic variations also modify thermal patterns in small areas. One of the best-known effects of this is the creation of an inversion that, at times, can create a distinctive air flow. On cool, still nights, air close to the surface becomes cooler than that above. If this cooling occurs in areas of uneven topography, the cold dense surface air tends to flow downslope and accumulate in the bottom of valleys and depressions. Such air flow is called a **katabatic wind**.

If the temperature of the ground is at or just below freezing, the collection of cold air in the valley bottom causes a localized frost. Such an effect in citrus growing areas can cause appreciable damage to crops. For this reason, apple and cherry orchards in the Appalachian Mountains are planted on slopes. The cold air flows through the orchards to the valleys, leaving the fruit orchard frost free. Topography also plays a role in air quality. Air descending a slope can cause particulates and other airborne pollutants to collect in valleys creating either an air pollution event or chronic poor air quality. If the pool of cold air becomes deep enough, any effluent carried into the inversion layer is trapped in it. Should the inversion remain in place for more than a few days, the stagnant air can become highly polluted.

LOCAL AND MESOSCALE CLIMATES

Local climates consist of a number of microclimate environments because they comprise a larger area. There is, of course, overlap in the organization framework, but a further differentiation can be made because local climatic studies stress horizontal rather than vertical differences in climate. There are many examples of local climate studies, and here the climates of forest areas are used to illustrate this level of analysis.

Forest Climates

Forest climates differ from those of surrounding nonforested areas. The **boundary layer** of the forest is its canopy, and it is at this level that energy exchanges occur. Some insolation is returned directly to space, and the amount depends on the albedo, or reflectivity, of the canopy layer. In some forests, the quantity varies enormously from season to season. Some energy is trapped within the canopy layer and some penetrates to the forest floor. The amount of penetration is characteristically low, and varies with both the state of the sky along with amount and type of foliage that exists.

Compared with the flow of air over open areas, wind inside a forest is reduced. The amount of decrease depends on the type and structure of the forest. For example, in a deciduous forest, the wind velocity is reduced by as much as 60% to 80% at a distance of 30 m (100 ft) inside the forest. Similarly, in a Brazilian forest, the wind speed has been found to decrease from about 8 kph (5 mph) to 1.6 kph (1 mph) with the same distance. The flow of air is, of course, highly complex and varies in the vertical and horizontal dimensions. Studies have shown that winds from 8 to 24 kph (5 to 15 mph) above the canopy are often less than 3.2 kph (2 mph) at the surface.

The forest environment also modifies local moisture conditions. Evaporation from the forest floor is relatively low because of reduced insolation and wind velocity. This effect is counterbalanced by the fact that, with the profuse vegetation, high transpiration occurs. The humidity within a forest depends on the density of the forest and the rates of transpiration that occur. It is generally found that the relative humidity in the forest may be from 2% to 10% higher than that of nonforested areas, with the highest humidities occurring during the high-sun season. Note that comparisons of humidity using relative humidity values are not always meaningful because the modified thermal environment directly influences the water-holding capability expressed by using relative humidity.

The thermal differences that occur result from a combination of the factors already outlined. These variations include shelter from direct rays of the sun, heat modification through water transfer, and the blanket effect of the canopy. The essential result of the interaction of these factors is that temperatures inside a forest are moderated. The maximum temperature is lower and the minimum higher than those of nonforested areas experiencing a similar climate regime. The amount of variation is seasonal with the main difference in summer being as much as 2.8°C (5°F) within a low-altitude, midlatitude forest. Exceptions to such moderating influences do occur. The Forteto oak forests of the Mediterranean, for example, experience higher temperatures than neighboring, nonforested areas. Such trees transpire slowly; therefore, the usual hydrologic and thermal conditions of the forest are modified.

Mesoscale Climates

Mesoscale climates are frequently identified with a distinctive geographic region. In such a region, the physical controls of climate are similar and not modified by major differences within the region. Thus, a mesoscale climatic study might concern the climates of areas such as the Central Valley of California, the lands in the vicinity of the Great Lakes, or the Mississippi Delta lowlands.

Mesoscale climatic studies have often concerned scales of motion rather than scales of area. This development corresponds to the meteorological concept of mesoscale phenomena, which includes analysis of such features as severe storms and mountain–valley winds. The climatological equivalents are seen in studies ranging from the cause of drought in the Sahel to precipitation variations resulting from circulation patterns in the Midwest of the United States.

MACROSCALE CLIMATES

Although the distinction among various scales of climate is not always clear, the identification of **macroscale climates** is aided by the concept of **filtering**—the averaging of circulation features over successively longer time periods. As the time scale becomes longer, the smaller scale patterns are filtered out and only the general characteristics remain. Thus, a mesoscale feature such as a hurricane or land–sea breeze would not appear on a macroscale representation.

The climatic analysis of macroclimates can follow two methods of approach. First, the analysis can consider the major surface climatic characteristics, ranging from temperature and precipitation patterns to statistical analysis of other climatic elements. Such an approach provides the descriptive climatology of large climatic regions or even continents. Second, the analysis can deal with the dominant circulation patterns that influence the climate and provide the key to understanding its cause. This process is accomplished through analysis of the state of the atmosphere as inferred by pressure maps showing patterns and winds at various levels of the atmosphere. Alternatively, the patterns can be depicted as cross-sections through the troposphere and stratosphere. This approach obtains a synopsis, or condensed view, of the atmosphere at a given time and is referred to as the *synoptic climatology*.

As noted, one approach to describing the climates of the world involves the descriptive climatology of large regions. Historically, the boundaries of such regions have been defined in terms of the surface conditions, and a number of classification schemes have been implemented. Such classification procedures are considered in the next section. The treatment is in terms of both the traditional descriptive regional climatology, in which the major surface climatic characteristics are used, and the dynamic analysis that uses the principles of synoptic climatology. Of necessity, the scale used to provide an overview of the entire climate of the world is at the macro level.

CLASSIFICATION OF WORLD CLIMATES

The classification of climates on a formal basis began with macroscale climatic regions, which cover large geographic areas, some being subcontinental in size. To produce a useful classification of any data, it is necessary to first group together those items that present the greatest number of common characteristics and then subdivide those groups on a uniform basis until a satisfactory degree of subdivision is reached. In most areas of science, attempts have been made to produce classifications based on the most fundamental characteristics possible rather than on elements that might be more easily observed but of less intrinsic importance. Sometimes items do not lend themselves easily to grouping or classifying. Attempting to group similar regional environments on the earth is much like trying to group students on the basis of height or weight. There are no clear-cut divisions.

The climatic elements of a region do not distinguish a region by their presence or absence, but by the differences in their character. Change from place to place is a basic assumption in regional study. This is not to imply that the differences from place to place are not without order. There is a systematic variation in climatic elements from place to place, sometimes rather abruptly and in other cases over considerable distance. Because change with time and place is an integral part of the earth's environment, it is appropriate that a classification of climates should be based on the patterns of variation that occur.

Throughout this book, there have been repeated references to the annual and diurnal **periodicities** that exist. The motions of the earth in space give rise to periodic climatic, hydrologic, biological, and geological events. If the basic pattern of energy fluctuation is considered, we find that it is responsible for the largest share of the periodicity and spatial variation in the earth environment. The changing intensity and duration of solar radiation bring about changes in the atmosphere and seas, influence migratory and hibernation habits in animals, and set the life cycle of much of the biota, including humans. The changing intensity of solar radiation between the northern and southern hemispheres produces a shifting of the general circulation of the atmosphere, which in turn is responsible for periodic changes in the hydrologic cycle over the earth. Thus, an energy flow that varies systemically through time and space results in an environment that also varies systemically through time and space. It is the purpose of this chapter to consider the major regions that result from periodic patterns of energy and moisture and to show how other environmental variables are related to these seasonal patterns.

Attempts to classify regional differences date back thousands of years. During the period of the ancient Greek civilization, the earth was divided into three broad temperature zones: torrid, temperate, and frigid. The use of the word *temperate* to describe midlatitude weather might not have been a particularly good choice, but nevertheless the classification persisted through the centuries. Since the recording of that very early division of the earth, classification schemes for organizing regional differences on the earth have appeared with ever greater frequency.

The division of the earth into three temperature zones as formulated by the Greeks centuries ago is appropriate. The tropics are essentially winterless areas as they are not directly affected by the outreach of cold polar air. Instead, the tropics are dominated by air currents that originate in the warm areas between 30° N and 30° S. The polar regions are areas that lack a warm summer, and they are without the incursion of tropical air currents. The *midlatitudes* are characterized primarily by the very marked summer and winter seasons because these areas are dominated seasonally by tropical and polar air currents.

A second element that varies markedly over the earth is the seasonal pattern of moisture. Some areas have a nearly equal probability of rain every day of the year. Most of the earth is subjected to a seasonal probability of rainfall as the primary circulation shifts back and forth with the solar energy supply. Some desert areas have nearly equal probability of having no rain on any day of the year. This breakdown gives perennial precipitation regimes, seasonal precipitation regimes, and dry regions. Such basic divisions of global climate using temperature and precipitation form the basis of a number of well-known classification systems.

Approaches to Classification

Of basic importance in the classification procedure is the selection of the variables used in the delineation process. Much of the early work in classification was limited to the use of temperature and precipitation. Furthermore, much of the early work was carried out by plant physiologists and plant geographers who found a correlation between vegetation and the temperature and moisture characteristics of an area. Mid-19th century researchers were influential in the development of climatic classification. It is not surprising, in view of their botanical training, that the distribution of climate and natural vegetation should be treated simultaneously. As an example of the influence of plant geographers on climate classification, many climatic regions are still identified by plant association. This procedure is still followed by some writers. It is not unusual to find a climate type described as savanna, taiga, or tundra. The correlation between climate and vegetation is still prevalent, and climates of the world are still described in terms of natural vegetation distributions.

It is evident that, in relating the distribution of natural vegetation to the distribution of climate, the *effect* of climate is being measured, instead of the climate. It is assumed that a given climate gives rise to a distinctive vegetation association. To identify the climate type, it is first necessary to determine the vegetation and then infer the climate. If climate can be so identified—that is, by expressing it as the cause of the distribution of one selected component of the environment—then equally useful climate designations can be made through other measures of the climate's effects. It becomes possible to devise climate schemes using factors ranging from the human response to climate to the effects of climate on rock weathering. Such systems would be based on the observed effects of climate, and the criteria used to delimit their boundaries established by best-fit statistical properties. Systems derived through such a method are collectively termed *empiric classifications*. The use of the qualifying term *empiric* connotes identification through the observed effects of climate.

Because it is possible to observe the effects of climate on a whole range of environmental phenomena, there are many bases that can be used in the formulation of an empiric system. The following list illustrates some of the innumerable interrelationships that can be examined:

- The human response to climate
- Climatic requirements for crop growth
- Water needs and precipitation effectiveness related to vegetation
- Study and identification of climatic analogs (e.g., agricultural analogs)
- Vegetation distribution related to climatic controls
- Geomorphic processes acting under different climatic conditions
- Climate and soil-forming processes
- Synoptic conditions and satellite imagery

A little thought could provide many other relationships, and for each there is probably a climatic classification available.

In recent years, the availability of extensive databases and electronic computing technology has seen the development of what is termed **numerical classification**. This method uses many variables and numerical procedures, such as correlation and cluster analysis, to classify climates. Numerical classification is becoming an increasingly important method, especially in special purpose classifications, and it is anticipated that most future systems will be of this type.

THE KÖPPEN SYSTEM

Of the many classifications that have been devised, it is inevitable that one would develop into what might be termed a *standard system*. Such a classification becomes standard as a function of wide usage. It follows that the most widely used systems are those that facilitate an orderly description of world climates. One such system that has developed along these lines is that formulated by Wladimir Köppen. In its various forms, it is probably the most widely used of all climatic systems.

Köppen made one of the most lasting and important contributions to the field of climatic classification. He was trained as a botanist and a meteorologist. In the early stages of his work, he was strongly influenced by the writings of botanists. The systems he formulated range from a highly descriptive vegetation zonal scheme to a classification in which boundaries are defined in relatively precise mathematical terms.

Beginning with his doctoral dissertation in 1870 and continuing to his death in 1940, Köppen proposed, modified, and remodified his system. By 1951, it had become so established that F. Kenneth Hare reported to the Royal Meteorological Society of Canada that some regarded the system "as an international standard, to depart from which is scientific heresy." Such interpretation probably was not intended by Köppen, to whom the scheme was never completely satisfactory. The evolution of the system shows that Köppen was not as concerned with the precise boundaries as he was with attempting to use simple observations of selected climatic elements to provide a first-order world pattern of climates.

Köppen's early work was completed at a time when plant geographers were first compiling vegetation maps of the world. His early publications (1870 and 1884) were concerned with temperature distribution in relation to plant growth, and it was not until 1900 that any of his publications were really concerned with world climatic classification. The 1900 system, which did not get much notice, is a highly descriptive scheme making use of plant and animal names to characterize climate. In 1918, Köppen produced a system that is substantially the one in use at the present time. Boundary values have changed and new symbols have been introduced, but the framework of the present system was clearly evident. The scheme demonstrates Köppen's major contribution to the systematic treatment of the climates of the world. He recognized patterns underlying world climate regions and introduced a quantitative method that allows any set of data to be categorized within the system. The classification is considerably enhanced by the introduction of a unique set of letter symbols that eliminates the necessity for long descriptive terms.

The classification is based on the distribution of vegetation. Köppen's assumption was that the type of vegetation found in an area is very closely related to the temperature and moisture characteristics of the region. These general relationships were already known at the time Köppen's classification was produced, but he attempted to translate the boundaries of selected plant types into climatic equivalents. The Köppen system is based on monthly mean temperatures, monthly mean precipitation, and mean annual temperature.

Köppen recognized four major temperature regimes: one tropical, two midlatitude, and one polar (Table 15.1). After identifying the four regimes, he assigned numeric values to the boundaries. The tropical climate is delimited by a cool month temperature average of at least 18°C (64.4°F). This temperature was selected because it approximates the poleward limit of certain tropical plants. The two midlatitude climates are distinguished on the basis of the mean temperature of the coolest month. If the mean temperature of the coolest month is below −3°C (26.6°F), it is **microthermal**. If the temperature is above −3°C (26.6°F), the climate is **mesothermal**. The fourth major temperature category is the polar climate. The boundary between the microthermal and polar climates is set at 10°C (50°F) for the average of the warmest month, which roughly corresponds to the northern limit of tree growth. A fifth major regime, the dry climates, was based not on temperature criteria, but lack of moisture. Dry climate boundaries are obtained using derived formulas. Figure 15.6 shows the distribution of the major climate types.

TABLE 15.1 Köppen's Major Climates

A	Tropical rainy climates
B	Dry climates
C	Midlatitude rainy climates, mild winter
D	Midlatitude rainy climates, cold winter
E	Polar climates

Principal Climate Types According to Köppen's Classification

Af	Tropical rainy	Cw	Midlatitude wet-and-dry, mild winter
Aw	Tropical wet-and-dry	Cf	Midlatitude rainy, mild winter
Am	Tropical monsoon	Dw	Midlatitude wet-and-dry, cold winter
BS	Steppe	Df	Midlatitude rainy, cold winter
BW	Desert	ET	Tundra
Cs	Mediterranean	EF	Ice cap

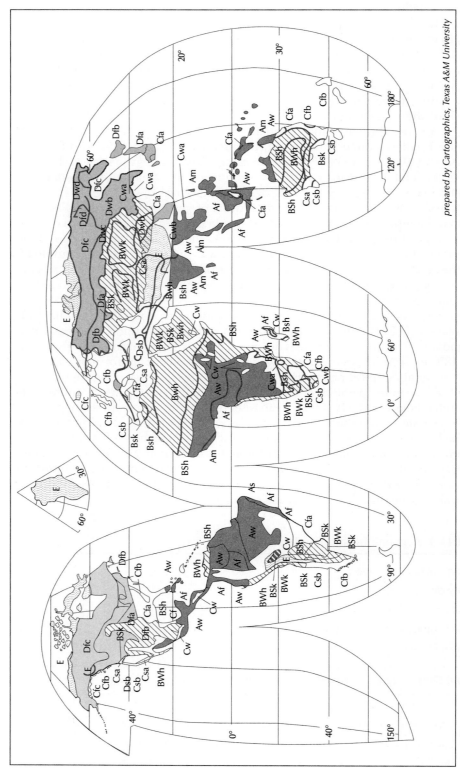

FIGURE 15.6 The Köppen classification of climate. (From Griffiths J. F. & Driscoll D. M., *Survey of Climatology,* © 1982, Prentice Hall, Upper Saddle River, N.J.)

prepared by Cartographics, Texas A&M University

The system has been subjected to criticism from two aspects: (1) There is no complete agreement between the distribution of natural vegetation and climate. This is to be expected since factors other than average climatic conditions, such as soils and topography, affect the distribution of vegetation. (2) The system is also criticized on the basis of the rigidity with which the boundaries are fixed. Temperatures at any site differ from year to year, as does precipitation, and the boundary based on a given value of temperature changes location from year to year. Despite the criticisms and empiric basis of the classification, it has proved quite useful as a general system.

CLIMATE AND THE DISTRIBUTION OF VEGETATION

Given that early climate classifications were often based on vegetation distribution, it is appropriate to briefly examine the relationship between climate and the distribution of natural vegetation. Among the factors that determine the distribution of flora are moisture availability, radiant energy, soils, parent material, slope, and other biota. Clearly climatic factors are a major determinant in the distribution of individual species and communities. At the **biome** level, the implications of climate are most readily visible. A biome is a climatically and geographically defined area of ecologically similar climatic conditions such as plant and animal associations and soil types. Because the entire world can be classified into 10 major types of biomes, the biome is far larger than an ecosystem. Where the climatic conditions are fairly similar from year to year in terms of temperature and moisture, a fairly distinct mature plant formation has evolved. In the classification of climates, four such regions were identified: tropical rainy, tropical desert, polar rainy, and polar desert or ice caps.

A recognized biome associated with the tropical rainy climate is the tropical rain forest. The relationship between the rain forest and climate is close enough so that, until recent years, the distribution of rain forest has been used to map the areas with a tropical rainy climate. It is now clear that the association between the two is not that perfect. In the tropical deserts is found the desert biome, which covers large areas on the earth's surface. Associated with the polar rainy climate is to be found the coniferous forest (taiga). In the vicinity of the ice caps, little or no vegetation is found as a result of very harsh conditions. In the transition areas between these four very different biomes, classification is more difficult and often confused by the intermix of communities and species. In these regions, the biomes present transitions from tropical to polar and humid to dry. For instance, some of the different biomes found in the tropical seasonal rainfall regime include seasonal forest, woodland, savanna, and steppe. Since the communities form a continuum over the landscape, placing the boundaries is generally difficult. A brief description of some of the more widely recognized biomes follows.

Forests

There are many definitions of *forest* because of the many varieties of forest communities. Here forests include those formations in which trees are a prevalent plant form. As already suggested, this classification covers a multitude of different ecosystems, and here it is possible to mention but a few. Rain forests are formations in which broadleaf evergreen trees are dominant and the canopy is more or less continuous. Rain forests exist where moisture is abundant, if not on a year-round basis, at least during the greater part of the year. Most of the vegetation in these communities is found in the uppermost canopy. The foliage of the trees is concentrated in the crowns; hence, the canopy and foliage of the lianas, epiphytes, and parasites are concentrated in the canopy. The understory consists largely of young trees of the dominant species and a sparse ground cover of shade-tolerant or shade-demanding shrubs and lower forms of vegetation. These forests are very limited in extent.

The majority of the world's forests occur where there is a dry season long enough to affect a seasonal change in the forest community. The seasonal forest may include evergreen, semi-deciduous, deciduous trees, or some combination of these. Where a mixture exists, it is not usually a random mixture of individuals of each species, but mixed stands of one type or the other. Local differences in soil or other site characteristics often determine which community persists. Since the seasonal forests exist where there is seasonal precipitation, the character of the forest is closely associated with the length of the rainy season. As the length of the rainy season decreases, the density of the canopy decreases. The most dense forests, or true jungles, are found in the seasonal forests. Here the dry season is long enough to spread the canopy and allow sunlight to reach the ground, but there is not such a long dry season that edaphic, or soil, desiccation can occur with any regularity.

Deserts

A desert is characterized by a discontinuous plant cover. The total vegetal coverage is usually small. Some desert plants, such as the creosote bush, produce their own population control devices. They produce hormones through their roots or leaves that inhibit the sprouting of other individuals within the proximity of the parent plant. This controls spacing of the plants and nutrient resources in a fashion that increases the probability of survival of some individuals of the species.

Many of the animals found in the deserts are the same as those found in the grasslands, but in greatly reduced numbers. There are some animals, including the kangaroo rats, which have adapted specifically to the deserts, but such species are relatively few in number.

Grasslands

The world's grasslands consist of formations made up of communities in which the herbs and shrubs are dominant. The communities are dominated by grasses and legumes, some of which reach considerable size. Grasses ranging upward of 3 m (10 ft) in height are not uncommon in the more humid grasslands. Grazing animals live in their greatest numbers in the grasslands. Among their attributes are their tendencies to live in groups and depend on speed for defense against predators. Some creatures have developed the abilities to leap high above the grasses for easier traveling and to see over the grass to watch for predators. The jack rabbit of the American and Australian plains is a good example. Many small animals burrow for shelter and concealment. Among this group is the prairie dog of the Great Plains of North America, which exhibits both group social structure and the construction of extensive underground towns.

Savannas are tropical grasslands with scattered trees or clumps of trees. Isolated trees are often found right at the desert edge. It is this scattering of drought-resistant trees that gives the tropical grasslands a distinctly different appearance from the mid-latitude grasslands. Fires are a recurrent phenomenon in the savannas, and both the grasses and trees are fire-tolerant. The varieties of species of trees and grasses are few compared with the tropical forests. Although the species are fewer because of the necessary adaptation to drought and fire, they are also very hardy and respond rapidly after a disturbance. Acacia and baobab trees are among the species that spread through the savannas.

Very few areas of natural grassland remain on the face of the earth. The grasslands have proved to be the most useful of the biomes when measured in terms of agricultural purposes. The natural communities of the grasslands have been either burned off or plowed up and replaced by the simpler communities of the domestic cereals such as corn, rice, wheat, and barley.

The grasslands are found where there is a seasonal moisture regime. They occur on all of the continents except Antarctica, and convergent adaptation has led to similar species in each and strange forms in some. Grasslands have been subdivided by secondary structural characteristics. Grasses have been divided between steppe and prairie, tall and short, and sod and bunch grass. The grasslands border forests, seasonal forests, woodlands, or deserts.

Colder Realms: Taiga and Tundra

Taiga is a word used to describe the great northern, or boreal, forests. The dominant trees of this forest are the needle-leaved evergreen trees. The members of the spruce, pine, and fir families are most common. Under a mature coniferous forest, there is very little understory as a result of the dense shade. The ground is often covered with a fairly thick layer of undecomposed and partially decomposed needles. This forest is associated with cool, moist conditions poleward to the limits of tree growth and extending equatorward considerable distances along mountain ranges, where favorable temperature and moisture conditions prevail.

Vegetation of the **tundra** consists largely of grasses, sedges, lichens, and dwarf woody plants. The tundra is associated with the seasonal rainfall areas around the Arctic Ocean and at high altitudes in mountains. It exists where there is a short summer season that is too cool for trees to thrive. The vegetation of the tundra has distinct characteristics that allow it to survive the cold temperatures, wind, and very long physiological drought. Almost all tundra plants are low growing and compact to escape the wind, reduce evaporation, and conserve heat. Perennials predominate, and many reproduce asexually by runners, bulbs, or rhizomes. Being perennial, the plants store food through the winter, and the buds are sheltered either underground or close to the surface.

Disturbed Formations

Some scientists maintain that the grassland, savanna, and brush formations are disturbed formations that have persisted because of repeated razing by fire. The sharp boundaries that exist, the variety of formation boundaries, and the lack of woody plants have led to this hypothesis. At this time, the debate seems to be a long way from resolution. The suggestion that grasslands are a product of disturbance does not in any way change the fact that disturbances are a very real factor in formation structure. Disturbance affects formations in several ways. It tends to sharpen the boundary between formations, at least in some areas, and often favors the intrusion of one formation into another. Local disturbances within a formation may cause an alternative formation to become established. The effects of a disturbance vary depending on the type of formation. Some communities in midlatitude grasslands may completely reestablish a mature community in less than 50 years. The tundra, where ecological processes are very slow, may not recover from a disturbance for centuries.

Summary

Climates can be studied at different scales—from the smallest microclimate through local climate and mesoclimate to the largest macroclimates. The study of microclimates usually requires instrumentation that is placed in the location under study. From the many unique microclimatic studies, a number of generalized observations have been made. It is shown that surface cover plays an important part in determining microclimate

characteristics. The types of study associated with local and mesoscale climates are seen in forest climates where the canopy layer often serves as the boundary layer.

To study climates at the global scale, a classification system is required. The widely used Köppen classification originally based on vegetation distribution is an example of an empiric system. Natural vegetation of the earth is closely related to climate. A useful level of equating the two is at the biome level. The forest biomes consist of the rain forests and the seasonal forest with the latter occurring because of a seasonal change of climate. Deserts, tropical grasslands (savannas), midlatitude grasslands, coniferous forests (taiga), and the tundra are the other identified biomes. The various climates that give rise to different plant associations provide a practical approach to climate classification.

Key Terms

Biome, *261*	Katabatic wind, *254*	Microclimates, *250*	Periodicities, *257*
Boundary layer, *255*	Laminar, *252*	Microthermal, *259*	Savanna, *262*
Canopy layer, *253*	Local climate, *255*	Numerical	Taiga, *263*
Climatic sheath, *253*	Macroscale climate, *256*	classification, *258*	Tundra, *263*
Filtering, *256*	Mesothermal, *259*		

Review Questions

1. Describe the four scales at which climate may be studied.
2. Why does the study of microclimatology necessitate field work?
3. How does topography influence wind?
4. How does a forest modify temperature and moisture?
5. What is filtering in relation to climatic studies?
6. Outline the basis of the Köppen climate classification system.
7. Both the rain forest and taiga are forest climates. Outline the main differences between them.
8. What is a biome?
9. What are the main differences between tropical and temperate grasslands?
10. Give examples of three biomes.

Tropical Climates

Tropical regions have long had a certain mystique for midlatitude peoples. The rich flora and fauna of certain tropical areas have encouraged some to envision a utopian economic development. Diseases such as malaria and yellow fever, fears of the debilitating effects of the tropical climate, and fears of ferocious insects, animals, and people have retarded development. Many of the diseases are under control. The climate is tolerable and, with air conditioning, comfortable indoors.

RADIATION AND TEMPERATURE

There are some distinctive attributes that characterize tropical climates in general. One of the most important of these is the energy balance. There are several aspects of the energy balance that distinguish these climates from those farther toward the poles. One is that the influx of solar energy is high throughout the year, although it does vary with the seasons. In this sense, there are indeed climates without winter. Intensity of solar radiation is high all year. There is very little variation in the length of the day from one part of the year to the next. The **photoperiod**—the relative lengths of day and night to which plants must adjust—varies between 11 and 13 hours from winter to summer. Contrary to popular notion, although solar radiation is relatively high all year, sometimes it is not as high as it is in midlatitudes, particularly in summer. In the equatorial lowlands, for instance, clouds can and usually do reflect more than half of the total solar radiation. In addition to the large amount of sunlight reflected by clouds, the high humidity and smoke from widespread burning further reduce solar radiation near the ground. At no time during the year do the tropical regions receive as many hours of sunlight as midlatitude locations receive in the summer.

Annual temperatures average about the same throughout the tropical regions. There are slightly higher averages in the drier areas due to more intense radiation at the surface. The annual range in temperature depends on the length of the dry season. Where there is no dry season, the annual range in mean monthly temperatures may be as little as one or two degrees. Where there is dry weather in the winter months, the mean temperatures drop and the annual range increases.

Another significant aspect of the energy balance of tropical climates is that the primary energy flux is diurnal (Figure 16.1). Tropical regions are not places of continuous high temperatures. Nights can be rather cool. The diurnal radiation and temperature cycles are more important than the annual temperature cycle as a regulator of life cycles (Figure 16.2).

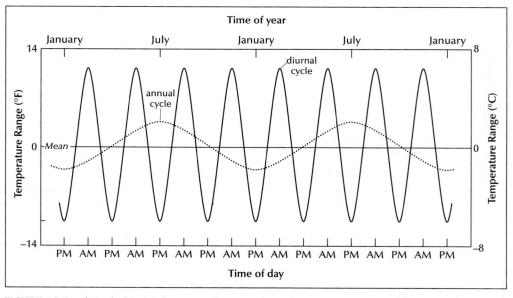

FIGURE 16.1   The relationship between the annual and diurnal energy periodicities in tropical regions.

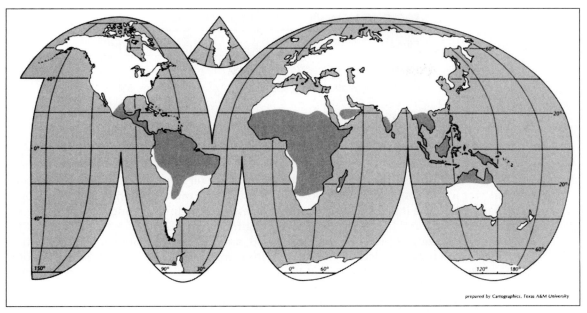

FIGURE 16.2   Shaded land areas experience a greater diurnal energy change than annual energy change.

PRECIPITATION

In tropical areas, rainfall rather than temperature determines the seasons. It is the amount and timing of rainfall that form the chief criteria for distinguishing the various climates. Contrary to popular belief, only a very small portion of the tropical regions has a year-round rainy season. For the continent of Africa, for instance, the area with substantial rainfall in each month is less than 10% of the total land area. The largest portion of tropical environments has a marked seasonal regime of rainfall that governs the biological productivity of the system. The remaining areas are deserts, where rainfall is incidental throughout the year. In fact, it is the seasonal moisture pattern that distinguishes the major tropical environments—*rain forest*, *savannas*, and *desert*—from each other. The notion that the major tropical environments are monotonous, seasonless regions is far from the truth. Unlike the midlatitudes, tropical areas lack snow storms, ice storms, and tornadoes. Change is an attribute of the tropical regions as much as it is anywhere else, and to the resident it as much a conversation item.

THE TROPICAL RAINFOREST CLIMATE

Tropical **rainforests** exist along the equator in Asia, Africa, and South America. These forests cover about 7% of Earth's land area. Extensive tracts of rainforest are in the Amazon River Basin, the Congo River Basin, and the East Indies. In 1990, half of the earth's tropical forests were located in four countries: Brazil, Indonesia, Peru, and Congo. Smaller tracts exist at other sites. There are two different varieties of tropical rainforests: equatorial rainforest and tropical wet rainforest. Two-thirds of the rainforests are classified as equatorial rainforests. The tropical wet rainforests are found on the margins of the equatorial rainforest and have a distinctly short dry season. The rainforest is also restricted to low elevations, usually below 1000 m (3300 ft), because at higher altitudes temperatures are considerably lower. As altitude increases, the forest changes character. Trees become shorter in height, number of species decreases, and highland vegetation replaces the rainforest.

TABLE 16.1 Classification of Tropical Climates

Climate	Köppen Type
Climates dominated by mT air masses	Af
Climates with alternating seasons of mT and cT air masses	Aw, Am, BSh
Climates dominated by cT air masses	BWh

The rainforest climate refers to the luxurious evergreen forest typical of the tropical wet lowlands. The distribution of this climate is more limited than most people imagine because it is confined by a rather limited set of climatic conditions. The region has fairly even and high temperatures averaging between 20° and 30°C (68° and 86°F), a high frequency of precipitation year round, and total rainfall of over 200 cm (80 in.). The climate is dominated by tropical maritime air masses (Table 16.1). So much solar radiation is reflected and scattered that a light-skinned person has difficulty getting a tan. The ultraviolet radiation is largely filtered out.

Part of the reason for the restricted geographic area is that the region must lie within the tropical convergence zone throughout the year. This does not happen over an extended amount of land because the convergence zone is continually shifting north and south. This migration is due to the passage of the overhead sun, as schematically illustrated in Figure 16.3.

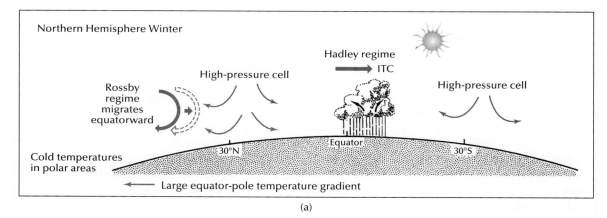

(a)

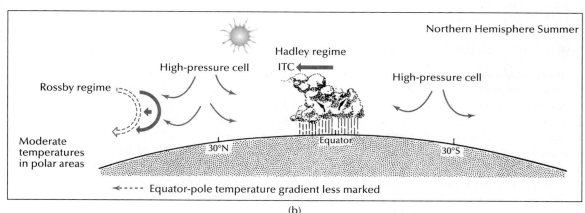

(b)

FIGURE 16.3 Schematic diagram showing the extent and migration of the Rossby and Hadley regimes in (a) winter and (b) summer.

The Rainforest Biome

The abundant precipitation and high-intensity radiation year round in this region results in an evergreen forest of tremendous productivity and diversity. The tropical rainforests have been in existence for as long as 45 million years. In this time, millions of species of plants and animals have evolved to take advantage of every niche in the system.

The monthly average temperatures vary from 24 to 30°C (75 to 86°F), with an annual range of only 3°C (5°F) or so. Belem, Brazil, for instance, experiences ranges from the warmest to coolest month of only 1.6°C (3°F). Because there is greater variation in radiation within the day than from day to day, the diurnal range may be two to three times the annual range. The fairly even temperatures of the rainforest are due primarily to:

- Equal or nearly equal periods of daylight and darkness
- Uniformly high intensity of radiation throughout the year
- Constantly high humidity

Associated with the convergence of tropical maritime air are high humidity and cloudy skies. Because humidity is so high in the daytime, when nocturnal cooling occurs, early morning fogs and heavy dew are common. Dew-point temperatures are in the range of 15 to 20°C (59 to 68°F).

The primary circulation determines the regional pattern of atmospheric circulation and the seasonal regime, and thunderstorms and easterly waves account for most of the day-to-day weather. Due to the degree of surface heating, local thermals, cumuloform clouds, and thunderstorms predominate. The cumulus clouds build during the daytime hours into towering cumulus and thunderstorms when the air becomes unstable. This process provides a distinctive diurnal rainfall pattern, such as that shown in Figure 16.4. Annual totals range up to 2 m (6.6 ft). The highest amounts occur on mountain slopes such as Mount Waialeale, Hawaii, where the annual average is 11.8 m (39 ft).

Average annual rainfall is not as important a factor as the frequency of precipitation. Precipitation occurs on more than 50% of the days. Duitenzorg, Java, averages 322 days a year with thunderstorms. Since the precipitation is largely from thunderstorms, it is of high intensity and short duration. Rainfall amounts of 50 to 65 mm (2.0 to 2.5 in.) per hour occur at a return period of less than 2 years. Hourly rainfalls of 95 to 120 mm (3.7 to 4.7 in.) are recorded in the East Indies. Although rainfall intensities for 1-hour periods are not significantly higher in the tropics than in midlatitudes, sustained periods of high-intensity rainfall are significantly more common in tropical regions. The maximum 24-hour rainfall on record occurred at Cilaos, La Reunion, on March 16, 1952, where 1870 mm (73.62 in.) of rain fell (Table 16.2).

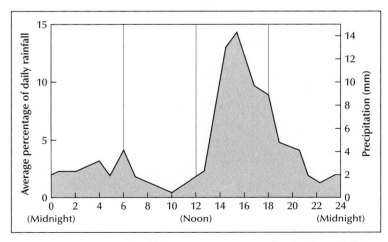

FIGURE 16.4 Diurnal distribution of rainfall at Kuala Lumpur, Malaysia. The midafternoon maximum is typical of many tropical wet locations.

TABLE 16.2 Precipitation Extremes Recorded in Tropical Wet Climates

Highest 12-hr total	Belouve, La Reunion Feb. 28–29, 1964	1340 mm (52.76 in.)
Highest 24-hr total	Cilaos, La Reunion Mar. 16, 1952	1870 mm (73.62 in.)
Highest 5-day total	Cilaos, La Reunion Mar. 13–18, 1952	3854 mm (151.73 in.)
Highest number of rain days in a year	Cedral, Costa Rica 1968	355 days

Hail is fairly infrequent, but it does occur on occasion. In 1922, Mount Cameroon erupted in West Africa. The eruption produced a storm that left a snow cover on the summit, and at Debundscha at the foot of the mountain 14.53 m (572 in.) of rain were recorded during the year. Not all rainfall is convectional as there are periods of prolonged showers.

These forests produce a substantial part of the global oxygen supply and in the process remove carbon dioxide. They act to modify the climate in the regions where they are found and also influence the climate of the entire planet. An interesting aspect of this region is that the forest plays a significant part in the wetness of the climate. During the morning and early afternoon hours, this forest transpires huge quantities of water into the lower atmosphere. This added moisture aids in condensation of water in building cumulous clouds and afternoon and evening showers, and thunderstorms may drop copious amounts of precipitation back onto the canopy layer of the forest. Much remains there to start the process over again the next day. The regularity of these showers in some areas is such that one can almost set one's clock by them.

Species Diversity

The tropical rainforest is a multilayered forest of largely evergreen species. Most of the leaf matter in the rainforest is found in the canopy. The canopy is represented by the crowns of the dominant tree species. In an area of approximately 1 hectare (2.5 acres) of equatorial rainforest, there may be as many as 200 tree species. Within the canopy is where most of the other vegetation and animal life are found. The canopy within a true rainforest is a mat of nearly continuous vegetation. Epiphytes such as orchids are most likely found in the canopy as are birds, monkeys, and tree snakes. This canopy may be over 30 m (100 ft) above the ground. So thick is this layer that little light reaches the forest floor. The intensity of light on the forest floor may be as little as 10% of that on top of the canopy. Extending above the canopy are the true giants of the rainforest whose crowns appear well above the surrounding canopy and are called *emergents*.

Tropical forests are rich in species diversity. It seems certain that over half of all species found on Earth are found only in the rainforests, and this number may be as high as 90%. Half of Earth's bird species and 90% of primates inhabit the tropical rainforests. Half of the plant species that provide the basic food grains for humans originated in the tropical forests. These include rice and maize. Of an estimated 3 to 4 million species of organisms thought to exist in the tropical forests, only about 15% are yet classified.

Although there is little difference in the length of the photoperiod during the year, the variations that exist are important biologically. Even very small differences in length of day are enough to stimulate flowering or dormancy in plants. Many tropical plants are more sensitive to slight changes in day length than are plants found at higher latitudes. The common poinsettia is a tropical shrub for which reduced day length triggers flowering. In their natural habitat, these shrubs do not bloom at Christmas but much later in the year. We force them to bloom before Christmas by controlling exposure to light and reducing the exposure artificially in early autumn.

Tropical Deforestation

The past, current, and projected destruction of this forest with all its species may be the most destructive act of humanity on the planet. These forests, which have been in existence for tens of millions of years, could effectively be destroyed to the place where they can no longer continue their present function by the year 2025. Destruction of the tropical rainforest is occurring rapidly. Some estimates suggest the annual loss to be an area equal in size to the state of West Virginia. In many tropical areas, 1 to 2% of the rainforest is cut or burned each year. This is equivalent to removal of 20 to 50 hectares (49 to 124 acres) each minute. The area of tropical rainforest in the Ivory Coast dropped 30% from 1956 to 1966. From 1980 to 1990, some 154 million hectares (380 million acres) of forest land was cleared for other use. Studies now show that the area of degraded and fragmented tropical forest may be greater than the area deforested. This is extremely important in terms of loss of biodiversity.

The pressure to cut these forests comes from several sources. The need for land for farming and ranching is the major reason for removal. Second is the demand for tropical wood for lumber. Between 10 and 15 million hectares (25.4 and 37 million acres) of land are cleared of forest in the tropics each year for agriculture and pasture. Brazil and Indonesia account for about 45% of global rainforest depletion. In India between 1972 and 1982, about one-fourth of the forest was cleared.

Deforestation in the Amazon River Valley

The Amazon rainforest is the largest continuous stand of forest left on the face of the earth. In 1980, the area of the remaining forest was some 5 million km^2 (2 million mi^2)—an area nearly half the size of the United States. Brazil contains about 30% of Earth's tropical forest. The forest covers about 40% of the land in Brazil. It extends into Bolivia, Columbia, French Guyana, Guyana, Peru, Surinam, and Venezuela.

For several decades, the government of Brazil envisioned the economic development of this vast region. To begin development, the 2500-km (15,500-mi) Trans-Amazon highway project was begun in 1970 and completed in 1974. The highway was funded by the World Bank. The highway, known as BR-324, runs from eastern Brazil to the western state of Rondonia. The highway is part of a scheme to develop the northwest section of the rainforest. Included in the plan was the clearing of 410,000 km^2 (160,000 mi^2) of forest. The road was initially a 1400 km (900 mi) dirt road of poor construction. It often washed out during the rainy season. Even in its poorest form, it cut the travel time from the eastern part of the country to Rondonia from weeks to days. The road was paved in 1982, and a flood of people moved along the road, clearing land as they went. They built the highway partly to move people out of the overcrowded northeastern part of the country. The result of constructing this road and the thousands of kilometers of feeder road was disaster for the forest. A population explosion of humans and livestock followed the construction of the road. Between 1966 and 1978, settlers cleared 80,000 km^2 (31,000 mi^2) forest. Most of the land was planted in grass for cattle ranching. Some estimates place the amount of original forest already cut or burned as high as 30%.

The government of Brazil encourages deforestation. To encourage development, the government provided incentives to foreign investors to explore for minerals and harvest the timber. The rate of deforestation has increased rapidly since 1980. Brazilian leadership recently announced a controversial new plan to harvest an additional 40 million hectares (95 million acres) of timber. At this rate of removal, the entire forest in South America will be gone by 2050.

The Galapagos: An Equatorial Dry Region

As with most other generalizations, there are exceptions to the rules about where certain types of climate are actually found along the equator. Some extremely dry environments exist where we expect to find the humid tropics. Herman Melville wrote the following in *The Encantadas* (Galapagos Islands):

> But the special curse, as one may call it, of the Encantadas, that which exalts them in desolation above Idumea and the Pole, is that to them change never comes; neither the change of seasons nor of sorrows. Cut by the equator, they know not autumn and they know not spring; while already reduced to the lees of fire, ruin itself can work little more on them. The showers refresh the deserts but in these isles, rain never falls. . . . Nowhere is the wind so light, baffling, and in every way unreliable, and so given to perplexing calms, as at the Encantadas.

Melville wrote of one island that used a system for collecting dew to provide fresh water:

> And here was a simple apparatus to collect the dews or rather doubly distilled and finest winnowed rains, which in mercy or in mockery, the night skies sometimes drop upon these blighted Encantadas. All along beneath the eave, a spotted sheet, quite-weather-stained, was spread, pinned to short, upright stakes, set in shallow sand. A small clinker, thrown into the cloth, weights the middle down, thereby straining all moisture into a calabash placed below. This vessel supplied each drop of water ever drunk upon the site by the Choles. Hunilla told us the calabash would sometimes, but not often, be half-filled over-night. It held six quarts, perhaps.

Global Warming in the Tropics

Just as in most other regions of the earth, the effects of a warming planet are apparent in the tropics. Throughout the tropics, there are mountains that support glaciers. Many of these mountains are famous as destinations for tourists and in literature. Melting is now taking place more rapidly here than in the past. In Africa, Mt. Kenya, which lies just a few miles north of the equator, supported a large glacier on its peak, and it is now all but gone. Just to the south of the equator on the border between Kenya and Tanzania lies Mt. Kilimanjaro. Famous for its majestic shape with a snow- and ice-capped crown, it is now also undergoing change, as the snow and ice are melting at an unprecedented rate. Around the base of the mountain and partway up the flanks is an intensive agricultural region based on the rich volcanic soil found there. Meltwater from the snow and ice provides a steady supply of irrigation water. As the snow and ice retreat to higher elevations, more of the precipitation occurs as rain. The rain runs off almost immediately and so is lost for irrigation purposes. The livelihood of thousands is in jeopardy.

In South America, the Quelccaya ice cap in the central Andes is melting at a more rapid rate than in the past. The same is true in Irian Jaya, where three glaciers have experienced a 50% increase in the rate of retreat of the glacier fronts since the 1920s.

THE TROPICAL WET AND DRY CLIMATE

The distinguishing feature of the tropical wet and dry climate is the pronounced seasonal moisture pattern. Atmospheric humidity, precipitation, soil moisture, and stream flow change through the year in a rhythmic pattern. These environments are found next to the tropical rainforest environments and in most cases poleward of them. They occupy much of the area from the boundary with the tropical rainforest to the Tropics of Cancer and Capricorn. They occupy a much larger area than that of the rainforest. There are areas with this type of climate north and south of the Amazon Valley in South America, across Africa north and south of the Congo Basin, and in much of Southeast Asia and parts of the Pacific islands. Their location in the general circulation is such that they experience the kinds of weather found in both the equatorial convergence zone and the subtropical divergence zone.

The tropical wet and dry climate has seasons dominated alternately by maritime tropical and continental tropical air masses (Table 16.3). The seasonal pattern of moisture is due to the migration of the tropical convergence zone. The rainy season is concurrent with the high sun and the presence of the convergence zone. The dry season is a product of the more stable air from the subsidence in the subtropical highs. Seasonality increases away from the equator. In a traverse away from the equator, the low sun, or winter, precipitation begins to decrease first, with the high sun precipitation remaining as high as in the tropical convergence zone. Between the equatorial convergence zone and the tropical desert, the winter precipitation drops to near zero. From that point poleward, the high sun precipitation declines until it is no longer significant, and we find the tropical desert.

The tropical wet and dry climate has the most pronounced seasonality of precipitation of any climatic type. Rangoon provides an example of the extremes of precipitation that occur in these regions. Rangoon, Myanmar (formerly Burma), has a 3-month winter average of 25 mm (1 in.) and a three-month summer average of 1880 mm (74 in.). The extremes increase northward to Akyab, where the averages are 3.8 and 426 cm (1.5 and 167 in.). The greatest extremes in the Asian region are at Cherrapunji, India. In two winter months, they receive an average of only 2.5 cm (1 in.) of precipitation, but in the two summer months, they get 528 cm (207 in.) of rain. Cherrapunji recorded a five-day total of 405 cm (159 in.) in August 1841 and a one-month total of 915 cm (30 ft) in July 1861. In one year, from August 1860 to July 1861, they measured 26 m (85 ft) of rain.

The average annual precipitation varies so much from place to place that it cannot be used as a criterion for distinguishing the region. On the wet margins where topography is favorable for orographic increase in rainfall, totals run over 1000 cm (400 in.). On the dry margins, it drops to less than 250 mm (9.8 in.). In fact, it is quite likely that the highest annual average precipitation totals on Earth occur where there is a strong seasonal pattern of precipitation. Variation in annual rainfall is higher here than in the tropical rainforest. The precipitation total in any year is subject to the extent of migration of the general circulation and the length of time the zone of convergence stays over an area. The farther a site is from the heart of the convergence zone, the lower the annual average precipitation, the shorter the rainy season, and the greater the annual variation. The weather is, of course, very different from season to season. During the rainy season, it is warm and humid, and there are frequent rainstorms. During the dry season, more-or-less desert conditions exist.

The lag of the rainy season behind the migration of the sun produces three seasons in many parts of the wet and dry tropics: a cool season, a hot season, and a rainy season. The cool season is during the winter months, when solar radiation is at a minimum. The hot season follows, when temperatures rise as the sun moves higher in the sky and solar radiation increases. The onset of the wet season brings an increase in cloud cover, slightly reduced radiation, and cooler temperatures.

Where moist air of oceanic origin exists, precipitation occurs. Where the dry continental air is usually present, precipitation is largely absent. Often there is a sharp boundary between the two types of air. The **Hadley cells** north and south of the equator and the convergence zone between them shift north and south, following the migration of the vertical rays of solar energy. This migration produces the seasonal pattern of precipitation that characterizes so much of the tropical region. The system moves through 20° of latitudes through the year from about 5° S to 15° N.

TABLE 16.6 Comparison of Temperature and Relative Humidity in Tropical Deserts and Coastal Deserts

	Average of Warmest Month			Average of Coolest Month			Mean Annual Temp. (°F)	Annual Temp. Range (°F)
		R.H. (%)			R.H. (%)			
	T(°F)	0700H	1400H	T(°F)	0700H	1400H		
Tropical deserts								
In-Salah (Sahara)	98	36	25	56	63	37	77	42
Riyadh (Arabian)	93	47	31	58	70	44	75.5	35
Laverton (Gr. Australian)	87	36	24	52	60	43	74.5	35
West coast deserts								
Arica (Atacama)	72	74	61	60	83	74	66	12
Walvis Bay (Kalahari)	66	91	73	58	83	65	62	8
Villa Cisneros (Coastal Sahara)	72	88	63	63	75	51	67	9

east of the city. These are unusual climatic areas and do not cover large areas of the earth's surface.

DESERT STORMS

Severe storms are not frequent in the deserts primarily because of the lack of moisture to supply the energy. However, deserts are very windy, particularly in the afternoon and during the hottest months. Common features of the air over deserts are a steep lapse rate and instability in the air near the surface. **Dust devils** are the most visible form of the turbulence. Developing in clear air with low humidity, they become visible due to the debris they carry. They can reach a height of nearly 2 km (1.2 mi). On occasion they will sustain velocities high enough to blow down shacks or blow screen doors off their hinges. These dust devils are very common in deserts, and it is not unusual to see a multitude of them at one time under the right atmospheric conditions. The dust devil results when there is an intense thermal at or near the ground surface. Surrounding air moves inward to replace the rising air. As the air moves in toward the thermal, the radius of curvature decreases and velocity increases.

Other storms of the tropical deserts are the dust storms and sand storms. Dust storms occur as a result of **deflation**. One of the readily visible aspects of much desert surface is the lack of fine sediments. The wind keeps the surface swept clean. Winds blowing out of the deserts are often dust laden. On the south side of the Sahara, they call the dry wind the **harmattan**. Because it is dry wind, it is cooling and welcome, although it may limit visibility and leave household furnishings covered with dust. These same winds blow out of the north side of the Sahara Desert. Occasionally, they travel across the Mediterranean Sea, bringing disaster to agricultural crops. They are important enough in the climate of Mediterranean countries to warrant local names such as **sirocco** in Italy and Yugoslavia and **leveche** in Spain.

In the sandy parts of the deserts, it is the wind's ability to move sand that is significant. Sand dunes cover only a small part of the tropical deserts. Sand dunes cover something less than 30% of the Sahara Desert and only about 2% of the Sonoran Desert of North America. Sand storms only occur when wind velocities reach a high enough velocity to move the sand. The velocity at which sand begins to move is the threshold velocity and depends on the size of the sand particles, wetness of the sand, and other less important variables. For medium-sized dry sand (0.25 mm), the threshold velocity is 5.4 m/s (18 fps) and is exceeded some 30% of the time. **Sand drift** increases rapidly as wind velocities increase above threshold velocity. Measurements indicate an average rate of 12.0 L/m width for winds 5.5 to 6.4 m/s and a rate of nearly 10 times that for wind velocities of 10 meters per second. The increase is rapid because above the threshold velocity the power of the wind increases as the cube of the wind velocity.

Two kinds of sand storms develop in the tropical deserts. One is the result of surface heating and the resultant turbulence. It is not unusual for sand temperatures at the surface to reach 85°C (180°F). This type of sand storm is chiefly a daytime phenomenon and is most pronounced in the hottest months. At Al-Hofuf, Saudi Arabia, the percentage of time in which sand drift occurs between 0600 and 1800 hours ranges from 76% in February to 91% in June. The total number of hours per day when winds exceed the threshold velocity also has a seasonal pattern, increasing from 3 hours per day in February to 9.4 hours per day in June. As a result of the increasing number of hours of high-velocity winds in summer, sand movement increases in summer. In the Nafud, it increases from 116 L/m/day in February to 406 L/m/day in June. In summer, the high frequency of afternoon sand storms makes travel and other outdoor activity particularly difficult.

The second type of sand storm is brought about by low-pressure disturbances passing through the area. These storms often generate higher velocity winds than does diurnal heating, and they last for longer periods. One such sand storm in the Nafud lasted for 43 hours, during which sand blew constantly. The wind averaged 10.7 m/s (24 mph) and exceeded 15.6 m/s (35 mph) for hours at a time.

Both types of sand storms have the ability to move huge amounts of sand. In the Nafud, the wind moves an estimated 80 m^3 of sand across each meter width of the sand field each year. The dunes move at fairly high speed. In the Nafud, dunes with an average height of 10 m (33 ft) move at a rate of 15 to 19 m (50 to 63 ft) each year.

DESERTIFICATION

In many parts of the world, particularly on the margins of the deserts, a process of environmental degradation called **desertification** is taking place. The process consists of the breakdown in the vegetal cover and soil until the land is no longer productive, and erosion by water and wind produce relatively sterile land. Desertification does not involve the outward expansion of the climate conditions responsible for the core areas of the world deserts. Rather, it is the alteration of the land to desertlike character by human activity.

The nature of the ecosystems in much of the Sahelian zone of Africa is such that they will not support many people on a subsistence basis. Dry-land cultivation practices are mostly both primitive and damaging to the land. They clear areas and farm them for several years, during which the soil gradually deteriorates and loses fertility. When the soil is exhausted, they abandon it on the assumption that it will rejuvenate with time. If the amount of land cultivated in any given year is small enough and the number of livestock units on the grassland is below the carrying capacity, the ecosystem will suffer little long-range damage.

Desertification takes place when the numbers of people and livestock exceed the capacity of the precipitation to support them. It is not a new problem, but the rate at which it is taking place is unprecedented. Before colonization of Africa, there were intermittent periods of environmental degradation. Whenever a dry cycle began, there was too much grazing and cultivation, which was followed by loss of vegetation and soils. Population increased during wet periods and decreased during drought (Figure 16.5).

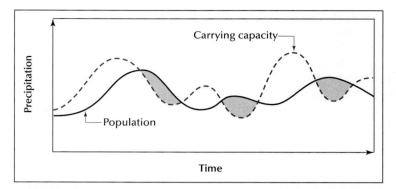

FIGURE 16.5 Population and carrying capacity in the precolonial era. Population size followed the precipitation cycles. When favorable periods occurred, population expanded; when dry periods followed, the population died back. The shaded areas represent times of major environmental degradation.

If an area became unproductive, the population either died back or moved away. The introduction of colonial administration brought some measures that altered the balance between the population and environment. They reduced intertribal warfare and introduced public health measures. These measures resulted in rapid population growth. In the year 2000, several of the Sahelian countries had the highest population growth rates in the world. Parallel with the increase in human numbers is an increase in animal populations.

The result of the rapid population growth in the face of varying rainfall is that the demands on the ecosystems become greater than the carrying capacity more frequently and for longer periods. As shown in Figure 16.6, the demand ultimately exceeds the carrying capacity on a continuing basis.

The result of the increasing pressure on the land is the expansion of the Sahara Desert southward into more humid areas. If human activity is not responsible for this phenomenon, it certainly accelerates it. Degeneration into desert usually occurs in scattered patches of bare ground from a few meters to several kilometers across. It is not desert expansion in the form of an even front of desert surface advancing across the landscape. Hence, it is not easy to measure. Desertification is taking place at a steady rate in many areas of the earth's grasslands and occurs on every continent except Antarctica. It becomes particularly rapid during periods of major drought. One example of how fast the process can take place is that of southeastern Australia in the first decade of the 21st century.

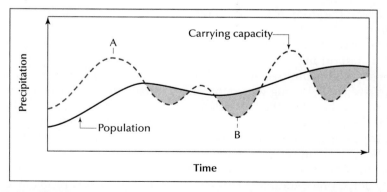

FIGURE 16.6 Population and carrying capacity following colonization when conditions changed to greatly increase population growth rates. At Point A, when the carrying capacity begins to decline, various forms of aid to the distressed areas keep down the death rate. Thus, when precipitation increases again at Point B, a much larger carryover population remains from which to continue growth. The result is a population greater than the environment can support, even under more favorable conditions. Desertification becomes even more accelerated.

When the plant cover falls below the minimum required for protection of the soil against erosion, the process becomes irreversible. The soil holds less water. It dries out. The temperature increase speeds the collapse. Bare soil appears and wind and water remove the fine soil particles. This loss of topsoil is large and very important. The movement of dust by wind has existed throughout historic times, and is by no means a recent phenomenon. Winds move the dust picked up from the Sahara and adjacent lands for great distances. The amount of dust carried by the atmosphere is large. Some of this dust moves across the Atlantic Ocean to the Americas. The total drift westward off the African coast may be as much as 60 million metric tons each year. It is extensive enough to be measurable in the Caribbean Sea.

Wind erosion is not restricted to dust-sized particles. When large patches of bare ground form, the soil is exposed to erosion of the larger sand-sized particles, humus, and mineral salts. Barren deflation pans then dot the surface. During the height of the Sahelian drought, satellite images showed thousands of square kilometers of surface obscured by blowing sand. Blowing sand and silt further destroy the standing vegetation by stripping leaves or burying it under dunes.

Summary

Tropical climates cover a large portion of the earth's surface between latitudes 25° N and 25° S. In most of this region, temperatures change as much or more from day to night as from summer to winter. Radiation intensity is relatively high all year. Within the tropical realm, the precipitation regime defines the climatic regions. Some places have a high chance of precipitation every day of the year, and others have an equally low chance of precipitation on any given day of the year. Between these extremes are the vast areas with a rainy season and a dry season. In the wet and dry tropics, there is a wide range in both the length of the rainy season and in total precipitation received during the rainy season. The monsoon regions of Asia represent the extreme example of seasonal precipitation, and it is there that some of the highest amounts of annual precipitation occur. The driest areas of the world are in tropical regions. The coastal margins of the tropical deserts include stations with the lowest known annual total precipitation.

Deforestation of the tropical rainforests is a critical problem facing the human species today. The tropical forests not only affect the regional climate, but provide humanity with many crucial medicines and other products. If the forests are destroyed, a major planetary biome will cease to function. In the tropical wet and dry regions desertification if equally devastating.

Key Terms

Deflation, *278*	Hadley cell, *273*	Photoperiod, *266*	Sirocco, *278*
Desertification, *279*	Harmattan, *278*	Rainforest, *267*	
Dust devils, *278*	Leveche, *278*	Sand drift, *279*	

Review Questions

1. Sand drift in deserts occurs under two different sets of conditions. What are these two different sets of conditions?
2. What is the primary reason that coastal deserts average less rainfall than interior deserts?
3. Why is the intensity of solar radiation high throughout the year in the tropics?
4. What factors determine the difference between the various tropical climates?
5. Of the three main types of tropical climate, which is the smallest in geographic area?
6. Does most precipitation occur in the summer or winter season in the wet and dry tropics?
7. In which tropical climates do the highest temperatures occur?
8. What is the difference between a dust devil and a sand storm?
9. In which of the tropical climates is desertification most common?
10. On which continents is desertification taking place?

17

Midlatitude Climates

The climates found in midlatitudes differ from those of tropical regions in two major respects. First, in tropical regions, the variation in radiation and temperature is greater from day to night than it is from season to season. In midlatitudes, the seasonal change is greater than diurnal variation (Figure 17.1). Second, in midlatitudes, the primary circulation is quite different from that of the Tropics. The subtropical highs over the oceans and polar highs act as major source regions for air moving toward the convergence zone centered in the latitudinal range of 50° to 60°. These two source areas are very different in character. One is very warm, and the other quite cold. Although this is enough to produce different air streams in midlatitudes, differences in moisture further distinguish the air masses. Thus, in the tropical convergence zone, the converging air streams are quite similar in temperature and moisture, but such is not the case in midlatitudes.

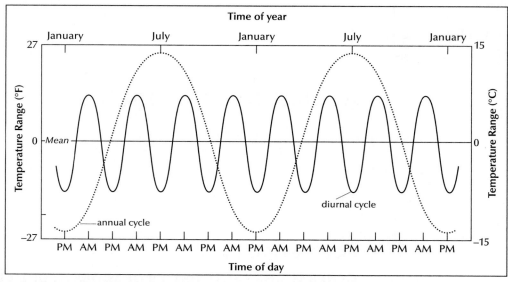

FIGURE 17.1 The relationship between the annual and diurnal energy cycles in midlatitudes.

GENERAL CHARACTERISTICS OF MIDLATITUDE CLIMATES

Midlatitude Circulation

Westerly upper air flows of the Rossby regime dominate the circulation in midlatitudes. These are located beneath the **subtropical jet stream**. Equatorward of this axis, the Hadley circulation operates.

The temperature gradient between the equator and the poles determines the location of the subtropical jet stream. As polar areas become warmer in summer, the gradient lessens and the strength of the westerlies diminishes. Surface circulation patterns are closely related to the flow of the upper-air westerlies in midlatitudes. The migration of the subtropical jet and the weakening of the westerly circulation provide very different conditions from season to season. The movement and relative dominance of air masses varies, as does the frequency and path of traveling low-pressure systems. Figure 17.2 illustrates the main source regions for North American air masses, and Table 17.1 provides some of the attributes of these air masses.

Summer

Summer in midlatitudes has many of the characteristics of the tropical climates. At the time of the summer solstice, radiation intensity at 47° N and 47° S is as high as it is at the equator. Not only is the radiation intensity high, but the duration of radiation is longer than it is at the equator on this date (Figure 17.3). Inequalities in length of day and night increase with latitude. This is added to the variation in solar intensity in such a way that it compounds the seasonal differences. During the summer months, solar intensity and hours of daylight are high. The result is greater radiation reaching the surface in summer in midlatitudes than in the equatorial zone. During summer, temperatures average in the 20 to 25°C (68 to 77°F) range at most stations. Along the equatorward margin, the warm-month temperatures may average 26 to 30°C (78 to 86°F)—comparable to rainy tropics. July averages differ only slightly from the Gulf

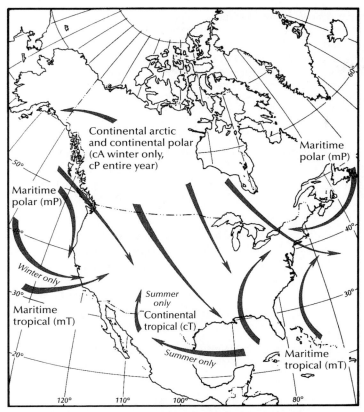

FIGURE 17.2 Source regions and paths of air masses affecting North America.

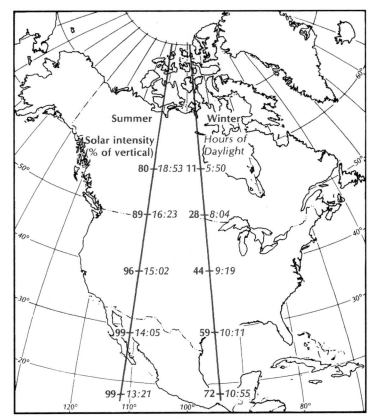

FIGURE 17.3 The length of daylight in summer and winter, and the relative intensity of solar radiation compared to a vertical solar beam.

TABLE 17.1 Weather Characteristics of North American Air Masses

Air Mass	Source Region	Temperature and Moisture Characteristics in Source Region	Stability in Source Region	Associated Weather
cA	Arctic basin and Greenland ice cap	Bitterly cold and very dry in winter	Stable	Cold waves in winter
cP	Interior Canada and Alaska	Very cold and dry in winter, cool and dry in summer	Stable entire year	a. Cold waves in winter b. Modified to cPk in winter over Great Lakes, bringing "lake effect" snow to leeward shores
mP	North Pacific	Mild (cool) and humid entire year	Unstable in winter, stable in summer	a. Low clouds and showers in winter b. Heavy orographic precipitation on windward side of western mountains in winter c. Low stratus and fog along coast in summer; modified to cP inland
mP	Northwestern Atlantic	Cold and humid in winter, cool and humid in summer	Unstable in winter, stable in summer	a. Occasional nor'easter in winter b. Occasional periods of clear, cool weather in summer
cT	Northern interior Mexico and southwestern U.S. (summer only)	Hot and dry	Unstable	a. Hot, dry, and clear, rarely influencing areas outside source region b. Occasional drought to southern Great Plains
mT	Gulf of Mexico, Caribbean Sea, western Atlantic	Warm and humid entire year	Unstable entire year	a. In winter, it usually becomes mTw, moving northward, and brings occasional widespread precipitation or advection fog b. In summer, hot and humid conditions, frequent cumulus development and showers or thunderstorms
mT	Subtropical Pacific	Warm and humid entire year	Stable entire year	a. In winter it brings fog, drizzle, and occasional moderate precipitation to northwestern Mexico and southwestern United States b. In summer it occasionally reaches western U.S., providing moisture for infrequent conventional thunderstorms

Source: Lutgens F. K. & Tarbuck E. J., *The Atmosphere,* 2nd ed., © 1982, p. 203. Reprinted by permission of Prentice Hall, Inc., Upper Saddle River, N.J.

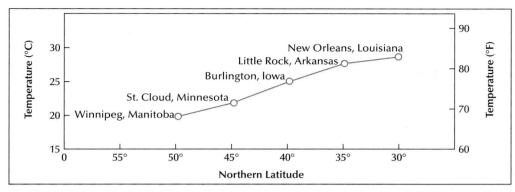

FIGURE 17.4 Mean July temperature along a north–south transect from New Orleans, Louisiana, to Winnipeg, Manitoba.

Coast to the Canadian border (Figure 17.4). The warm-month temperature means drop slowly in a poleward direction. Only at latitudes greater than 45° will the summer averages drop below 20°C (68°F).

Precipitation decreases toward the poles and toward the interior of the continents. The primary reason for this decrease is the increasing distance from the source of mT air. Areas on the west sides of the continents that receive their precipitation from mP air normally have less total precipitation than areas receiving precipitation from mT air. This is because the mP air is cooler and contains less water vapor. The frequency of thunderstorms decreases rapidly from south to north. Cyclonic precipitation is most frequent at the polar margins and is more frequent in all sections of the midlatitudes in the winter. Hurricanes provide another precipitation mechanism, and some of the heaviest rains along the coastal areas result from hurricanes.

Winter

Winter truly sets the midlatitude zone apart from the tropics. All areas in the midlatitudes experience a range of 47° in the angle of the solar beam between the summer solstice and the winter solstice. This varying angle produces changes in radiation intensity at the surface. In winter, solar intensity is low and the hours of sunlight are short; temperatures reflect these conditions. Mean winter temperatures also decrease markedly as latitude increases. January temperatures average above 10°C (50°F) closer to the equator and drop to below –18°C (0°F) toward the pole. The range between mean July and mean January temperatures increases from about 8°C (14°F) in the warmer areas to about 40°C (72°F) in the Canadian Arctic. The coldest winter temperatures have a steep poleward gradient also, but not as steep as the average temperatures. Temperatures as low as –15°C (5°F) occur along the equatorward margin, and lows from –31 to –46°C (–25 to –50°F) characterize the more northerly locations. **Diurnal ranges** in winter are slightly greater than the summer ranges because the relative and absolute humidity are somewhat lower. The winter controls are, in essence, low solar radiation, short days, and moderate atmospheric humidity.

The equatorward migration of the polar front results in the dominance of polar air masses over much of the midlatitudes (Figure 17.2). Periodically, the invasion of very cold Arctic air gives rise to cold waves. These are any sudden drop in temperatures within a 24-hour period such that measures to prevent extreme distress to humans and animals may be necessary.

Equally cold outbreaks of Arctic air have occurred in Europe and Asia. In A.D. 763, a cold wave froze both the Black Sea and the Dardanelles Strait in Europe. In 1236, the Danube River froze solid. In 1468, the winter in Flanders was so severe that wine rations issued to soldiers had to be cut with hatchets and distributed in frozen chunks. In 1691, crazed by the lack of food in the countryside, wolves invaded Vienna, Austria,

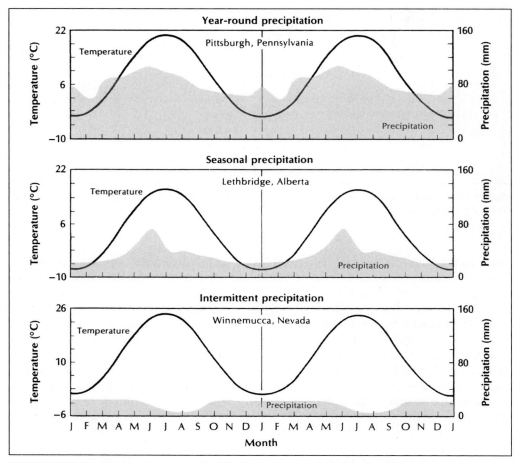

FIGURE 17.5   The three major types of midlatitude climates, based on the seasonal pattern of precipitation.

killing humans for food. However, with the recent warming of winter temperatures associated with climate change, severe cold waves have become less frequent.

CLIMATIC TYPES

Tropical climates are distinguished from each other on the basis of the seasonal pattern of precipitation. All midlatitude locations have a seasonal temperature regime, stronger in some places than in others. Midlatitude climates are also distinguished from each other on the basis of the seasonal pattern of precipitation (Figure 17.5 and Table 17.2).

TABLE 17.2 Midlatitude Climatic Types	
Climate	**Köppen Type**
Climates dominated by mT and mP air	Cf, Df
Winter-dry with seasonal mT and cP air	Cw, Dw
Summer-dry with seasonal mP and cT air	Cs, Ds
Climates with seasonal cT and cP air	BWk

Midlatitude Wet Climates

The midlatitude wet climates are associated with the subpolar lows over continental areas. They tend to stretch nearly across the continents in the northern hemisphere near 60° N and 60° S. They reach equatorward as far as 25° or 30° along the eastern margins of land masses. In North America, this type of climate extends from the Pacific coast of Canada eastward in a crescent to the Atlantic coast. In the United States, it includes the area east of the Mississippi River and the first row of states west of the Mississippi. In Eurasia, this climate extends from the offshore islands of Great Britain and Ireland eastward into the Soviet Union. It also occurs along the Pacific coast of Asia in latitudes 30° to 60°. The climate also exists in South America, Africa, and Australia. South America has two such areas: southern Chile, including the southern end of the central valley. The other includes the Pampas of Argentina and Uruguay and extends southward to include part of Patagonia. The area in Africa is very limited, consisting of a small area on the very southeast tip of the continent. Both Australia and New Zealand have sectors with this type of climate. Table 17.3 provides sample climatic data.

Summer conditions are very similar to those found in the rainy tropics. The primary differences are two: cyclonic storms that move across the region and temperatures that average from 21° to 26°C (70 to 79°F). The tropical margins reach as high as 29°C (84°F), somewhat warmer than the humid tropics.

The summer extremes exceed those of the tropical wet climate. Few stations in the humid tropics have recorded temperatures in excess of 38°C (100°F). Most stations in the midlatitude climate have recorded maximums this high or higher. The summer diurnal range is typically 8 to 11°C (15 to 20°F), again quite comparable to the tropical humid climates. In the three summer months of June, July, and August, the temperatures of the southeastern United States average 1 to 2°C (2 to 3°F) higher than in Belem, Brazil, and the rainfall is about the same. Nighttime temperatures are quite often within a few degrees of the daytime temperatures because the nights are short. Summer temperatures in midlatitude humid climates are controlled mainly by the high solar intensity, long days, and high atmospheric humidity.

Precipitation is frequent throughout the year. The annual total is quite variable depending on the latitude and continental position. It varies from as little as 510 mm (20 in.) upward to 1780 mm (70 in.). Mobile, Alabama, is among the wettest of U.S. cities, with a mean annual precipitation of 1.73 m (68 in.). The variability of the annual precipitation is similar to that of the humid tropics, normally less than 20%. The seasonal distribution is fairly even, but some areas have more in one season or the other. Frequency of precipitation varies. Bahia Felix, Chile, averages 325 days a year with rain. Thus, there is a 90% probability of rain on any and every day of the year. This is not typical of these climates, of course, but it does show to what extent the precipitation mechanisms can persist.

The precipitation is more varied in form in the midlatitudes than in the tropics. During summer and on the equatorial margins, convectional rainfall is the primary mechanism of precipitation. The U.S. Southeast averages 40 to 60 days per year of thunderstorms. The frequency of thunderstorms decreases rapidly from south to north. Cyclonic precipitation is common on the polar margins and is more frequent in all sections of the region in the winter. Hurricanes provide another mechanism for inducing precipitation, and some of the heaviest rains along coastal areas result from hurricanes.

Although most of the annual precipitation is in the form of rain, snow is a factor to some extent in all places. In the United States, the mean snowfall varies from a trace along the Gulf to an average of about 1 m (40 in.) through the Great Lakes district. In January 1977, even Miami, Florida, had snow.

Humidity is generally high in the midlatitudes. On occasion it drops quite low, but usually for short periods. High humidity and high temperatures make the sensible temperatures of summer quite high. As in the tropics, radiation fogs are common on clear nights in summer and fall.

TABLE 17.3 Monthly Data for Midlatitude Wet Climates

	Jan.	Feb.	Mar.	Apr.	May	June	July	Aug.	Sept.	Oct.	Nov.	Dec.	Yr.
							New Orleans, LA 29°59′ N, 90°04′ W: 1 m						
Temp. (°C)	12	13	16	20	24	27	28	28	26	21	16	13	20
Precip. (mm)	98	101	136	116	111	113	171	136	128	72	85	104	1371
							Montreal, Quebec 45°30′ N, 73°35′ W: 60 m						
Temp. (°C)	–11	–9	–4	5	13	18	21	19	15	8	1	–7	6
Precip. (mm)	83	81	78	72	72	85	89	77	82	78	85	89	971
							Buenos Aires, Argentina 34°20′ S, 58°30′ W: 27 m						
Temp. (°C)	23	23	21	17	13	9	10	11	13	15	19	22	16
Precip. (mm)	103	82	122	90	79	68	61	68	80	100	90	83	1026
							London, England 51°28′ N, 0°00′ W: 5 m						
Temp. (°C)	4	4	7	9	12	16	18	17	15	11	7	5	10
Precip. (mm)	54	40	37	38	46	46	56	59	50	57	64	48	595

The summer in the midlatitudes is much like the rainy tropics. Temperatures are high, humidity is high, and convectional showers are common. Summer storms such as tornadoes, thunderstorms, and hurricanes, as well as heat and cold waves bring variety to summer. Cold waves result from equatorward flow of summer continental polar air from the interior of high-latitude land masses. These cold waves may have temperatures above 10°C (50°F), but they represent much cooler air than is normal.

Within the wet climate area, there are differences in temperature that provide further divisions in the climate. Köppen distinguished between areas that have hot summers and mild winters (Cf) and those with cool summers and mild winters (Cbf). The latter region is mainly on the western side of the continents, where there are cooler summer temperatures due to the onshore winds from the oceans.

Marine West Coast Climates

The midlatitude wet climates are perhaps best developed on the west sides of the continents from about 45° to 65°. Here the westerlies from the oceans bring frequent mP air masses onshore, resulting in a high percentage of cloud cover, high humidity, and a high frequency of precipitation. The largest area of this type of climate is in Europe. Here it spreads inland a fair distance because there is no north–south mountain range to block the flow of the westerlies from the Atlantic Ocean. In North America, the zone extends from northern Oregon to Alaska. In South America, it extends from 40° S to the tip of the land mass. Mountain ranges parallel to the coast restrict this type of climate in both North and South America. Small areas with this type of climate exist in South Africa, Australia, and New Zealand.

Winter is fairly long but mild, and summer is cool. The winter maritime air masses are usually at least as warm as the offshore ocean. The winter air is damp and chilling. Fogs are common because of the high humidity. The storm systems often become occluded, bringing extensive periods of rain and drizzle. For this reason, London, England, is regarded as a foggy, misty, and damp city. London averages 164 days each year with precipitation. This translates into a 45% probability of rain on any and every day. Yet the total annual rainfall in London is only 625 mm (25 in.), so the amount of rain received on each rainy day averages less than 4 mm (0.15 in.). Cloudy weather, fog, and drizzle are indeed typical.

Very cold weather is not common along these coasts. Cold cP air masses develop over land, and the westerlies carry the air masses south and east most of the time. Only when an extremely large mass of cold air spreads out and expands in a westerly direction do these cold waves affect coastal areas. Significantly, the warm ocean currents increase winter temperatures.

In summer, precipitation is less frequent because of the stabilizing effect of the subtropical highs as they move northward. Sites near sea level do not receive high total amounts of precipitation primarily because of the coolness of the air and cyclonic activity that produces the precipitation. Summers are subject to cool weather as the result of the influence of the adjacent oceans. The oceans in these latitudes never warm very much, and a cold current flows equatorward along most of these areas. Where highlands exist along these coasts, even summer precipitation is frequent, and rainforests evolved containing huge trees and dense surface vegetation interspersed with the trees. The famous redwoods of California and Oregon typify these forests. The Olympic Peninsula of Washington contains a magnificent forest of Douglas fir. These forests extend well into Canada. The green meadows of Ireland are also due to the frequency of the moist Atlantic air masses.

Midlatitude Winter-Dry Climates

The midlatitude winter-dry climates (Table 17.4) have a strong seasonal pattern of both temperature and moisture. The general location of this climate is the interior of

TABLE 17.4 Monthly Data for Winter-Dry Climates

	Jan.	Feb.	Mar.	Apr.	May	June	July	Aug.	Sept.	Oct.	Nov.	Dec.	Yr.
Denver, CO 39°22' N, 104°59' W: 1588 m													
Temp. (°C)	0	1	4	9	14	20	24	23	18	12	5	2	11
Precip. (mm)	12	16	27	47	61	32	31	28	23	24	16	10	327
Bismarck, ND 46°46' N, 100°45' W: 507 m													
Temp. (°C)	–13	–10	–5	6	13	18	22	21	14	8	–2	–8	5
Precip. (mm)	11	11	20	31	50	86	56	44	30	22	15	9	385
Ulan Bator, Mongolia 47°55' N, 106°50' E: 1311 m													
Temp. (°C)	–26	–21	–12	–1	6	14	16	14	9	–1	–13	–22	–3
Precip. (mm)	1	2	3	5	10	28	76	51	23	7	4	3	213

the continents in midlatitudes. This continental location provides the characteristic feature of a large average annual temperature range.

Specific areas with this type of climate are the Great Plains of the United States and Canada, the grasslands of Eurasia, and small areas of Australia, Africa, and South America. These are continental locations lying where there is seasonal reversal of wind systems. There is frequent convergence during the summer and thus frequent precipitation. In winter, there is more divergence and a drier atmosphere. This is the most variable of all the climates. These areas receive maritime tropical air masses and occasional tropical continental air masses from the adjacent deserts in summer. In winter, they have frequent outbreaks of polar continental air masses and an occasional maritime polar air mass. Each type of air stream brings its particular variety of weather.

A major difference in the response to solar input in this region is the result of lower atmospheric humidity. There are more clear days, and so there are higher rates of incoming and outgoing radiation. Both summer and winter extremes are greater as the result of more frequent dry air masses. Extreme high temperatures average from 38 to 44°C (100 to 110°F) in the wet and dry regions. They range up to 49°C (120°F) in the warmer sections.

In the midlatitude winter-dry climates, winter is the dry season. Dry weather increases radiational cooling during the long nights with the result that mean winter temperatures are usually a few degrees lower than in the humid climates. Extreme low temperatures are 3 to 5°C (5 to 9°F) cooler. This climatic region experiences the lowest temperatures in the northern hemisphere. In North America, the temperature dropped to –52°C (–62.8°F) at Snag, Yukon, on February 3, 1947. At Tanana, Alaska, –51°C (–60°F) was recorded in January 1886. The lowest official temperature of the northern hemisphere is –57°C (–71° F) at Verkoyansk, Russia, and an unofficial –61°C (–78°F) was recorded at Oimekon, Russia. As a result of less humid conditions and excessive Earth radiation, the average annual range is up to 34°C (67°F) in extreme cases and the absolute range up to 100°C (186°F). A winter characteristic of these areas is rapidly changing temperatures. At Browning, Montana, on January 23-24, 1916, the temperature dropped from 6.7 to –48.9°C (44 to –45°F)—a 55.6°C (99°F) drop in 24 hours.

To measure the effect of a continental land mass on climate (i.e., a minimum impact of oceans), climatologists use the concept of **continentality**. As early as 1888, it was suggested that by measuring the average annual range of climate and adjusting for latitude, continentality could be quantified. This idea is illustrated by the formula used by Gorzynsky:

$$K = [1.7(A/\sin\phi)] - 14$$

where K is continentality, A is annual thermal amplitude (°C), and ϕ is latitude.

Many modifications of this formula have been suggested including the formula proposed by Conrad in 1946 where continentality is given by:

$$[1.7A/\sin(\phi + 10)] - 14$$

Thorshaven, in the Faroe Islands, has a continentality index of 0 using this formula.

While maritime climates may be expressed by low values of continental indices, oceanicity indices have also been proposed. The first, introduced by Kerner in 1905, is given by:

$$O = 100[(T_o - T_a)/A]$$

where O is oceanicity, T_o and T_a are the mean monthly temperatures for October and April, and A is mean annual range of temperature (all in degrees Celsius). Another, developed by Kotilainen in Norway, also includes precipitation:

$$K = N \, dt/100\Delta$$

where N is precipitation in millimeters per year, dt is the number of days with mean temperature between 0 and 10°C, and Δ is the difference between the warmest and coldest month. Although the scientific validity of the empiric measures has been questioned, the measures do provide a general comparison of the oceanic and continental influences.

These climates are subject to the effects of a monsoon. During summer, the excessive heating of land in the northern hemisphere causes the subpolar low to expand equatorward. This forms a low trough or cell in the middle of the continent. Converging air brings moisture to the continent, producing a summer rainy season. During the winter cooling of the land mass, the subpolar trough splits into cells over the oceans and a ridge of high pressure develops between the polar high and the subtropical high. This reduces the probability of precipitation.

Winds are a significant factor in the weather. With the seasonal change in pressure is a seasonal change in wind direction. During the high-sun season, onshore winds prevail. During the winter, there is a switch of 90 to 180° in prevailing wind direction. The degree of wind shift depends to a large extent on the continental position. The more interior the station, the greater the wind shift. Wind velocities are high and persistent. In the United States, Oklahoma City has the highest average wind velocity of any city, with a mean of 22 kph (13 mph). Chinook winds are a winter phenomenon that produces extremely rapid changes in temperatures. They occur in the Great Plains from Alberta to Colorado. On January 22, 1943, Spearfish, South Dakota, observed a 2-minute rise of 27°C (49°F), from –20 to 7°C (–4 to 45°F).

Atmospheric humidity and precipitation are seasonal in character, and cloud cover varies with humidity. Precipitation decreases with distance from the equator and the sea. Although it is difficult to delimit annual precipitation, most areas receive between 300 and 900 mm (12 and 35 in.). Of the annual total, about 75% falls during the summer half-year. The annual total is highly variable and subject to long-term cycles as yet unexplained. Several factors are responsible for the variable amount of rainfall. Nearly all rainfall east of the Rocky Mountains comes from midlatitude cyclones. Winds from the Gulf of Mexico take a curved path across the eastern United States, moving east of the Great Plains. The dry tropical air masses that move northward across the Great Plains develop in Mexico and contain little moisture.

In the Great Plains, the cold fronts are often very steep as they move to the southeast, and extremely severe rainstorms develop along them. The unofficial heaviest 12-hour rain fell at Thrall, Texas, on September 9, 1921: 810 mm (32 in.) of rain fell. At Holt, Missouri, on June 22, 1947, an unofficial rain of 42-minute duration totaled 300 mm (12 in.). The frequency of hailstorms is high here as a response to drier air. The largest recorded hailstone fell in Potter, Nebraska, on July 7, 1928. It measured 137.4 mm (5.4 in.) in diameter.

The winter-dry climate is well defined over the Eurasian land mass. Here the largest contrasts between summer and winter weather conditions occur, with the possible exception of the two polar regions. Much of Asia is subject to seasonal reversal of pressure, wind direction, and precipitation.

Summers are hot and humid with intense summer convectional storms. Along the equatorward fringe as much as 70% of the annual precipitation occurs in the summer months; toward the interior, this increases to as much as 80%. The summer precipitation results from convectional instability, and severe rainstorms are frequent. A significant portion of the annual rainfall may come in a single storm. Total annual rainfall may be as much as a meter along the coastal areas, but this decreases inland to less than 675 mm (25 in.).

In winter, cold, dry, continental air from the Siberian high-pressure zone brings a majority of clear days, with short periods of cyclonic disturbances. In the interior, winter temperatures drop below −34°C (−30°F). The frost-free growing season is short—less than 100 days in most years, particularly on the northern fringes. Since most of the precipitation occurs in the summer, winter snowfall is not particularly abundant. Snow stays on the ground for as much as 200 days in the northern interior.

The extreme of this climate occurs in northeastern Siberia. Summers can be warm and winters severely cold. At Oimekon, the range in temperature between January and July is 67°C (115°F). The range between the highest and lowest temperature recorded is 104°C (187°F). Winter is at least seven months long. Snow often remains on the ground until May, when temperatures rise rapidly.

Permafrost appears in the coldest parts of the climatic zone. Since there is often little snow to cover the soil, the ground freezes down to a depth of as much as 50 m (150 ft). In summer, the top few centimeters thaw, leaving a wet, soggy soil and extensive swamps.

Midlatitude Summer-Dry Climates

One area with a seasonal rainfall regime differs from the rest. In this climate, the rainy season comes in the winter rather than in the summer as illustrated by data in Table 17.5. This type of climate occurs on the western margins of the continents between 30° and 40°. It normally does not spread into the continents very far, and thus is limited in extent. There are three main areas of summer-dry climates. The largest is around the Mediterranean Sea and extends eastward into Iran. The second is the west coast of North America from near the Mexico–United States boundary northward into the state of Washington. It exists along this coast between the Pacific Ocean and the Sierra Nevada and Cascade Mountains. The third major area is across Southern Australia. Smaller areas are in the Capetown district of South Africa and in central Chile. The climate is often called the Mediterranean climate for the area associated with the Mediterranean Sea. In total, only about 2% of the land area of the earth experiences this type of climate.

Much of the earth's land area receives most precipitation in the summer or spread over the year. In this climate, there is a pronounced winter maximum of precipitation. At Perth, Australia, 85% of the annual precipitation occurs in the winter. At Istanbul, Turkey, some 70% of the yearly total falls in the winter half year. San Diego, California, receives 90% of its annual total from November to April. Santa Monica, California, has recorded only traces of precipitation in the three months of June, July, and August. The six months of April through September average only 25 mm out of a total for the year of 381 mm. The occurrence of the precipitation in winter and the relatively low intensity of the cyclonic precipitation permit more efficient plant growth than is the case in the climates where the rainy season is in summer. Although mean winter temperatures average above freezing, occasional snow falls even along the equatorward margins. It rarely stays on the ground very long as daytime temperatures rise high enough to melt the snow. Total annual precipitation averages less than 750 mm (30 in.) and tends to increase poleward. Table 17.6 provides the annual rainfall for several cities along the west coast of North America. The data show the rapid increase in annual total that takes place in a poleward direction.

Several climatic and topographic factors give prominence to these regions as resort areas. In summer, the coastal areas are sunny but not as hot as other areas as the cool ocean waters and the sea breeze offset the intensity of the sunshine. Winter temperatures are mild, with a lot of sunny weather and short periods of rain. The subsidence associated with the subtropical high keeps the sky relatively clear so it has an unusually bright blue color. The Mediterranean Sea is noted for its brilliant blue color. This is partly due to sky color and partly due to the low organic content of the water.

Like all other climates that have a rainy season, these environments have two distinct seasons, each associated with a different air mass and a different part of the general circulation. During summer, these areas are under the influence of stable

TABLE 17.5 Monthly Data for Summer-Dry Climates

	Jan.	Feb.	Mar.	Apr.	May	June	July	Aug.	Sept.	Oct.	Nov.	Dec.	Yr.
						Santiago, Chile 33°27′ S, 70°40′ W: 512 m							
Temp. (°C)	19	19	17	13	11	8	8	9	11	13	16	19	14
Precip. (mm)	3	3	5	13	64	84	76	56	30	13	8	5	360
						Los Angeles, CA 33°56′ N, 118°23′ W: 37 m							
Temp. (°C)	13	14	15	17	18	20	23	23	22	18	17	15	18
Precip. (mm)	78	85	57	30	4	2	T	1	6	10	27	73	373
						Rome, Italy 41°48′ N, 12°36′ E: 131 m							
Temp. (°C)	8	8	10	13	17	22	24	24	21	16	12	9	15
Precip. (mm)	65	65	56	65	51	30	25	17	66	78	96	97	711

TABLE 17.6 Latitudinal Variation in Annual Precipitation in the Summer-Dry Climates

City	Precipitation (mm)
San Diego	269
Los Angeles	380
San Francisco	510
Portland	900

oceanic subtropical highs, giving them essentially tropical desert weather conditions. In winter months, the anticyclonic circulation moves equatorward, allowing the westerlies to bring moisture to the region.

Midlatitude Deserts

The midlatitude deserts are basically in the interiors of continents, although they merge with tropical deserts on the west sides of the land masses. In North America, there is desert from southern Arizona northward into British Columbia between the Sierra Nevada and Rocky Mountains. In Eurasia, they are in the trans-Eurasia **cordillera** or lie on the flanks of these mountain masses. All of the central Asian countries have areas of deserts within their boundaries. The southern hemisphere has a much smaller area of desert in midlatitude locations. The basic reason for the small size of the deserts south of the equator is a lack of land area in the latitudes where the deserts are most extensive.

These deserts have the highest percentage of possible sunshine of any of the midlatitude climates. Arizona averages 85% possible sunshine through the year, with an average of 94% in June and 76% in January. Temperatures reflect the clear skies that dominate the weather. Summer temperatures are the highest of the midlatitude climates (Table 17.7). The summer averages are 5 to 8°C (9 to 14°F) higher than in humid areas, and the maximum temperatures are also higher. During summer, when the days are long and the sun high in the sky, temperatures will reach above 50°C (122°F). Death Valley in California recorded an official shade temperature of 56.7°C (134°F). This is only 1°C (1.8°F) below the highest sea level temperature ever recorded on Earth. Winter temperatures compare with those of the humid climates at the same latitude primarily because the coldest temperatures anywhere in midlatitudes occur with outbreaks of cold Arctic air.

This climatic type is the most difficult to place in the pattern of the general circulation. Several factors contribute to the aridity of these regions. Each is not enough to produce extreme drought, but all the elements together result in aridity. Lying between the subpolar low and the subtropical highs, these deserts maintain dry conditions due to low moisture supply and intermittent anticyclonic circulation.

The midlatitude deserts are not as arid as the tropical deserts. During the winter months, the general circulation shifts equatorward far enough to allow occasional passage of weak cyclones associated with the polar front. Precipitation totals are, of course, very low. Phoenix, Arizona, averages 180 mm (7 in.) of precipitation per year, and Ellensburg, Washington, 230 mm (9 in.). There is little seasonal pattern to precipitation. Phoenix receives a trace of snow, and Ellensburg, Washington, averages 780 mm (31 in.) of snow each winter. Moisture efficiency is very low throughout these deserts. In some areas, potential evaporation exceeds the precipitation by as much as 10 times.

TABLE 17.7 Monthly Data for Midlatitude Deserts

	Jan.	Feb.	Mar.	Apr.	May	June	July	Aug.	Sept.	Oct.	Nov.	Dec.	Yr.
Yuma, AZ 32°40' N, 114°36' W: 62 m													
Temp. (°C)	13	15	19	22	26	31	35	34	31	25	18	14	24
Precip. (mm)	10	9	6	2	T	T	6	13	10	10	3	8	77
Winnemucca, NV 40°50' N, 117°43' W: 1312 m													
Temp. (°C)	–3	0	3	8	12	16	22	20	15	9	2	–1	9
Precip. (mm)	27	24	21	21	24	19	7	4	9	21	20	24	221
Alice Springs, NT, Australia 23°38' S, 133°35' E: 546 m													
Temp. (°C)	28	27	25	20	16	12	12	14	18	23	25	28	21
Precip. (mm)	27	45	18	10	18	15	14	10	6	25	23	29	240
Ashkhabad, former U.S.S.R. 39°45' N, 37°57' E: 230 m													
Temp. (°C)	2	5	9	16	23	29	31	29	24	16	8	3	16
Precip. (mm)	22	21	44	38	28	6	2	1	3	11	15	19	210

WARMING IN NORTH AMERICA

Melting of sea ice, glaciers, and the accompanying rise in sea level produce drastic and rapid changes on coastlines and island nations. In Alaska, the average temperature has risen approximately 4°C (7°F) since 1970. Here, whole villages are facing extinction. Coastal villages are succumbing to erosion by ocean waves and to melting permafrost. Shishmaref is one such village of approximately 600 Inupiaq people. It is located on a barrier island called Sarichef in the Chukchi Sea, just south of the Arctic Circle. In the past, sea ice remained near the shore even in summer. With warming temperatures, the sea ice melts progressively farther from shore and freezes for a shorter time each winter.

The reason for the increased erosion is primarily due to the melting and wind. Most waves are generated by wind. Waves do not readily form under ice, only in open water. When the sea ice is close to shore and there is little open water, the wind does not have a long trajectory over water and so does not generate high waves. The net result of more open water near the island is higher waves reaching the shore. In Shishmaref, the waves are rapidly eroding the shoreline. Houses have disappeared into the sea, and others have been moved back from the shoreline. This kind of erosion was unknown until recent decades. To try to protect the village, nearly $10 million was spent on sea walls, but they were ineffective in stopping the erosion.

Melting permafrost has also become a problem with the warming. Buildings tilt and settle, and some collapse. Water and sewer pipes break, and roads crack, buckle, and collapse. Until recently, two-thirds of Alaska was covered by permafrost. It is melting and retreating poleward rapidly. The swamps that result from the melting make land travel by any means difficult or even impossible. The thawing of the permafrost has led to exploding insect populations. Mosquitoes now inhabit Alaska north to the Arctic Sea. This was not the case in the recent past. The federal and state governments buried large quantities of radioactive waste and other toxic chemicals beneath the frozen ground to get rid of it. Now as the permafrost melts, the waste gases and liquids are being released into the atmosphere and onto the surface. In addition, as the permafrost melts, methane gas—one of the most important greenhouse gases—is released into the atmosphere.

In Shishmaref, houses have been destroyed by melting permafrost. Between the erosion and melting permafrost the village will disappear. The only alternative is to move the village of Shishmaref. It is not the only village in this predicament. A 2004 government study determined that 184 native villages out of a total of 213 were already being damaged by the increased temperatures. Some coastal villages will be gone by 2020. Moving the villages inland and building new homes would run into millions of dollars for each village.

North America is being affected by climate change in the same ways as the rest of the world. One sample of sites in North America that perhaps illustrates the results of warming is America's national parks and monuments. The United States pioneered the idea and development of national parks. The parks were set aside in part to preserve unusual or exceptional natural areas for the people of the country. Many of America's unique parks and wildlife areas are already being altered by global warming. In some cases, the very basis for establishing the park is being destroyed. The following are some of the areas affected and the manner in which they are being affected:

- *Glacier National Park:* The park has lost 73% of its glacial ice since 1850. Some 40% of the loss has taken place since 1966. Forecasts indicate that all the glacial ice in the park will be gone by 2050 or even 2030. The snow fields in recent years have been melting weeks earlier in the spring than in the past. Warmer temperatures are pushing smaller animals that live in cooler environments higher and higher in the mountains.
- *Blackwater National Wildlife Refuge:* The water level in Chesapeake Bay has increased 0.3 m (12 in.) over the past century. One-third of the wildlife sanctuary

has been inundated since 1938. Scientists believe that one-third of this rise is due to global warming.

- *Rocky Mountain National Park:* The stunted pines that inhabit the tree line are growing taller and more rapidly. The forests below the tree line are becoming denser, crowding out alpine tundra plants.
- *Yosemite National Park:* Warming and drying are expected to change some of the meadows into desert-like landscapes, with sagebrush predominating.
- *Joshua Tree National Park:* Joshua trees depend on winter freezes to flower and seed. Warmer winter temperatures are killing off the trees. Freezing temperatures are occurring less frequently in winter, and the trees are not reproducing.
- *Channel Islands National Park:* Rising sea level will inundate the habitat for seals and sea lions. Also, the temperature of the California Current, a fertile current of water flowing southward along the coast, has risen nearly 1.6°C (3°F) since 1951. As a result, the amount of zooplankton is down 80%. This has caused very rapid decline in species that depend on these microorganisms.
- *Seney National Wildlife Refuge:* Migratory birds now arrive an average of 21 days earlier in the spring than they did in 1965.
- *Yellowstone National Park:* Less snow and more forest fires will reduce populations of grizzly bears, bison, and other animals.
- *Arctic National Wildlife Refuge:* The Arctic has grown steadily warmer since the 1840s. Permafrost is beginning to melt along the southern edge. Spring thaw is occurring earlier.

Summary

Midlatitude climates have marked differences in temperature through the year in response to changes in the intensity and duration of insolation. Winter is accentuated by the incursion of cold air from the polar regions, and it is winter that distinguishes these climates from the tropics. The convergence of very different air streams results in the development of actively moving low-pressure systems, which often produce severe storms. These changing atmospheric conditions provide a distinct pattern of summer, fall, winter, and spring. Like the tropics, midlatitude climates subdivide on the basis of the distribution of precipitation through the year. In midlatitudes, the wet and dry climates are further subdivided into summer-dry and winter-dry regimes. Different types of climate result depending on whether the rainy season occurs in winter or in summer.

Key Terms

Continentality, *293*

Cordillera, *297*

Diurnal range, *287*

Permafrost, *295*

Subtropical jet stream, *284*

Review Questions

1. How does the variation in solar radiation differ though the year between the tropical and midlatitude climates?
2. In the summer months, why do the 48 contiguous states in the United States receive more total radiation than is received at the equator?
3. Why is the precipitation more efficient for plant growth in the summer-dry climates than in the winter-dry climates?
4. What impact does the freezing of lakes and soil moisture have on the water balance?
5. What effect does continentality have on temperatures?
6. Why do the coldest winter temperatures in North America occur in the interior of the continent?
7. Why does annual precipitation decrease from north to south along our Pacific Coast?
8. Why does California receive most of its precipitation in the winter months?
9. The Great Plains tend to be very windy. Why does this region have higher average wind velocities than the states east of the Mississippi River?
10. The state of Washington has the fewest thunderstorms of any of the contiguous 48 states. Why is this the case?

Polar and Highland Climates

The distinguishing feature of the polar climates is cold weather brought about by low levels of solar radiation. The radiation balance differs from that of the tropics and midlatitudes in both the diurnal and annual pattern of solar energy. In the tropics, the major energy change is from day to night. Daily changes in radiant energy and temperature are greater than from season to season. In midlatitudes, there are large fluctuations in both the diurnal and annual periods. The farther poleward a site is located, the larger the seasonal fluctuation in radiation and temperature relative to the diurnal changes. Like the tropics, the polar climates are dominated by a single periodic fluctuation in energy. In the tropics, the major fluctuation is diurnal, but in polar areas it is the annual fluctuation that is most important (Figure 18.1).

Every place on the surface of the earth receives nearly the same number of hours of sunlight during the year. However, there is a big difference in the distribution through the year. In the tropics, the radiation is evenly spread throughout the year, with about 12 hours of insolation each day. At latitudes higher than the Arctic and Antarctic circles, most of the energy comes in one continuous dose during the summer

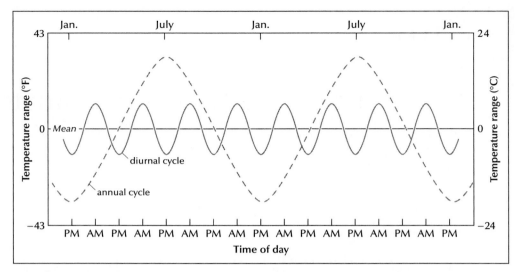

FIGURE 18.1   Relationship between the annual and diurnal energy cycles in polar climates.

half year. The other half year receives very little energy. At the time of the summer solstice in the northern hemisphere (June 22), the north polar axis tilts toward the sun 23.5°. This is the longest day of the year in the northern hemisphere. On this date, the sun does not drop below the horizon in the zone from 66.5° N to the north pole. At Murmansk (69° N), the continual daylight of summer lasts for 70 days. At Spitzbergen (80° N), there are 163 days of continuous light, and at the north pole there are 189 days of continuous daylight.

At the time of the fall equinox in the northern hemisphere, the days are 12 hours long everywhere on the planet. The vertical rays of the sun are over the equator, sunset is occurring at the north pole, and sunrise is at the south pole. The equinoxes are far more important to the polar regions than are the solstices because they mark major times of change in energy. The equinoxes represent times when solar radiation either becomes significant or drops to near zero.

At the time of the winter solstice, the days in the northern hemisphere are the shortest. The north pole is in the midst of a single period of darkness lasting 176 days. At Spitzbergen, the period of darkness is 150 days long, and at Murmansk continuous darkness lasts for 55 days—from November 26 to January 20.

During the winter months in polar regions, direct solar radiation may be absent. Some radiation, including some in the visible range, continues to find its way to the surface. Total darkness seldom exists. When the sun is below the horizon, refraction and scattering of the rays bring some sunlight to the surface until the sun is 18° below the horizon (**astronomical twilight**). Thus, at some latitudes, there may be several months with no direct sunlight but almost continuous twilight. The stars, moon, and **auroras** are also sources of light for the poles, so the darkness is not as intense as it might be.

Another important element in the polar climates is that the intensity of radiation is never very high for any length of time compared with midlatitude or tropical regions. Even on the equatorward margins, the sun rarely reaches more than 45° above the horizon. Although summer insolation persists for many hours a day, temperatures do not get very high. This is because radiation intensity is low, and much of the energy goes to melt snow and ice.

Although there are elements of similarity between the two polar areas, there are also basic differences that influence the climate of each. There is one important and obvious difference between the Arctic and Antarctic. The Antarctic is a relatively

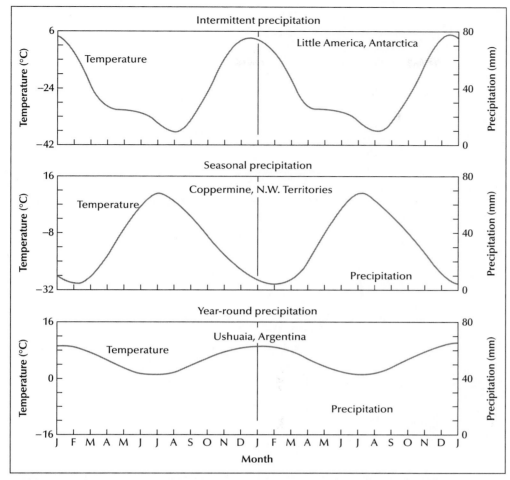

FIGURE 18.2 The polar environments: All three environments have a low solar energy input year-round. The heat from atmospheric water and the sea keeps the average temperature at Ushuaia above freezing.

large and high land mass surrounded by water, whereas the Arctic is a sea surrounded by land. The three climates of the polar regions are the polar wet, wet and dry, and dry (Figure 18.2). The polar wet and dry climate exists primarily in the northern hemisphere, and the other two climates occur mainly in the southern hemisphere.

THE ARCTIC BASIN

The Arctic Ocean consists of the entire sea surface north of the Arctic Circle, whether covered by ice or open water (Figure 18.3). The central part of the Arctic Ocean is frozen most of the year but does have occasional open leads. Until recent years the central ice pack covered 11.7 million km^2 (4.5 million mi^2) in winter and melted back to 7.8 million km^2 (3 million mi^2) in the warmest months. In the past decades, the summer ice has melted back much further due to regional warming. The ice drifts from east to west due to Earth's rotation. In the process of moving, leads open and close so some unfrozen areas are always present. The ice ranges in age up to 20 years and in thickness up to 5 m (16.5 ft), with the oldest ice being the thickest. In the summer, the ice melts away from the edge of North America and

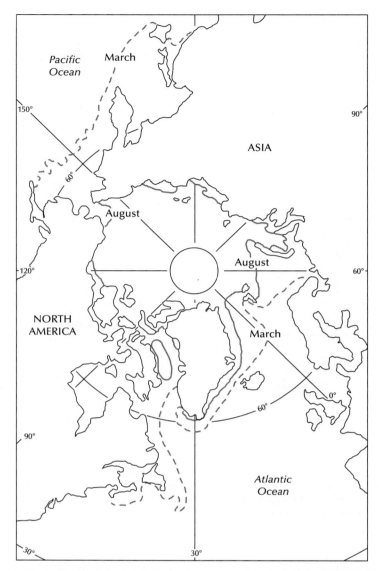

FIGURE 18.3 Map of the Arctic Basin, showing the average location of the ice pack in March and August.

Eurasia so there may be up to 2 months when there is mainly open water with patches of drifting ice. Because of the heat in the Gulf Stream, the Norwegian Sea and Barents Sea remain open all year at latitudes where the sea is frozen in other areas (Figure 18.4).

The earth–space energy balance in the polar areas is such that the poles are major energy sinks. There is much more energy emitted by the polar regions than received from solar radiation. In the Arctic Basin, radiation reflected and radiated away as long-wave radiation is 60% greater than direct solar radiation. For this to happen, a supplementary heat source must make up the difference. This energy source is heat carried poleward by air and water currents. It is also this steady flow of energy poleward that maintains Arctic temperatures at a level much higher than insolation would allow. For the Arctic Basin, the primary source of heat is the inflow from the sea and atmosphere. This energy is nearly double that of solar radiation absorbed at the surface. Warm ocean currents move much of this heat into the Arctic.

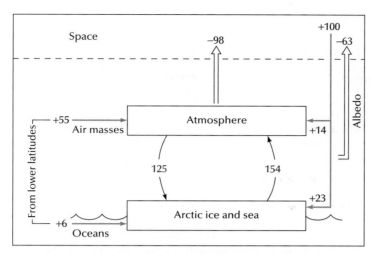

FIGURE 18.4 Energy balance of the Arctic Basin. Note that because of the influx of heat brought into the Arctic Basin by the Gulf Stream and atmospheric storms, the radiation loss to space is much greater than the radiant energy received from space.

The water releases heat to the atmosphere by radiation, conduction, and evaporation. The amount of open water compared with ice is of extreme importance. The heat transferred to the atmosphere from a water surface is more than 100 times that of an ice surface.

Sea water acts differently from fresh water as it cools toward the freezing point. In fresh water, the maximum density is at 4°C (39°F), and freezing takes place at 0°C (32°F). In sea water, the maximum density is just near the fusing point. This is about –2°C (28°F), depending on salinity. One result of this phenomenon is that surface water in polar seas may cool up to 6°C (11°F) more than would otherwise be the case before ice forms. The colder polar sea water is denser than either fresh water at the same temperature or warmer sea water. As it chills to near the freezing point, it sinks to the bottom and spreads out, settling in the deeper parts of the ocean basins. This settling cold water helps promote circulation in the ocean, which in turn helps equalize the energy between the equator and poles. This circulation also explains why most of the water in the ocean is so cold. The average temperature of water in the sea is about 3.5°C (38°F). Warm surface water in tropical regions and along beaches in midlatitudes is the exception and not the rule.

The atmosphere over the Arctic Basin never becomes really cold. The sea, even when frozen, modifies atmospheric temperatures. Whenever the air temperature over the ice drops below –1.6°C (29°F), there is heat transferred through the ice and from the open water. Over much of the Arctic Basin, temperatures over the ice range between –20°C (4°F) and –40°C (–40°F), with the lowest temperature yet recorded near the surface being –50°C (–58°F). The influence of the Arctic Ocean as a heat source is such that there is a very abrupt atmospheric change along the coastline.

Polar Lows and Arctic Hurricanes

It has been known for many years that intense and short-lived low-pressure systems develop along the edge of the continents in the North Pacific and North Atlantic Oceans. Tor Bergeron used the term *extratropical hurricane* in 1954 to describe severe storms in the Arctic. More recently, they have been referred to as **Arctic hurricanes**. These polar lows are small compared with midlatitude lows. The polar lows are less than 1000 km (620 mi) in diameter, and sometimes a fully developed system is only 300 km (186 mi) across. Wind speeds may be sustained at velocities of 40 kt (46 mph) and there may be gusts to 60 kt (69 mph). The central pressure drops to 970 mb or less. The same type of low pressure system forms around the Antarctic continent year-round.

In some respects, these storms are similar to tropical hurricanes. They are circular, have a relatively clear eye, and sustain high wind velocities. They also have symmetrical spiral bands of deep cumulus clouds. Like tropical cyclones, there is a high-level outward flow of air and the storms reach to the tropopause. The outflow of moist air aloft produces a broad and thick layer of cirrus clouds. They dissipate rapidly when they reach land.

These storms differ from tropical cyclones in important respects as well. They develop to full strength in 12 to 24 hours and may have a total life span of only 36 to 48 hours before they strike land. They form under very different conditions than tropical lows. They form in cold air north of the polar front. They travel at speeds up to 30 kt—nearly twice as fast as the average tropical hurricane. They have gale-force winds and heavy snow and sleet. In the northern hemisphere, they form near the ice pack and move eastward toward land. They form where there are extreme differences in the merging air streams along an Arctic front and where there is an existing low-pressure disturbance along the edge of the ice pack. This low draws in the cold air from off the ice sheet and relatively warm moist air from the subpolar ocean. Very cold air develops over the pack ice in winter. Temperatures may be as low as –40°C (–40°F). The temperature contrast between the air masses may be more than 40°C (72°F). The sea-surface flux of heat plays a major role in intensifying and maintaining the mature storm.

THE POLAR WET AND DRY CLIMATE

The polar wet and dry climate, found primarily along the shores of the Arctic Ocean, has cold winters, cool summers, and a summer rainfall regime. Specific areas experiencing this climate are the North American Arctic coast, Iceland, Spitzbergen, coastal Greenland, the Arctic coast of Eurasia, and the southern hemisphere islands of McQuarie, Kerguelen, and South Georgia. The land areas with this climate occupy only about 5% of the land surface of the earth. These regions generally sustain tundra vegetation (Table 18.1).

A sample of five tundra weather stations in the northern hemisphere shows a January average temperature of –30°C (–22°F). Mean annual temperatures are below freezing, but summer temperatures go above freezing at least for a few weeks. The summer period, if measured by above freezing temperatures, averages two to three months but may be as long as five months. Mean temperatures in the warmer months seldom go above 5°C (41°F). High temperatures range from 21 to 26°C (70 to 79°F).

TABLE 18.1 Classification of Polar Climates

Climate	Köppen type
Climates with seasonal flux of mP and mT air masses	ET
Climates dominated by cP air masses	EF
Climates dominated by mP air masses	DF

Pressure gradients are weak in the tundra, so it is not a stormy region. In fact, it is one of the least stormy areas of the earth's surface. High-pressure systems with cold, still, dry air are frequent, particularly in winter. The North American Arctic is even less stormy than other areas. Because of the mountainous character of Alaska, the cyclones of the Pacific Ocean do not move inland very far or very often. Spring and fall are the seasons with the most active weather, as in most high-latitude locations. Winds are light in velocity and variable in direction. The average velocity is less than 8 kt (9 mph) and less than 5 kt (6 mph) in the Canadian tundra.

Atmospheric humidity is high in the tundra during the summer months, averaging 40 to 80%. During the winter it is less, averaging 40 to 60%. Annual precipitation averages less than 250 mm (10 in.) for most stations; more than three-fourths of the annual total falls during the summer half year. Winter snowfall is only about 50 mm (2 in.) of water because the snow is dry and compact. A deep snow cover is absent over most of the tundra, as winter snow is not excessive and tends to drift easily. Thus, exposed areas are often free of snow, although the sheltered areas along streams may have a deep snow cover. Cloudiness varies through the year in the tundra. In the summer months, cloud cover may average 80%, but during the winter it drops to around 40%.

THE ANTARCTIC CONTINENT

The Antarctic continent (Figure 18.5) is the coldest area on the earth's surface. It is about 14.2 million km² (5.5 million mi²) in size. It is nearly twice the size of the United States and nearly three times larger than the Arctic Basin. The harshness of the environment around this land mass is such that even the existence of the continent was not known until extremely late in history. Throughout much of historic time, it showed on

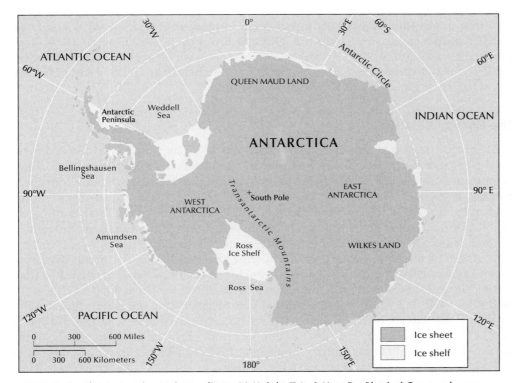

FIGURE 18.5 The Antarctic continent. (From McKnight T. L. & Hess D., *Physical Geography: A Landscape Appreciation*, © Prentice Hall, Upper Saddle River, N.J.)

maps as *Terra Australis Incognito* ("unknown southern land"). Nathaniel Palmer first discovered it on November 17, 1820. The first expedition commissioned by the United States set out in 1838, under Lieutenant Charles Wilkes.

The outstanding feature of the Antarctic land mass is, of course, the huge mass of ice surmounting and surrounding the continent. A sheet of ice that is more than a mile thick covers 95% of the land mass. In some places, the ice is more than 4000 m (13,000 ft) thick. This ice sheet is divided into two separate pieces: the West Antarctic Ice Sheet and the East Antarctic Ice Sheet. The West Antarctic Ice Sheet is thicker and at higher elevations than the East Antarctic. The surface ranges from 2400 to over 3500 m (8000 to over 10,000 ft) in elevation. The eastern ice sheet is firmly attached to the ground beneath, but the western sheet is fairly unstable. Approximately 86% of the planetary ice is on this land mass. The pole of inaccessibility, or the center of the ice mass, is about 4000 m (13,200 ft) elevation, and it is 670 km (400 mi) away from the geographic pole.

The polar ice cap is not the thickest where it is the coldest. The ice reaches its maximum depth where air temperatures average a little below freezing. The accumulation of ice is quite slow in some areas. Near Byrd Station, at a depth of about 300 m (1000 ft), the ice formed from snow that fell some 1600 years ago. The ice mass moves seaward over much of the continent. The rate of movement of the ice seaward is highly variable. In some cases the ice moves seaward as much as 1 km (0.6 mi) per year and in some locations becomes buoyant and extends far out to sea.

There are several locations where the ice sheet moves out onto the sea as a large floating ice shelf. One of these, the Ross Ice Shelf, extends almost 800 km (500 mi) out to sea and covers an area of some 520,000 km^2 (280,000 mi^2). This floating ice is as much as 270 m (890 ft) thick, with roughly 40 m (130 ft) above water. Icebergs breaking off from this shelf ice are tabular in shape and may be very large. A huge iceberg broke away from the Ross Ice Shelf in early October 1987. It was about 154 km long and 36 km wide (92.4 × 21.6 mi). The surface area was about 4750 km^2 (1834 mi^2). It is known to be at least 250 m (825 ft) thick and may be more than that in some places.

The oldest known large drifting iceberg formed from an ice shelf is known as C-2. It was first spotted by satellite in 1978 and broke up in 1990. It was a fairly large iceberg. It was about 30 km (18.6 mi) in diameter and over 60 m (198 ft) thick. For over 4 years, it was grounded in 60 m of water. It traveled nearly around the Antarctic continent for a distance of about 10,000 km (6200 mi).

The Energy Balance

The Antarctic region is a major energy sink. Almost all the stations on the continent have a net radiation loss on an annual basis. Most, in fact, have a net radiation loss even in the summer. The winter is six months long, and there are really only two months of summer. The heat loss to space is greatest over the ocean surrounding the land mass. In July, in the midst of the southern hemisphere winter, there is radiative heat loss to space over the entire area from the pole halfway to the equator. In January, the area of net loss shrinks southward to as far as 85° in some places.

Antarctic temperatures average 16°C (30°F) below those of the Arctic. There are several reasons for the colder regime. The area is mainly a land mass rather than water. The average elevation is high, with parts of the ice sheet 3000 m (10,000 ft) above sea level. The south pole is 3 million miles farther from the sun during its winter than is the Arctic during its winter. This distance reduces solar radiation 7% below that of the Arctic winter. Temperatures reflect both elevation and proximity to the ocean. Mean annual temperatures range from –19°C (–2°F) to as low as –57°C (–70°F) at Vosktok. Minimum temperatures in the winter drop to –57°C along the coast. They decrease with elevation and distance from open water. Vostok, a Russian Antarctic station located inland at an elevation of 3950 m (13,000 ft), has recorded the coldest

TABLE 18.2 Climatic Data for Amundsen–Scott Station (South Pole)

Item	Dates	Datum
Mean annual temperature	20 years (1957–76)	–49.3°C
Coldest year	1976	–50.0°C
Coldest day of record	July 22, 1965	–80.6°C
Coldest month of record	August 1976	–65.1°C
Coldest half year	1976 (April–September)	–60.7°C
Warmest day of record	January 12, 1958	–15.0°C

temperature at the surface of the earth, –88.3°C (–127°F). The geographic south pole is at 3000 m (10,000 ft) elevation. Table 18.2 provides illustrative data of the temperatures that occur.

In the Antarctic, the coldest temperatures lag far behind the solstice as a result of the long absence of direct solar radiation. At Vostok the coldest days occur in their spring after the sun has come above the horizon for a day or two. The sun comes above the horizon there on August 22, and the coldest temperature yet observed occurred two days later, on August 24. In winter, the changes in weather from day to day are brought about by the advection of air masses with differing amounts of moisture and cloud cover. These changing systems are mostly above the inversion layer.

As stated earlier in the chapter, the intensity of radiation reaching the surface of the polar regions in summer is low. The total amount of radiation received at the surface however, is large. In fact, stations in the Antarctic hold the record for the most solar energy received in a monthly period. Stations on the Antarctic continent also hold the records for the longest continuous period of direct sunlight and the most hours of daily sunshine in a month. These aspects of the energy balance do not change the fact that the coldest temperatures on the planet are found here.

In summers, the temperatures range upward toward the freezing mark but rarely reach it. Over most of the Antarctic, it does not get warm enough in summer to melt the surface snow, although there are a few areas that do shed their snow cover. Temperatures above the ice cannot rise much above the freezing point because the surface does not get above freezing.

The equinoxes are periods of great change in the Antarctic. Around the spring equinox (October), there is a burst of solar energy into the polar atmosphere. It is this burst of energy that begins the mixing process that causes the chlorine compounds to interact with the other gases to remove ozone. At the fall equinox, temperatures drop rapidly over the ice sheet resulting in very strong temperature, pressure, and moisture gradients around the edge of the continent. Intense cyclones develop and move around the perimeter. These are the Arctic lows discussed earlier in the chapter. These storms send large quantities of relatively warm moist air inland above the temperature inversion. This influx both retards cooling in the fall and provides heat to the surface throughout the winter.

By far the dominant feature of the Antarctic is the persistent cold. Second to the cold is the prevalence of a temperature inversion above the ice. It exists over the entire continent in winter and much of the continent in summer. The inversion ranges up to 25°C (45°F) in the lower 4 km (2.5 mi). On July 2, 1960, there was an inversion of 30°C (54°F) at Vostok. The amount of the inversion increases away from the coast in winter, reaching 25°C (45°F) or more at the cold pole. The strong inversion is brought

TABLE 18.3 World Climatic Extremes Recorded on the Antarctic Continent

Condition		Record
Solar Radiation		
Longest continuous period of direct sunlight	December 9–12, 1911	60 hr
Highest monthly mean (hours)	Syowa Base (69° S 40° E)	14 hr/day
Highest mean monthly intensity	South Pole (December)	955 Langley/day
Temperature		
Lowest mean monthly diurnal range	South Pole (December)	2°C (3.5°F)
Lowest average temperature (warmest month)	Vostok	–32.5°C (–27.6°F)
Lowest mean annual temperature	Cold Pole (78° S 96° E)	–58°C (–72°F)
Lowest mean monthly maximum	Vostok	–66.5°C (–88°F)
Lowest monthly mean	Plateau Station (August)	–71.7°C (–97°F)
Lowest mean monthly minimum	Vostok	–75°C (–103°F)
Lowest daily maximum	Vostok	–83°C (–117°F)
Lowest absolute temperature	Vostok	–88°C (–126.9°F)
Humidity		
Lowest dew point	South Pole	–101°C (–150°F)
Wind Velocity		
Highest annual mean	Cape Denison	19.3 m/s (43 mph)
Highest monthly mean	Cape Denison (July 1913)	24.5 m/s (55 mph)

about by the high altitude and clear atmospheric conditions, which permit rapid radiation loss from the surface.

Diurnal temperature ranges are small, less than 5°C (9°F), as a result of the small difference in insolation during the day. The absence of the normal pattern of day and night and the high heat capacity of the snow and ice reduce fluctuations in temperature on a diurnal basis. Table 18.3 contains data to show the extreme nature of the temperature regime of the Antarctic.

The extreme cold of the Antarctic continent influences the climate far from the ice cap. Cities that lie on the southern margins of the southern hemisphere land mass have average temperatures 3°C (5°F) cooler than cities at the same latitude in the northern hemisphere. Several of the islands in the South Seas support glaciers, including Kerguelan Island and Heard Island. In South America, mountain glaciers reach the ocean at latitudes nearer the equator than do those in the northern hemisphere.

The Windy Continent

Winds are a predominant factor in polar deserts' weather. Westerly winds are common at the surface from about 30° poleward to around 65°. Poleward of 65° they give way to low-level easterly winds that extend on to about 75°. Cyclonic storms develop over the oceans in the westerlies and move around the Antarctic from west to east. They normally do not penetrate far inland, and they account for much of the precipitation and weather along the coast. High-velocity winds that move snow and produce blizzard conditions occur at Byrd Station about 65% of the time. They have high enough velocity to produce zero visibility about 30% of the time. A slight increase in wind velocity brings a large increase in blowing snow. The amount of snow moved by the wind varies as the third power of the velocity. It is unfortunate for those working in the Antarctic, but the most accessible locations are also the windiest. Mawson's Base at Commonwealth Bay experiences winds that are above gale force (12 m/s [28 mph]) more than 340 days a year. At Cape Denison, the mean wind velocity is 19.3 m/s (43 mph), and during July 1913, it averaged 24.5 m/s (55 mph).

Strong gravity winds develop and blow off the land masses. These **katabatic winds** are the flow of cold, dense air down a topographic slope due to the force of gravity. They result from the extreme cooling of the air over the ice. Where the slopes are steep, the katabatic winds may regularly flow in a direction other than that of the pressure gradient. The strongest winds occur when the pressure gradient and topographic slope are coincident. The katabatic winds are the prevailing winds over much of the Antarctic, particularly where there is a steep drop from the interior highlands to the coast. Where these winds are fairly constant, they will produce formations of waves in the snow and ice surface called **sastrugi**. These frozen waves grow as high as 2 m (6.6 ft) and make surface travel extremely difficult. Strong winds, blizzards, and rising temperatures often occur together. This is because the usual temperature inversion disappears, and warmer maritime air will move inland.

Even the interior can be very stormy, particularly in winter. On July 27, 1989, a six-member international team of men and dogs set out to cross Antarctica by dog sleds and skis. It took them 7 months to make the crossing, the first on foot. They encountered 40 m/s (90 mph) winds and temperatures as low as –45°C (–47°F). One of them got lost in a blizzard one evening only a matter of meters from the camp. He dug himself a trench in the snow and lay down. Snow soon buried him. When morning came, they started looking for him again. When he heard them calling, he crawled out of the snow, cold but alive and unharmed. There were only rare interludes without howling wind and blowing snow. Visibility was often zero, in which case they could not travel. Sometimes they were forced to stay put for as much as 3 days at a time. One blizzard lasted for 17 days, and another stormy period lasted 60 days. Summers are not so stormy. Temperatures go up as high as –29°C (–20°F), and winds drop to around –9 m/s (20 mph).

Precipitation

The ice sheet covering Antarctica is comparable to the tropical deserts in amount of precipitation. The extremely cold air is incapable of holding much water vapor. Moisture data are more difficult to obtain for the polar deserts than are temperature data. Data show that relative humidity, absolute humidity, and cloud cover decrease inland. Relative humidity often drops as low as 1%. Humidity and cloud cover are also lower in winter as a result of the stronger subsidence and the surface inversion. The cloud decks that appear over the ice sheets produce warmer temperatures as they provide a heat source that radiates heat to the ground. Precipitation may be fairly frequent, but it is light. It probably does not exceed 10 mm (0.4 in.) at any time.

Average annual precipitation probably does not exceed 200 m (8 in.) at any place on the ice sheet. The total ranges from as little as 50 mm (2 in.) on the cold plateau to as much as 510 mm (20 in.) at some peninsular locations. Most of the precipitation occurs as snowfall, averaging 300 to 600 mm (12 to 24 in.) per year. At Halley Bay, there is snowfall about 200 days per year. Because of the high winds, drifting snow also occurs 200 days a year. Blowing snow reduces visibility to less than 1 km (0.6 mi) about 170 days a year. When there is no wind, loose dry snow will pile up to as much as 1 m (3.3 ft) deep. Some of the precipitation occurs in the form of ice pellets, and some moisture is deposited directly as condensation. Along the coasts, instances of steam fogs drifting inland have occurred when temperatures were as low as −30°C (−22°F). These steam fogs add to the accumulation of moisture.

THE POLAR WET CLIMATE

Surrounding the Antarctic is the planetary ocean. It is uninterrupted by land masses north to a latitude of 40° S except at the tip of South America. Over this ocean, the atmosphere has year-round high humidity, a high frequency of precipitation, and low temperatures. The only land areas are a series of islands scattered around the Antarctic continent. Table 18.4 contains data representative of this climate type.

In the polar wet climate, humidity is very high, averaging over 50%. Cloud cover is also extensive, averaging 80% during the winter and slightly less in the summer. Precipitation frequency is high in all areas, with at least a 25% chance of precipitation on any day of the year. The annual total varies, however, as the intensity of precipitation varies with latitude and distance from the ocean. The totals vary from 370 mm to 2.92 m (15 to 117 in.). The annual variation in precipitation is among the lowest found anywhere. Snow is a common form of precipitation. Most snows are very wet and last but a very short time.

The high humidity and cloud cover have a considerable degree of control over insolation and temperature. The summer averages range up to 10°C (50°F), and winter averages are between −6.7 and 0°C (20 and 32°F), so the annual range in temperature is not very large. Winter lows are exceptionally warm for a polar climate. The reason for the mild winter temperatures and low annual range is the marine location of these areas. Because the ocean does not freeze, it provides a constant source of heat for the atmosphere. The diurnal ranges are also quite low because of the high humidity. A frost-free season is nonexistent. Frost can occur any day of the year, and these areas average over 100 days annually with frost.

In winter, there is a much greater frequency of storms with high winds. This is the area named "the roaring forties, the furious fifties, and the shrieking sixties" by the sailors of the 16th century. They gave these names to the area when they started sailing around Cape Horn in South America.

TABLE 18.4 World Climatic Extremes Recorded in the Polar Wet Climate

Lowest mean annual diurnal temperature range	Heard Island, Indian Ocean (53°10′ S 74°35′ E)	3.3°C (between −0.5° and −2.8°C) 6°F (between 31° and 37°F)
	Macquarie Island, Indian Ocean (54°36′ S 185°45′ E)	3.3°C (between 2.8° and 6.1°C) 6°F (between 37° and 43°F)
Lowest mean annual hours of sunshine	Laurie Island (66°44′ S 44°44′ W)	500 hr
Lowest monthly percentage of possible sunshine	Argentine Island (June) (61°15′ S 64°15′ W)	5%
Highest mean annual relative humidity	Deception Island (63° S 61° W)	90%

This climate is distinctive in several ways. It is the cloudiest of all climates, with most of the area experiencing few clear days. No place can equal the South Atlantic zone for the frequency and severity of storms.

HIGHLAND CLIMATES

The tropical, midlatitude, and polar systems are distinguished from each other on the basis of periodic aspects of the energy balance. We subdivide each of these latitudinal regimes in turn on the basis of the seasonal pattern in the hydrologic cycle. Highlands, or mountain areas, exist in all of these different climatic regions and hence are subject to the same diurnal and seasonal patterns of solar energy and moisture as the surrounding low-lands. For example, the coast ranges of California are subject to a strong seasonal pattern of energy and a seasonal moisture regime consisting of a dry summer and wet winter.

The primary aspect of the climate of highland areas is the rapid change that takes place with elevation and orientation of the slopes to the sun or prevailing wind. Radiation, temperature, humidity, precipitation, and atmospheric pressure all change rapidly with height. Radiation intensity increases with height because of a steady decrease in thickness of the atmosphere. La Quiaca, Argentina, for instance, has the highest average annual radiation level of any known surface location (Table 18.5). The increase in radiation intensity with height is responsible for the sunburn that skiers or hikers often get in mountain resort areas. There is a sharp increase in ultraviolet radiation as the atmosphere thins. Ambient air temperatures decrease with height above sea level at about 1°C/100 m. As the density of the atmosphere decreases, the concentration of water vapor and carbon dioxide decreases. In the absence of these gases, the greenhouse effect of the atmosphere drops sharply, and the ground surface heats and cools rapidly. Higher diurnal temperature ranges result. El Alton, Bolivia, at an elevation of 4,081 m (13,468 ft), has a mean annual temperature of 9°C (50°F) and a mean daily range of 13°C (23°F).

The microclimate of a site in highlands depends on slope orientation. Orientation to the sun is significant in determining local heating characteristics. A slope facing the sun has much higher surface temperatures and resultant air temperatures than one facing away from the sun. This is a result of greater intensity of solar radiation (Figure 18.6).

Relative humidity increases with elevation up to a point. This is because the ability to hold moisture is a function of atmospheric temperature, and temperature decreases with height. For this reason, fog is more prevalent at higher elevations. Moose Peak in Maine averages 1580 hours of fog annually, and Mt. Washington, New Hampshire, averages 318 days a year with fog. The seasonal pattern of precipitation is similar to that of surrounding lowlands. The amount of precipitation changes rapidly with elevation and orientation to the wind. Since highlands provide a mechanical lifting mechanism, precipitation tends to increase up to a point and then decrease again as moisture is lost through precipitation. Orientation of slopes to the prevailing winds is just as significant as elevation in determining local precipitation. Windward slopes receive up to five times the amount of precipitation that leeward slopes receive. The rainiest areas found on the earth are where there is orographic lifting of onshore winds.

Because temperature decreases with elevation, a greater percentage of total precipitation occurs as snow in highland areas. The mean snow line for the earth is about 4550 m (15,000 ft). In the tropical regions, it rarely extends above 5450 m (18,000 ft). The snow line is currently moving higher due to atmospheric warming. The snow line is usually higher on the leeward sides of mountain ranges because there is less snow, more insolation, and Föhn winds melt the snow. Mountains in Africa, South America, and Asia all exceed this elevation, and so there are snow-capped mountains even along the equator. Extremely heavy snowfalls are common in some highlands. The 25 m (82.5 ft) of snow that fell at Paradise Ranger Station on Mt. Rainier in the winter of 1970–1971 serves as an extreme example.

TABLE 18.5 World Climatic Extremes Recorded in Highland Climates

Extreme		Amount Recorded
Rainfall		
Highest monthly mean	Cherrapunji, India (July)	2692 mm (106 in.)
Highest monthly total	Cherrapunji, India (July 1861)	9296 mm (366 in.)
Highest annual mean	Mt. Waialeale (Kauai, HI)	11.68 m (460 in.)
Highest total (4-month period)	Cherrapunji, India (April–July 1861)	18.74 m (737.7 in.)
Highest annual total	Cherrapunji, India (August 1, 1860–July 31, 1861)	26.47 m (1042 in.)
Snowfall		
Highest daily total	Silver Lake, CO (April 14–15, 1921)	1930 mm (76 in.)
Highest 6-day total	Thompson Pass, CO (December 26–31, 1955)	4420 mm (174 in.)
Highest 12-day total	Norden Summit, CA (February 1–12, 1938)	7722 mm (304 in.)
Highest monthly total	Tamarack, CA (February 1911)	9906 mm (390 in.)
Greatest depth on ground	Tamarack, CA (March 9, 1911)	11.53 m (454 in.)
Highest annual mean	Paradise Ranger Station (Mt. Rainier, WA)	14.78 m (582 in.)
Highest annual total	Paradise Ranger Station (Mt. Rainier, WA)	25.83 m (1017 in.)
Wind Velocity		
Highest 1-hr mean	Mt. Washington, NH	77.3 m/s (173 mph)
Highest 24-hr mean	Mt. Washington, NH	57.7 m/s (129 mph)
Radiation		
Highest mean annual	La Quiaca, Argentina (22° S, 3459 m elev.)	667 Langley/day

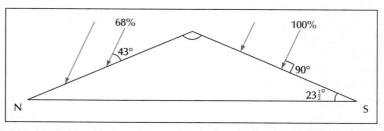

FIGURE 18.6   Slope orientation and radiation intensity. The figure represents the difference in radiation falling on a south-facing slope versus a north-facing one.

CHINOOK WINDS

The *chinook* is a down-slope breeze or wind found along the east side of the Cascade and Rocky Mountains of North America. The name comes from the Chinook Indians, who lived along the lower reaches of the Columbia River. The chinook is a relatively warm and dry wind. The wind develops fairly quickly with a sharp rise in temperature and a drop in relative humidity. These winds are most often identified with the sudden changes in temperature they bring. On January 22, 1943, the temperature rose from –20°C (–4°F) at 7:30 A.M. to 7°C (45°F) at 7:32 A.M. at Spearfish, South Dakota. This was a total rise of 27°C (49°F) in 2 minutes. These winds occur most often in a narrow zone about 300 km east of the crest of the Rockies from New Mexico north into Canada. They form on the leeward sides of mountain ranges.

The primary source of heat in the chinook is a result of compression. Chinooks form when there is a strong westerly flow of air across the mountains. When there is a strong wind, a trough or cell of low pressure forms on the east side of the mountains. The air is stable, so the low-pressure trough pulls the wind down the eastern slope of the mountains to its original altitude. The subsiding air heats at the dry adiabatic rate of 10°C/km.

The Chinook is strengthened if precipitation takes place in the air stream as it rises over the windward side of the mountains. Condensation adds heat to the air. When the air descends on the lee side of the mountains, it is warmer than at the same height on the west side. A cloud called a chinook wall cloud forms over the Rockies when a chinook is blowing. The cloud forms and remains over the mountain crest providing a visible sign of the wind system. As the air descends the eastern flanks of the mountains, the clouds evaporate in the warmer air and a clear sky often exists above the chinook.

Chinook winds are gusty and can reach velocities as high as 160 kph (100 mph). The chinook has the capacity to absorb much water. Because they are warm and dry, in the winter they may sublimate large amounts of snow. It is not unusual for these winds to remove 150 mm (6 in.) of snow a day. They have removed 0.5 m (19.5 in.) of snow in a single day. It is for this reason that these winds are called *snow eaters*. The same wind occurs in other parts of the world. It is the Föhn wind in the Alps Mountains of Europe. In Argentina, the wind is a **zonda**.

Summary

Cold is the distinguishing characteristic of the polar climates. Solar radiation of very low intensity is the primary factor in producing the cold in both polar regions. The Antarctic is colder than the Arctic because it is an elevated land mass. The Arctic Ocean continually supplies heat to the atmosphere even when covered with ice. For this reason, the northern hemisphere cold pole is some distance away from the geographic pole. The major continental ice sheets existing on the planet at present are in Greenland and the Antarctic. Both are related to moderate snowfall and cold temperature. The primary circulation of the atmosphere carries heat into the polar areas to raise the temperature above what it would otherwise be.

Highland climates are unique in that elevation and orientation to the winds are the primary variables associated with temperature and precipitation amounts. The seasonal distribution of precipitation and insolation is the same in the mountains as in the surrounding lowlands.

Key Terms

Arctic hurricane, *305*

Astronomical twilight, *302*

Aurora, *302*

Katabatic wind, *311*

Sastrugi, *311*

Zonda, *315*

Review Questions

1. How do the number of hours of sunlight received at the two poles over a period of a year compare with the number of hours of sunlight at the equator?
2. What is the primary factor resulting in the low temperatures at the two poles?
3. What is the fundamental difference in the physical environment at the north pole from that of the south pole?
4. Why do air temperatures not get as cold in the Arctic as they do in the Antarctic?
5. Temperatures do not change much during the normal 24-hour period in the Antarctic. Why is this the case?
6. Annual precipitation is low in the Antarctic. It is sometimes referred to as a desert. Why is precipitation so limited over the Antarctic continent?
7. Laurie Island in the south Atlantic Ocean averages the fewest hours of sunshine per year of any location on Earth. Why does this island have so little sunshine?
8. Mountain ranges are often places of extremes. Paradise Ranger Station on the southwest slope of Mt. Rainier averages more snow per year than any site in the United States. What factors contribute to this high amount of snowfall?
9. Why do Chinook winds occur along the Rocky Mountains but not along the Appalachian Mountains?
10. How can there be snow-capped mountains and glaciers along the equator when radiation intensity is so high?

PART IV

Applied Climatology

19

The Human Response to Climate

Throughout history, many scholars have attempted to show that the physical environment is the single most important determinant of human actions and activities. This concept led to a philosophy called environmental **determinism**, which can be defined as the philosophical doctrine that human action is not free but determined by external forces acting on human attributes and behavior.

It was only natural that climate, a major component of the environment, gave rise to climatic determinism. In the latter part of the 19th century and up to the 1930s, many books and papers purported to show how environment and climate determined all sorts of activities, ranging from religious beliefs to cultural activities. Deterministic writers often made generalizations, based on limited data, about human–environmental relationships and often ignored contrary evidence. Unfortunately, one outcome of the deterministic viewpoint was that, once it fell into disfavor, it led to a marked decrease in legitimate studies relating human activities to climate. However, the studies of such relationships today are based on better science and rigorous interpretation of data, and the current study of people and the climatic environment deals with many facets of human society. Climate has been related to activities ranging from crime to recreation, from artistic endeavor to the decline of civilizations. This chapter examines a few of these relationships.

THE PHYSIOLOGICAL RESPONSE

The body temperature of a healthy human varies a little around 98.6°F (37°C). Maintenance of that temperature calls for a balance between the body's heat loss and heat gain. Chemical processes within the human body (**metabolism**) are related to the intake of food, the production of energy (and waste), and the physical status of the body.

Energy production in the body is related to energy gains and losses to and from the surrounding environment. Gains of body heat from the environment occur because of absorption of long-wave radiation and conduction from the surrounding air, if the air temperature is higher than the skin temperature. Likewise, heat losses occur through radiation, conduction, and, most important, evaporation of moisture from the skin surface (Figure 19.1). If more heat is lost than is gained, the body temperature must fall. Conversely, if more heat is gained than is lost, then the body temperature must rise. Under conditions of imbalance, a number of physiological responses occur (Table 19.1). The term **hypothermia** refers to conditions that occur when the core body temperature is reduced beyond a manageable range.

As indicated in Table 19.1, cold conditions cause blood flow to the body periphery (e.g., hands, feet) to be reduced and less heat is lost from the body. Although this reduced outward flow conserves heat and maintains the core temperature of the body,

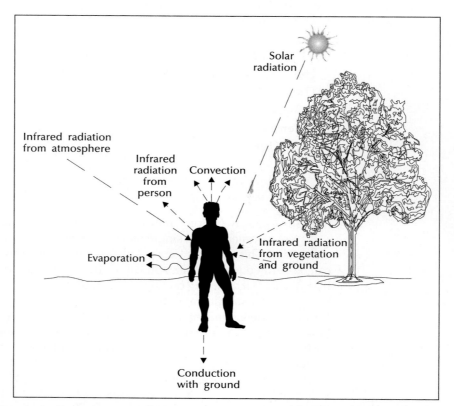

FIGURE 19.1 The energy exchanges that occur between the human body and the surrounding environment. (After Moran and Morgan, 1997.)

TABLE 19.1 Summary of Human Responses to Thermal Stress

To Cold	To Heat
Thermoregulatory Responses	
Constriction of skin blood vessels	Dilation of skin blood vessels
Concentration of blood	Dilution of blood
Flexion to reduce exposed body surface	Extension to increase exposed body surface
Increased muscle tone	Decreased muscle tone
Shivering	Sweating
Inclination to increased activity	Inclination to reduced activity
Consequential Disturbances	
Increased urine volume	Decreased urine volume. Thirst and dehydration
Danger of inadequate blood supply to skin of fingers, toes, and exposed parts leading to frostbite	Difficulty maintaining blood supply to brain leading to dizziness, nausea, and heat exhaustion. Difficulty maintaining chloride balance, leading to heat cramps
Increased hunger	Decreased appetite
Failure of Regulation	
Falling body temperature	Rising body temperature
Drowsiness	Hearing regulating center impaired
Cessation of heartbeat and respiration	Failure of nervous regulation terminating in cessation of breathing

Source: After Lee, 1958.

the reduced blood flow to the extremities can have dire effects. Under extreme conditions, frostbite can occur, resulting in frozen tissue and cell destruction.

Shivering is another physiological response to cold conditions. Its function is to increase the metabolic rate. Although this certainly occurs, the shivering function is not a highly efficient process. It brings more blood to the surface layers of the body, but, especially with uncontrolled shivering, it also increases heat loss by radiation and convection from the body to the surrounding environment. This causes difficulty in breathing, muscular rigidity, unconsciousness, and, eventually, death.

The response of the body to an excessive heat load is variable. An initial effect is dilation of surface blood vessels, which causes a rise in skin temperature. This increased temperature causes sweating, which is a cooling mechanism. The moisture on the skin surface is evaporated, and in the change of state that occurs, the removed heat cools the body surface. Such responses are not, however, without danger. Physical exertion in a hot, humid environment can result in heat stroke—a condition caused by cessation of sweating and consequent rise in core body temperature. In a hot, dry environment, excessive sweating causes salt depletion and dehydration. Both heat stroke and dehydration can prove fatal.

BIOMETEOROLOGICAL INDICES

To better understand the relationships between the human body and a set of climatic conditions, **biometeorological** (or bioclimatological) **indices** have been developed. These serve to allow prediction of various responses to the sensation of warmth and to assess the physiological strain imposed by combined atmospheric

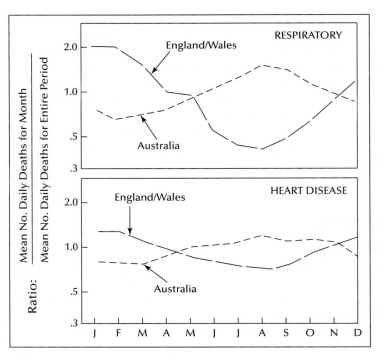

FIGURE 19.3 Comparisons of seasonal mortality for two causes in England/Wales and Australia.

direct and indirect. During cold weather, the cooling of the outer parts of the body places greater stress on the heart while increasing blood pressure. For persons with an impaired circulatory system, this proves highly stressful. Stress becomes much greater if physical activity is carried out, and heart failure while shoveling snow is not an uncommon event.

Despite the influence of modern methods on rates and seasonality of mortality, the role of weather is still very important. Figure 19.4 shows the deaths in the United States on a daily basis for a typical year. The upper graph reveals a departure from the number of deaths in winter and represents a sudden peaking in the number of deaths in summer. The lower figure presents an enlarged version of the peak and relates it to the daily temperatures at an east coast station—Philadelphia. There is a clear statistical relationship between the number of deaths and the heat wave temperatures. In this case, research shows that the elderly were mostly affected.

Similarly, the 1995 Chicago heat wave led to at least 600 deaths over a period of five days in mid-July. Most of the victims were poor and elderly individuals living in the downtown area, with no access to air conditioning. The worst case of heat-related deaths to date was the European heat wave that occurred in August 2003 and claimed 35,000 lives. Temperatures in France reached as high as 40°C (104°F) and remained exceptionally high for two weeks, resulting in nearly 15,000 deaths in that country alone.

CLIMATE AND ARCHITECTURE

Shelter, with food, is one of the mainstays of human life on Earth. The nature of the shelter required largely depends on the conditions of the environment, with the climate providing one base that determines the type needed. Primitive peoples of the world—primitive in the technological and preliterate sense—using the limited resources at hand often developed shelters that were in harmony with the climatic conditions under which they lived.

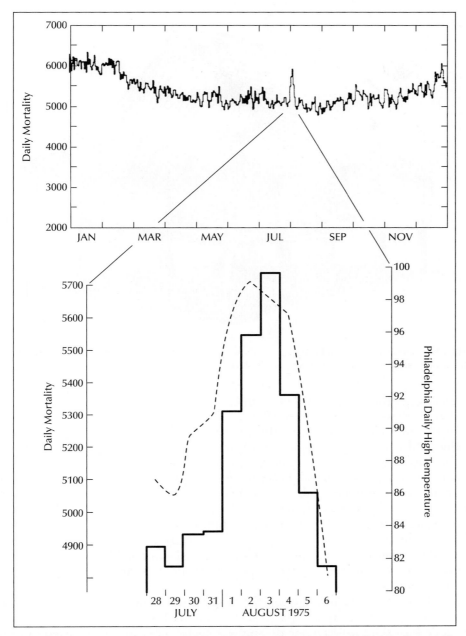

FIGURE 19.4   The upper graph shows the daily deaths in the United States in a single year. The peak in August of that year corresponded to a deadly heat wave in the northeast. The lower line graph shows how temperatures peaked in Philadelphia for that period.

Climate and Primitive Constructions

In the hot-wet tropics, climate is characterized by small annual variation of temperature, intense solar radiation, high humidity, and heavy rainfall. Thus, houses should provide maximum ventilation and shade, while the walls should be open or supplied with movable shades. Roofs, of necessity, must be waterproof. The profuse vegetation of such areas supplies ample raw material, with vines, poles, and bamboo readily available. In the permanent settlements, where agriculture or fishing are the livelihood,

FIGURE 19.5 The basic design of a traditional house in the wet tropics. Note the waterproof roof, the stilts, the shaded veranda, and the movable shutters.

houses are constructed on a skeletal frame, often stilted, with impervious thatched roofs that overhang to provide shade (Figure 19.5).

In the tropical deserts, climatic problems are related to the intense solar radiation and heat during the day and low temperatures at night. Humidity and precipitation are low and present no problem in housing comfort. Some vegetation is available as building material; however, for the most part, mud and straw are the basic materials. These materials are ideal because the high heat capacity of adobe helps maintain an even temperature in the building. The thick walls, with minimum window space, are admirably demonstrated in many desert dwellings. Figure 19.6 shows an example of this and further indicates how dwellings in dry areas are often built close together for maximum shade—a fact well exhibited in the narrow streets of many Arab communities. Modern expansive boulevards are not so well suited to such conditions.

Savanna areas, located between the desert and the humid tropics, experience seasonal rains of varying intensities. In such locations, homes are frequently dome or cone shaped to facilitate drainage during wet spells. They are constructed of grass, mud, and branches, although animal skins may be used. Such traditional homes are now being replaced over wide areas by cement block types.

The regions that abut the wet and dry zones present special problems of house design. The inhabitants are faced with the dual problem of seasonal climates. In the Mediterranean climatic zone, for example, it is necessary to combine the requirements of buildings of the arid regions with those of places that experience a cool, wet season. In effect, to cope with the cooler wet period, the houses of the Mediterranean climate must be more substantial than those designed for desert locations only. Older dwellings in such regions, ranging back to Roman times, are often identified by a central open courtyard. These are shaded during the day because even at the height of summer the sun is never directly overhead in these latitudes. At night they are effective as radiators. The microclimate might be further modified by fountains and pools, which were often located in the courtyards. During the winter, the substantial homes could be heated to combat cooler periods.

FIGURE 19.6 Sketch of buildings suited to hot, dry conditions. Note the close grouping of the houses to provide maximum shade.

Outside the tropical realm, primitive shelters were necessary to overcome the problem of cold. The best-known construction in this respect is the igloo—a house form that is rarely found today. Using extremely limited resources, the igloo represents a building that is well adapted to hostile environmental conditions. It is hemispherical, which both minimizes the proportion of area exposed and represents a streamlined building capable of withstanding high winds. It is constructed of dry snow blocks that are piled one on the other in an inward spiral. The snow blocks have a low conductivity to help conserve the interior temperatures (Figure 19.7). The interior temperature may be as high as 15°C (59°F)—a marked contrast to the temperature prevailing outside. Heat is generated by both oil lamps and body temperatures, and it is not unusual to find animal skins lining the interior of the dwelling. Igloos are winter dwellings. In summer, the moderate temperatures and long days, as the sun is in the sky continuously, require dwellings of different characteristics. The available materials—turf, earth, and driftwood—are used to construct sod-roofed dugouts.

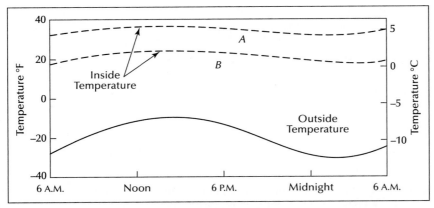

FIGURE 19.7 Comparison of temperatures inside and outside an igloo. *A* is air temperature, and *B* is floor temperature.

Midlatitude continental climates, with severe winters and warm moist summers, have been countered by some admirable dwellings. Nomadic people of such regions required a home suited to the extremes of climate, but that was also portable. Such structures are seen in the yurt (or Kazak) tent of Asia and the familiar Native American tepee. Human ingenuity and comprehension of the surrounding environment have given rise to dwellings of amazing diversity. Equal diversity is also found in modern buildings, but for quite different reasons.

Climate and Modern Structures

The construction of a home or business that uses its background climatic environment as a resource benefits in many ways. Some benefits are economic, such as decreased energy costs, and others aesthetic, as exhibited by landscape design. Beginning with site and eventually passing to construction materials, climate can be a significant component of building construction.

Of basic importance in the layout design of a building or building cluster is consideration of the effects of solar radiation. Direct solar radiation on an exterior surface and that which is passed into the structure via openings is, for the most part, a marked asset in cold climates and should be used to full advantage. In hot regions the effects of direct radiation should be avoided as much as possible. Because the sun's path in the sky and the resulting radiation intensity varies from location to location over the year, an understanding of earth–sun relationships is essential to formulating a rational design plan.

The position of the sun in the sky at any given latitude can be approximated using a simple representation. As in Figure 19.8, the apparent movement of the sun in

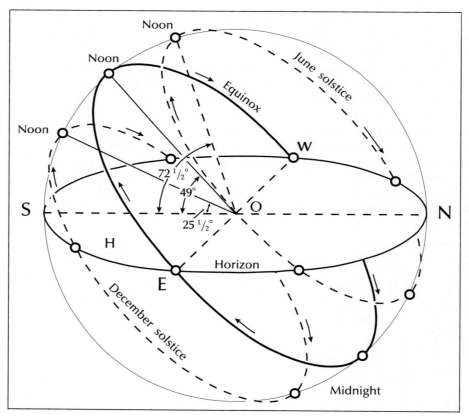

FIGURE 19.8 Apparent path of the sun in the sky at New York City. To the earth-bound observer, the earth's surface is a flat, horizontal disk. The sun, moon, and stars appear to travel on the inner surface of a hemispheric dome. (After Strahler, 1969.)

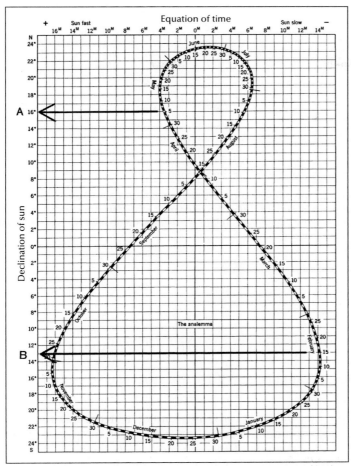

FIGURE 19.9 The analemma provides the declination of the sun for each day of the year. The examples show that (a) on May 5, the sun at noon is vertically overhead 16° N, and (b) on February 15, the declination of the sun is 13° S.

the sky can be illustrated by considering the observer (O) located at a given latitude (in this case, 41° N, the latitude of New York City) surrounded by a horizon (H). The large, surrounding circle (S) represents the celestial sphere across which the sun can be assumed to pass. At the equinoxes, the sun will rise in the due east, attain its highest elevation at noon, and set in the due west.

At any other time of the year, the vertical sun is overhead at noon at a latitude between the tropics. This, the solar declination, may be derived in a number of ways, one of which uses the **analemma**. This is illustrated in Figure 19.9, with an appropriate explanation. Accordingly, a general equation to describe this is:

$$\text{Solar altitude} = 90 - (\text{Latitude} \pm \text{Declination of sun})$$

where the minus sign is used when the sun and latitude are in the same hemisphere and the positive sign when they are in different hemispheres. So at the winter solstice, when the sun is overhead at 23.5° S, the altitude of the noon sun is given by:

$$\text{Altitude} = 90° - (\text{Latitude} + 23.5°), \text{ which for } 41°N \text{ is } 90 - (41 + 23.5) = 25.5°$$

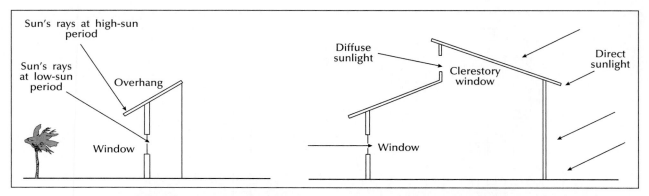

FIGURE 19.10 Correct use of overhangs (a) can allow penetration of sunlight when needed and exclude it at other periods, and (b) shows the principle of the clerestory. In this, scattered and reflected light is admitted while direct sunlight is excluded on opaque walls on the sunlit side of the building.

when the sun is overhead (Cancer, 23.5° N), in summer:

$$\text{Altitude} = 90° - (\text{Latitude} - 23.5°) \text{ so that at } 41° \text{ N is } 90° - (41° - 23.5°) = 72.5°$$

To fully consider the solar radiation load on a structure, it is also necessary to consider the changing lengths of day and night over the period of a year and the azimuth, the horizontal angle of the sun measured clockwise from a north–south reference line. With such facts, an architect is able to plan to best utilize the solar radiation load. Such information is used to determine shading requirements for best exposure to summer shade and winter sun. Perhaps the most practical adaptation is the use of an awning, which can be adjusted to the angle of the sun (Figure 19.10).

The thermal environment in a building is closely related to the solar radiation factor. Within any building, the object is to establish and maintain an internal thermal environment conducive to habitation in terms of the prevailing external climate. The modern approach to maintaining this is to rely, for the most part, on air conditioning and heating systems. Such systems are needed at times, but overreliance on them has often resulted in unbearable strains on power resources. The amount of energy needed to maintain a required internal temperature in a dwelling depends on both the external climatic conditions and the fabric of which the walls are made. A measure of the former are degree days and of the latter the construction materials used.

Heating and Cooling Degree Days

Heating degree days (HDDs) and **cooling degree days (CDDs)** are used as indicators of the amount of energy consumed for space heating and cooling. The idea of degree days was formulated by heating engineers and, in the United States, is computed using the Fahrenheit scale. The engineers found that when the outdoor mean daily temperature is below 65°F (18°C), most buildings require heating to bring their interior temperatures to 70°F (21°C). HDDs are accumulated by subtracting the average daily temperature from 65°F; for example, with a daily high temperature of 70°F and a low of 50°F, the average, 60°F, is subtracted from 65°F to give 5 HDDs. Suppose that the next day has a mean daily temperature of 58°F; then an additional 7 (65–58) HDDs will be accumulated. By keeping a running total of HDDs throughout the heating season, fuel distributors and power companies can anticipate the general amount of fuel oil to be delivered or power to be generated. Large differences are seen in these accumulated average annual totals. In central North Dakota, for example, some 9,000 HDDs occur, whereas in central Florida, HDDs amount to 500. Clearly, heating costs are much greater in the colder northern states.

CDDs are generally computed when the air temperature is 65°F or more, although sometimes a higher base temperature, often 70°F, is used. CDDs are computed in a similar way to HDDs, using the accumulated days above the selected base temperature. The higher the value, the greater the amount of energy used for cooling the interior of a structure. The locational energy needs are generally the inverse of the HDDs, with highest values in the United States occurring in the southern states.

Although HDDs and CDDs were specifically designed for assessing regional energy needs, they also have other uses. For example, by plotting the seasonal degree days over time, it is possible to detect warming and cooling trends or the relative periodicity of cold and warm years.

The Building Fabric

Both the adobe and igloo seem well suited to the environments in which they occur. These two examples clearly indicate the importance of construction materials under different climatic conditions. In these days of prefabricated houses or uniform building materials, the natural thermal properties of the material often are overlooked. This is in accord with the philosophy that, with artificial heating and cooling methods, most building problems can be overcome by manipulating input of energy. Nothing could be less correct, and the application of a few simple principles can provide much insight into optimum thickness and required conductance characteristics suitable for a given climate. Although prefabricated materials can be used in such diverse climatic regions as New England and the American Southwest, the manner in which they are arranged needs to be quite different.

To understand how a building reacts to a given external climate, it is necessary to estimate a number of thermal characteristics. Basically, the main concern is the flow of heat through the wall or roof in terms of both quantity and time. It is often convenient to express thermal properties as the R value—the heat resistance. It is found that the R value varies considerably. For example, a simple galvanized metal shed may appear to be a potential oven in a hot climate, yet simple modification of the structure and rational use of building materials can mitigate the conditions. The R value for galvanized metal over an open frame is 0.6; if a 1 in. insulating board is used, it becomes 3.6. With the addition of 4 in. of wood shavings under the metal, the R value becomes 10.5.

Once the relationship between building material and heat flow is derived, another important factor must be considered. The transfer of heat through a structure will not occur instantaneously, and there will be a time lag between the maximum outside and maximum inside temperatures. The time lag can vary from about three hours for a light wall made of two layers of sheet material separated by an air cavity to at least eight hours for a compacted earthen wall about 30 cm (11.7 in.) thick. Not only will time lag occur, but there will also be an interior dampening of the extremes found in

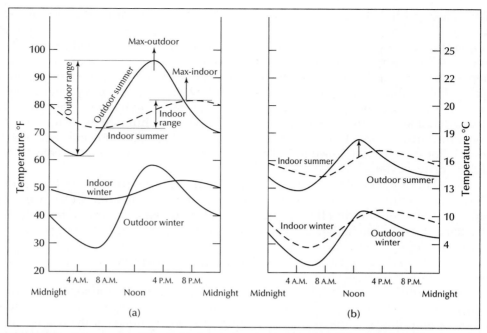

FIGURE 19.11 Comparison of the inside internal and external temperatures in (a) adobe in hot dry climates and (b) a wood structure in midlatitudes.

the exterior climate. As Figure 19.11 shows, the differences can be visualized as two cycles. The time lag causes them to be out of phase while the dampening (or decrement) causes the interior cycle to have less amplitude than that outside. Use of the correct wall materials and thicknesses can result in an ideal situation, where the phase and amplitude are designed to meet the needs of the climatic realm in question. In damp climates, building materials also need to act as vapor barriers. Lack of proper use of vapor barriers is seen in peeling surface paint, stains on inside walls of heated buildings, and damaged wood or metal structures. The resistance of material to moisture penetration is given by its vapor resistance.

Layout design also has a marked effect on air motion. Wind in hot moist climates is an asset and should be exploited as natural air conditioning for it tends to reduce temperatures and excessive humidity. In colder regimes, the main effect of the wind is to induce convective cooling and so lower the temperature sensation. Many of the climatic effects on buildings require both macro- and microanalysis. At the individual home level, ventilation is of considerable significance in climates where external conditions are not extreme. Ventilation maintains inside temperatures near those outside by promoting venting of internal heat (either from that produced inside or from solar loading). Very effective in this regard is cross-ventilation, which uses the natural effect of the pressure variation of the wind in relation to a building. On a windward side, the pressure of the wind against the house leads to a slightly higher pressure. On the leeward side, eddies form and create a slightly lower pressure. Careful placement of building outlets and inlets can harness this effect to use the wind to advantage by creating a flow of air through the building.

In many situations, local topographic effects give rise to highly specialized conditions. For example, high winds frequently occur on the windward sides of slopes in higher latitudes. If such situations are used for building, it then becomes necessary to

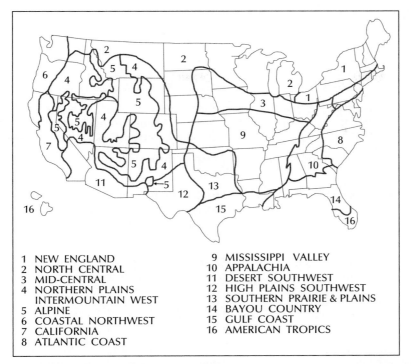

1 NEW ENGLAND
2 NORTH CENTRAL
3 MID-CENTRAL
4 NORTHERN PLAINS
   INTERMOUNTAIN WEST
5 ALPINE
6 COASTAL NORTHWEST
7 CALIFORNIA
8 ATLANTIC COAST

9 MISSISSIPPI VALLEY
10 APPALACHIA
11 DESERT SOUTHWEST
12 HIGH PLAINS SOUTHWEST
13 SOUTHERN PRAIRIE & PLAINS
14 BAYOU COUNTRY
15 GULF COAST
16 AMERICAN TROPICS

FIGURE 19.12 Regional climate variations identify zones applicable to housing styles. (After various sources, including Addison, 1994.)

design the structure in relation to windbreaks or natural topographic barriers. The channeling of wind through valleys also requires special design. The local building response to the cold mistral winds of the Rhone Valley provides a good example. Small windows, or even no windows at all, are characteristic of the northward-facing sides of buildings where windbreaks are commonly found.

A knowledge of prevailing wind direction is also of value in evaluating potential windborne pollution. Odor and particulate matter are common hazards as the odors generated by some industrial centers can prove quite overwhelming. Many homeowners, originally unaware of the problem, have been forced to relocate as a result of such pollution. A similar effect is seen in rural areas, but here lessons learned over time are more evident. It would be quite unusual, for example, for a farmer to locate his farmhouse on the downwind side of animal stalls and pigpens.

Regional House Design

Given the foregoing arguments, it is possible to identify the type of house design best suited for a selected regional area. The Department of Housing and Urban Development along with the Department of Energy and the American Institute of Architects have attempted such for the United States. Figure 19.12 shows the conterminous United States divided into 16 regions in which climates are best suited for particular climate-designed homes. Like any other climate classification, this grouping is somewhat arbitrary and debatable, but it does help identify regional characteristics. Each of the regions is described in Table 19.5.

TABLE 19.5 Region, Climates, and House Design

Region	Climate	Design
1. New England	Quick-changing seasons with a harsh winter and cool summers with low-to-moderate humidity.	Use compact shape and control wind and sun effects. Saltbox suits region well.
2. North Central	Cold winters with wide day-to-night (diurnal) temperature swings, summers are also severe; spring and fall are pleasant.	Balance sunshine, wind, and humidity carefully; high-mass, Earth-sheltered houses are effective and recall indigenous styles like log or sod cabins.
3. Mid-Central	Temperate, but with fierce windchill and humid summers.	Compact shapes with proper solar orientation will thrive.
4. Northern Plains and Intermountain West	Dry with cold, windy, stormy winters; wide diurnal temperature range.	Shading and thermal mass are priorities; evaporative cooling (adding water to dry air to cool it) can help in summer.
5. Alpine	Elevations above 6000–7000 ft are mostly cool to cold with wide diurnal variations.	Simple requirement—keep warm; nestle into mountain. Steeply pitched roof sheds snow. Insulate well.
6. Coastal Northwest	Mild with only 15°F seasonal and diurnal ranges.	Few constraints; compact, small-windowed houses above fog zone are best.
7. California	Chilly, rainy winters; warm-to-hot, dry summers; sunny with wide diurnal range.	Use evaporative cooling (e.g., fountains, pools, and house plants), maximize southern exposure, minimize or eliminate east- and west-facing windows.
8. Atlantic Coast	Relatively temperate with four distinct seasons; pleasant spring and fall; hot, humid summer; winter can be cold and stormy.	Insulate well and maximize ventilation—cover porch, raise first floor, use big windows.
9. Mississippi Valley	Hot, humid summers and cold winters similar to Atlantic Coast.	Shade and enhance ventilation; heavy materials with high thermal mass mitigate seasonal temperature range.
10. Appalachia	Similar to Atlantic Coast but with wider diurnal range and less humid summers; comfortable days are possible any time of year.	With many climate considerations, houses should be well-insulated, high-mass, shaded, and solar oriented.
11. Desert Southwest	Extremely hot summers, moderately cold winters, and an extremely wide diurnal temperature range.	Keep sun out; use high-mass, light-colored materials (e.g., adobe); and cool with evaporative techniques.
12. High Plains Southwest	A cousin to the Desert Southwest—sunny, dry, and wide diurnal temperature range.	Wind-swept west Texas has enhanced ventilation and evaporative cooling.
13. Southern Prairie and Plains	Both summers and winters can be uncomfortable; humid and windy summers, but humidity often falls by early evening; significant diurnal temperature range in spring and fall; often sunny in winter.	Single-story ranch houses have more access to Earth cooling.
14. Bayou Country	Flat, humid, rainy, with hot summers and some cold winters, though sea breezes and moderate diurnal temperature swings are reliable.	Cool with shading, ventilation, and open layout.
15. Gulf Coast	Mild, sunny winters, but otherwise hot and humid; persistent cloudiness limits diurnal temperature range, and sea breezes add some relief.	Emphasize cross-ventilation.
16. American Tropics	Sunny, warm, humid, but comfortable year-round.	Provide shading, ventilation, and light-colored building materials.

URBAN CLIMATES

Humans have a remarkable facility to alter the natural environment. Any change results in a modification of the ongoing natural processes at that site. Perhaps the most changes to an environment is the city, for the constructed environment of a city creates a totally different climatic realm from that which occurred prior to its founding. Consider the following:

- Concrete, asphalt, and glass replace natural vegetation.
- Structures of vertical extent replace a largely horizontal interface.
- Large amounts of energy are imported and combusted.
- Combustion of fossil fuels creates pollution.

These related factors modify the climatic process in the urban environment.

The Modified Climatic Processes

Figure 19.13 provides a graphic summary of the modified flows of energy in a city. Of prime importance are the energy characteristics of the asphalt/concrete/glass environment as compared with that of the vegetation-covered surfaces of rural areas. For the most part, the city surface has a lower albedo than nonurban areas and greater heat conduction and more heat storage. During the day, the city surfaces absorb heat more readily. After sunset, they become a radiating source that raises temperatures at night.

The energy flows are further modified by the geometry of the city buildings (Figure 19.14a). Walls, roofs, and streets present a much more varied surface to solar radiation than undeveloped areas. Even when the sun is low in the sky, a time when little energy absorption occurs on flat land, vertical city buildings feel the full impact of the sun's rays. In the early morning and late evening, the city is absorbing more energy than surrounding rural areas.

The different surfaces that make up the city modify energy availability by changing the water balance (Figure 19.14b). Rain falling on urban and nonurban areas is disposed of quite differently. On building-free rural surfaces, some of the water is retained as soil moisture. Plants draw on this source for their needs and eventually return the moisture to the air through transpiration. At the same time,

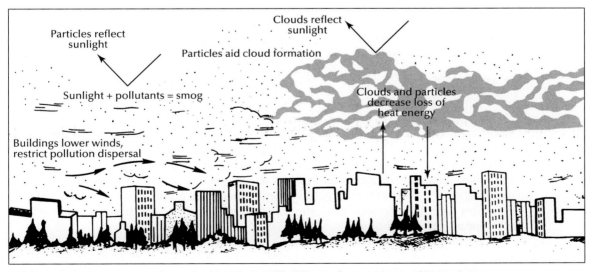

FIGURE 19.13 Schematic diagram showing the modified flows of energy in an urban environment.

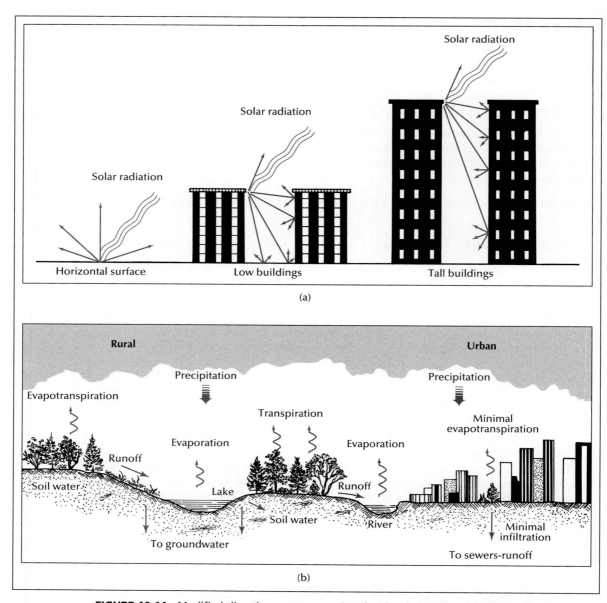

FIGURE 19.14 Modified climatic processes associated with urbanization: (a) effects of horizontal and vertical surfaces on incoming solar radiation and (b) disposal of precipitation on rural and urban surfaces.

standing water and soil moisture are evaporated. Solar energy is required for both transpiration and evaporation. In the city, pavements and buildings prohibit entry of water into the soil. With limited green areas, transpiration is minimal. Most of the rain water drains off quickly and passes to storm sewers, with the result that water availability for evaporation is greatly diminished. Since both evaporation and transpiration amounts are decreased, the solar energy available for this process provides additional surface heating.

As areas of concentrated activities, cities are high consumers of energy, and enormous amounts of energy are imported to maintain their functions. In New York City, for example, the amount of heat generated through the burning of fossil fuels in winter is 2.5 times the amount of heat energy derived from solar radiation for the same period. This high use of energy means that much *waste heat* from factories, buildings, and transportation systems passes to the atmosphere to add to the warmth of the city.

The actual combustion of fossil fuels results in cities having higher air pollution levels than their rural counterparts. Visible evidence of air pollution over cities is the dust dome that is often produced. This pall of smoke and smog acts in a number of ways to modify the urban climate. Some incoming solar radiation strikes particles in the air and is reflected back to space. Other particles act as a nucleus onto which water vapor condenses to form clouds. That cities tend to be cloudier places than rural areas is illustrated by data from London, England. A comparison of bright sunshine hours showed that the city received 270 hours less sunlight than that of the surrounding countryside over the period of a year.

The lower input of solar energy because of increased reflection from particles and clouds results in lower temperatures in cities. However, this lack of energy is more than counterbalanced by other effects. Some of the pollutants absorb rather than reflect energy, and the increased cloudiness reduces loss of long-wave energy to space.

The Observed Results

Of the many changes that occur in the climate of cities, the modified temperature regime is the best known and most closely studied. Cities tend to be warmer than the surrounding nonurban environment. The higher temperatures are best developed at night, when, under stable conditions, a **heat island** is formed. In passing from the center of the city toward the surrounding countryside, temperatures decrease slowly until, in rural areas, they remain about the same. The city is an island of warmth surrounded by cooler air. Figure 19.15a provides a schematic summary of this feature. It is possible to visualize the formation of an urban dome in which the normal lapse rate occurs because of the warmth of the city. This effect is in contrast to rural areas, where, during the night, a low-level inversion often forms. If a slight regional wind is blowing, the warmer air of the city is swept downwind as an **urban plume**, modifying the rural lapse rate conditions (Figure 19.15b).

The heat island effect has been measured in many cities around the world. Places of diverse size—from large cities such as Tokyo, London, and St. Louis to smaller ones such as Corvallis, Oregon, and San Jose, California—show similar patterns. Generally, when the heat island is best developed, temperature differences between city and country are as much as 5.5°C (10°F).

Wind speeds in a city are, on average, lower than those in open surrounding areas because of the increased roughness of the urban fabric. A comparison of highest expected wind speeds in the city and its airport (usually located outside the city in flat, unobstructed land) shows this to be true. In Boston, Massachusetts, the highest airport wind speed is 46 m/s (102 mph), and the value for the city is 32 m/s (71 mph). The same effect is found in Chicago (city 25 m/s, 56 mph; airport 31 m/s, 70 mph) and Spokane, Washington (city 23 m/s, 52 mph; airport 35 m/s, 78 mph).

The wind in cities can also be modified by the urban heat island. As noted, the heat island is best formed under calm conditions. In such instances, the city may create its own wind pattern. A small pressure gradient results when the warm city air rises and is replaced by that from surrounding areas. The uplift of air is not high, and a convective cycle results. This process means that the city air is recycled. If pollutants are being added to the air under these conditions, pollution levels can increase dramatically and give rise to a dust dome over the city.

The influence of cities on rainfall is still not completely understood. Generally, however, it is thought that a city gets about 10% more rainfall than surrounding rural areas. Intensive studies of this aspect of urban climates are part of the research project

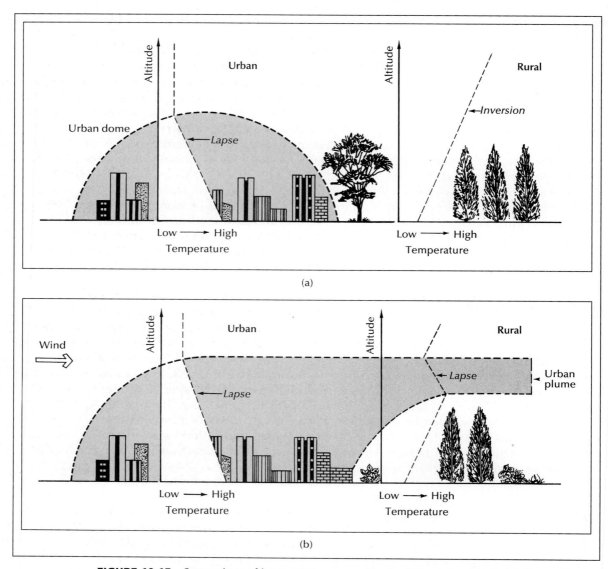

FIGURE 19.15 Comparison of lapse rates over rural and urban environments: (a) On a still night, normal lapse rate conditions occur in the city and an inversion in the surrounding rural areas; (b) with a slight regional wind, the rural conditions are modified by the creation of an urban plume.

called METROMEX (Metropolitan Meteorological Experiment), which is an in-depth study of the conditions in St. Louis. Many significant findings, such as enhanced convective precipitation and severe weather downwind of St. Louis, have resulted from this project, but the results cannot always be applied to other urban centers in different climatic regimes.

Table 19.6 lists some of the basic climatic modifications that occur within cities. Examples of changes associated with the elements outlined are given, together with other parameters. It is an imposing list that points to the amount of change that can inadvertently occur through modification of the environment by people. As the urban population of the world continues to expand, even more widespread results might be anticipated.

TABLE 19.6 The Characteristic Mix of Atmospheric Processes over Urban Areas

Radiation and sunshine	Greater scattering of shorter wave radiation by dust but much higher absorption of longer waves owing to surfaces and CO_2. Hence, more diffuse sky radiation with considerable local contrasts owing to variable screening by tall buildings in shaded narrow streets. Reduced visibility arising from industrial haze.

Summary

UV radiation	25 to 90% less
Solar radiation	1 to 25% less
Infrared input	5 to 40% more
Visibility	Reduced

Clouds and fogs	Higher incidence of thicker cloud cover in summer and radiation fogs or smogs in winter because of increased convection and air pollution, respectively. Concentrations of hygroscopic particles accelerate the onset of condensation.

Summary

Cloud	More in and downwind
Fog	More often

Temperatures	Stronger heat energy retention and release, including fuel combustion, give significant temperature increases from suburbs into the center of built up areas creating heat islands. These can be up to 8°C warmer during winter nights. Heating from below increases air mass instability overhead, notably during summer afternoons and evenings. Big local contrasts between sunny and shaded surfaces, especially in the spring.

Summary

Convective heat flux	About 50% greater
Heat storage	About 200% greater
Air temperature	1 to 3°C/100 years; 1 to 3°C/annual mean

Pressure and winds	Severe gusting and turbulence around tall buildings, causing strong local pressure gradients from windward to leeward walls. Deep narrow streets much calmer unless aligned with prevailing winds to funnel flows along them.

Summary

Turbulence intensity	10 to 50% greater
Wind speed	Decreased 5 to 30% in strong flow (10 m)
Wind direction	Altered 1 to 10°

Humidity	Decreases in relative humidity occur in inner cities owing to lack of available moisture and higher temperatures there. Partly countered in very cold stable conditions by early onset of condensation in low lying districts and industrial zones.

Summary

Evaporation	About 50% less
Humidity	Drier in summer daytime; more moist in winter and summer night

Precipitation	Perceptibly more intense storms particularly during hot summer evenings and nights owing to greater instability and stronger convection above built-up areas. Probably higher incidence of thunder in appropriate locations. Less snowfall and briefer covers even when uncleared.

Summary

Snow	Less (often turns to rain)
Total	More, mostly to lee of city
Thunderstorms	More

Summary

The study of the relationships between climate and people has a long history. Although it is clear that climate affects many aspects of human life and activity, the ideas of climatic determinism are not acceptable. However, the human body does react to changes in the surrounding atmosphere, and shivering and sweating provide two ways in which the body reacts to extreme temperatures. Biometeorological indices, such as the windchill index and the heat index, have been developed to measure the impacts of combined atmospheric elements on people. The role of climate in human morbidity and mortality has been extensively examined with the finding that seasonal impact is most clearly seen.

The nature of traditional primitive shelters largely depends on the prevailing climates. Homes in the wet tropics and desert areas provide fine examples. Modern structures are not always designed with climate, resulting in increased energy costs for heating and cooling.

The built environment of the city modifies climate. The urban climate is a result of altered energy flows and imported energy. All major climatic variables are modified, with the best known example being the urban heat island—the city being an island of warmth surrounded by a cooler rural region.

Key Terms

Analemma, *331*

Apparent temperature (AT), *322*

Biometeorological indices, *321*

Climatotherapy, *325*

Cooling degree days (CDDs), *332*

Determinism, *319*

Heat index (HI), *322*

Heating degree days (HDDs), *332*

Heat island, *339*

Hypothermia, *320*

Metabolism, *320*

Ultraviolet index, *324*

Urban plume, *339*

Windchill, *322*

Review Questions

1. What is climatic determinism?
2. How is the temperature of the human body maintained?
3. What are the two most widely used biometeorological indexes? Explain each.
4. Why has it become necessary to derive an ultraviolet index?
5. How does climate influence seasonal mortality?
6. Relate the nature of a traditional home found in the wet tropics to the prevailing climate.
7. What differences would you find between homes (designed with climate) in New England and the American Southwest?
8. What are HDDs and CDDs used for?
9. How does the geometry of a city influence energy flows?
10. Why is the Bowen Ratio lower in urban areas than in rural areas?

Climate, Agriculture, and Industry

This chapter relates climate to several economic activities. By considering both agriculture and industry, it is clear that only the major facets of each relationship can be discussed within the limits of a single chapter. However, in covering but some of the relationships, it becomes apparent that climate is a significant variable in the economic well-being of people.

AGRICULTURE

From its early origins in the development of civilization, agriculture has evolved into a highly technical field. Through selective breeding and trial-and-error techniques, specialization of cultivated crops took place over the ages, with the most dramatic changes occurring in the 20th century. Genetics, improved fertilizers, increased disease resistance, and the substantial input of energy into farming methods have all led to significant increases in productivity. However, the prevailing climate of a region still places distinct limitations on what crops can be grown and, in part, determines the hazards to which the crops are exposed.

Just as solar energy is essential for crop growth, agricultural systems are maintained through human addition of energy in the form of fuel for tractors, food for farm workers, and the expenditure of labor. To maximize the benefits of this energy supplement and to obtain the highest yield from all inputs, single crops are grown over wide areas. Such monoculture is well exemplified by the wheat fields of North America. Yet this monoculture is in marked contrast to mature ecosystems, where diversity is a key to the maintenance of the system. Due to a lack of biological diversity, crop failure sometimes leads to complete disaster, and most widely grown crops have suffered from disease on grand scales. The boll weevil problem experienced in the old "Cotton Belt" of the southern United States is an appropriate example as is the disastrous potato famine of Europe. In the 1840s, the complete failure of the potato crop because of blight led to some 2 million deaths due to starvation, and it is estimated that another 2 million people emigrated from the affected region.

Because of the weather-related risks involved in agricultural practices, many ways have been devised to ease the vulnerability. Some of the adaptations include:

- *Risk spreading:* Examples—Crop insurance and cooperative farming schemes.
- *Environmental manipulation:* Examples—Irrigation, pesticide, and fertilizer application.
- *Managed diversification:* Examples—Crop rotation and dual-purpose livestock.
- *Modified farming system:* Examples—Modified crop calendar and improved storage.

Modern farming is both technical and highly specialized.

Climate and Crop Distribution

The limit to the areal extent of a given crop often closely corresponds to a climatic limit because crop yields depend highly on climate. It is found that some crops simply will not grow in areas outside the climatic limits, whereas others will grow but will not flower. For example, bananas will not grow in Wisconsin, while deciduous fruit trees will grow in the tropics but will not bear fruit there; in agricultural terms, they have a negligible yield.

The study of a few widely grown crops provides much insight into the problems and potentials of climate in relation to agriculture. Consider the example of wheat. Wheat can be produced in quite dry climates through dry-farming methods. The optimum amount of annual precipitation, however, appears to be on the order of 76 cm (30 in.) because higher amounts of rainfall—particularly in cool conditions—inhibit its growth. The wide distribution of wheat means that yields vary enormously. Some insight into the most productive areas can be obtained by assessing the quality of the wheat, rather than its quantity, because high-quality wheat—wheat high in protein content—brings the best price to the farmer. The protein content of wheat is highest in the wheat belts of North America and the black-steppe area of the former U.S.S.R. This might be accounted for in a number of ways. Such areas are, of course, technologically advanced and use the latest advances in agricultural techniques. They also correspond to the midlatitude grassland biomes of the world. Highest productivity is associated with the biome in which grasses, to which wheat is related, comprise the natural vegetation.

Despite the apparently ideal conditions for growing wheat in the grassland biomes, such areas are not without climatic problems. The continental locations make such areas highly prone, for example, to summer thunderstorms with associated hail.

Hailstones can achieve considerable size and, when falling onto standing wheat fields, can cause enormous damage. For example, a single storm in Nebraska destroyed 3 million bushels of standing wheat. Losses of wheat crops from hail occur every year in the Great Plains.

Many agricultural plants are now produced in areas where they do not occur naturally. The banana was native to Southeast Asia, but the main production area is now tropical America. The potato, with its origins in America, is now widely grown in Europe, and sugar cane from Southeast Asia is widely cultivated in tropical America. The list could be considerably extended. In many cases, the relocation of agricultural crops often results in better yields. The reason for this is twofold. First, in the area in which they were native, plants probably became an integral part of the environment in which they were found. They were susceptible to pests and competition from other plants of that region. Upon being reestablished in other areas, many of the natural limiting factors were avoided. Second, and probably of more consequence, plants transported by people for the express purpose of production are treated with special care and afforded much attention. Were it not for human inputs, such plants might not be able to survive in their new environments.

Perhaps the finest example of a plant that is highly productive in an area remote from its original area is the rubber tree (*Hevea brasiliensis*). A tree of the equatorial rainforest, it was confined to the Amazon basin, and its spread was perhaps inhibited by the dissimilar climates that occur on either side of the biome. Problems of collecting latex from trees scattered throughout the extensive forest led, through somewhat underhanded means, to its introduction as a plantation crop in Southeast Asia. The intriguing story of how rubber caused a boom in South America and how it was smuggled from there to Kew Gardens, London, and, ultimately, to Asia makes fascinating reading.

Clearly, climate plays a significant role in determining what crops are grown in any region. Some of the climatic limitations placed on agriculture are summarized in Table 20.1. While such limitations exist, many ways to overcome climatic problems have been devised, and some crops are now grown in areas because of manipulation of that environment by humans.

Extending the Climatic Limits

If a crop is susceptible to **frost**, the obvious place for it to be grown is where frost does not occur. However, it is sometimes found that some plants grow exceptionally well in areas where only an occasional frost occurs. The farmer is thus faced with a problem: Should one take the risk and grow the crop while being aware that high returns are equally balanced against total loss if a frost should occur? In modern agriculture, large investments often require that the end product is guaranteed, and chances should not be taken. To meet this requirement, methods to overcome climatic limitations on crop growth have been implemented. These methods not only mitigate the danger of frost, but also address the problem of an inadequate water supply.

FROST PROTECTION To overcome the problem of frost damage, it is necessary to be aware of the ways in which frost can occur. Radiation frost results from rapid cooling of the air layer above the ground usually during the early morning hours. This typically follows cool, clear nights when there is little air turbulence. A second type of frost occurs through advection, when cold air flows into a region. Radiation frost is often localized, whereas advection frost can cause damage over widespread areas.

To protect against frost, farmers have a number of strategies. One that recognizes the nature of radiational frost causes frost-prone plants to be planted away from cold air drainage in valley bottoms, as shown in Figure 20.1a. The cold, clear nights characteristic of radiational cooling form layers of dense, cold air near the ground so that mixing the layers decreases the probability of a frost. This mixing is accomplished by

TABLE 20.1 Temperature and Precipitation Requirements for Selected Commercial Crops

Crop Type	Temperature	Precipitation	Notes
Cocoa	Because temperatures between 10 and 15.6°C may be harmful, the crop cannot be profitably grown in regions where mean maximum of coldest month falls to 13.9°C or where absolute minimum of less than 10°C occurs.	Tree not resistant to dry weather, so generally restricted to areas where dry season does not exceed four months.	A tree of humid tropical lowlands—mostly grown within 20° of equator, below 457 m.
Citrus fruits	Little or no growth where temperature is below 15.6°C. Dormancy in cooler months of subtropical climates. Temperatures slightly below freezing are highly damaging.	Requires high soil moisture content. Orchards often irrigated even in fairly moist areas.	Can be grown on a variety of soils with high humus content.
Coffee	A tropical crop whose temperature requirement varies with species. Generally, optimum temperatures are between 15.6 and 25.6°C.	Depending on temperature, optimum amounts vary from 127 to 229 cm per year. The distribution is important, with an ideal minimum in the flowering season. Too much water can promote tree disease.	Generally does best on a well-drained loam soil. Thus some species are highland varieties. Others do well in lowlands.
Cotton	Needs a frost-free growing season of 180 to 200 days. Does not grow below 15.6°C, and optimum temperatures are from 21.1 to 22.2°C during the growing season. Four to five months of uniformly high temperatures are beneficial.	Can tolerate a wide range in annual precipitation; the distribution during the growing season is of critical importance. Frequent but light showers immediately following planting are advantageous.	Needs sunshine to ripen the boll in full maturity.
Rubber	For optimum growth and yield, the mean maximum should be between 24°C and 35°C.	Evenly distributed rainfall of more than 178 cm per year. Lengthy dry periods inhibit growth. Good drainage essential.	A tree of the equatorial rain forest (Amazonia) transferred to plantations in Southeast Asia.
Sugar cane	Susceptible to low temperatures. Little or no growth below 10°C, and the optimum is appreciably higher. Frost very dangerous to young cane.	During vegetative growth, requires a considerable amount of moisture and is sensitive to drought. In ripening period, should be relatively dry to maintain high sucrose level.	Often grown in cleared areas formerly occupied by tropical forests.
Tea	Optimum temperatures should not fall below 12.8°C nor exceed 32.2°C.	Can tolerate high amounts of rainfall (254 to 381 cm per yr) if rain fairly evenly distributed throughout year.	Often grown as an upland crop in tropical areas.

Source: Oliver (1973).

using fans or a heat source (a *smudge pot* as in Figure 20.1b), to create convective currents in the still and layered air.

Sprinkling and flooding are also used to combat frost. The objective in these methods is to reduce excessive cooling and increase thermal conductivity of the ground. Due to the release of latent heat, when water freezes, the temperature of the plants will not fall below freezing. The method does have limitations. Although it retards loss of heat during the night, it also limits heat gain during the day, and successive applications of water become less effective. Of the two approaches, sprinkling is probably better; however, problems occur in determining the amount and frequency

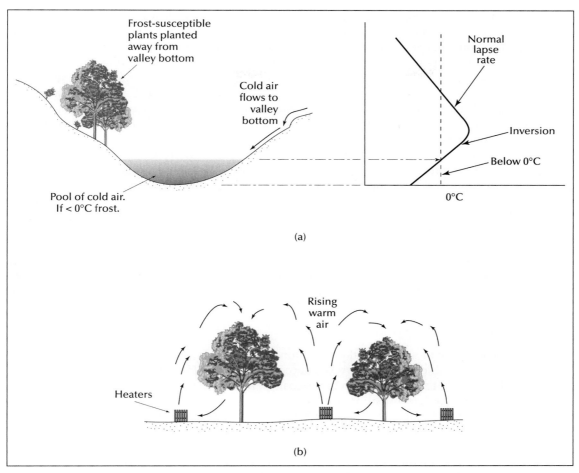

(a)

(b)

FIGURE 20.1 (a) Cold, dense layers of air close to the ground flow into the valley. Crops susceptible to frost are planted away from the valley floor. (b) Heaters in an orchard create air motion to decrease the likelihood of frost.

of water to be added. It has been found that areas not receiving water from the sprinkler are damaged, whereas loss also occurs in areas receiving limited added water.

Other methods to overcome the frost problem include brushing—adding a protective covering of paper—and placing individual covers (hot caps) over plants to reduce nighttime radiation loss.

IRRIGATION Water is essential for plant growth. It would therefore seem that agriculture must be restricted to areas in which water is available at the various stages of plant development. This is obviously not the case because people have long known that by transporting water to crops, agriculture can be successfully carried out. Indeed, irrigation is an integral part of the story of society's development of agriculture, and the great aqueducts and complex water diversion schemes of classical antiquity are well known. Some writers have suggested that the basic cultural aspects of large societies are partly a result of the human control of water.

It is possible to identify three types of regions in which precipitation must be augmented. There are the great deserts of the world in which a perpetual **drought** exists. There are regions in which the distribution of rainfall causes marked deficits of water to occur seasonally. Then there are the usually well-watered areas where, because of periodic droughts, water must be added to maintain the normal production. Therefore, it is possible to recognize perpetual, seasonal, and periodic drought.

Regardless of the problem of the definition of the drought, the major problem that faces the farmer using irrigation is how much water to add and when to add it. To attain an answer to this question, it is necessary to evaluate the amount of water required by a plant to function at its maximum capacity. It is generally assumed that this amount is equivalent to the potential evapotranspiration. Yet as already noted, there is no single method capable of deriving evapotranspiration with exactitude.

Irrigation agriculture has been the subject of many research efforts, and a great deal of literature exists on the topic. Addressed are such aspects as the methods of water application, economics of irrigation, the possibility of using seawater or desalinized water, and the possibility of disease vectors arising from irrigation schemes. The list is a lengthy one and contains much material beyond the scope of climatology. There are many studies concerning the way in which the creation of irrigation systems modifies the local climates. Large irrigation schemes mostly occur in dry climates for it is here that irrigation is needed most. In many of the schemes, large dams have been built and enormous amounts of water impounded behind them. In effect, people create oases in the middle of dry regions.

Climate and Crop Yields

The purpose of agricultural production is to produce as much foodstuff from a given area as possible. The line between success and failure in farming often depends on the yield per unit area in relation to the costs of maintaining that yield. Much research has gone into the determination of optimal conditions to produce maximum yields. Light, heat, and moisture conditions are of major importance in this research.

The conversion of radiant energy to chemical energy by plants, **photosynthesis**, is:

$$\text{Carbon dioxide} + \text{Water} + \text{Solar energy} = \text{Hexose sugar} + \text{Oxygen}$$

Clearly, sunlight is the basic requirement for the rate at which photosynthesis occurs. Plants also respond to varying lengths of daylight—**photoperiod**. Because seasonal climates of the earth are in part a response to differing lengths of day and night over the globe, photoperiodism is an important adaptation of plants during growth and development. Accordingly, plants are to be classed in one of four types:

Long-day plants	Flower only when daylight is greater than 14 hours.
Short-day plants	Flower only when daylight is less than 14 hours.
Day-neutral plants	Bud under any period of illumination.
Intermediate plants	Flower with 12 to 14 hours of daylight but not outside these limits.

This explains, for example, why there are limits to tobacco growth in areas of similar temperature regimes and why chrysanthemums are so colorful during the shorter days of autumn, while many other common flowering plants are limited to the longer days of summer.

Just as light plays a role in plant growth and development, so does temperature. The temperature gradient from the tropics to the polar realm has permitted identification of temperature zones over the earth. For example, it is generally assumed that tropical plants have temperature limits of 20 to 30°C (68 to 86°F), whereas midlatitude plants have limits of 15 to 20°C (59 to 68°F). A number of climatic classifications, such as Köppen's, use temperature as a parameter of climate types. These are practical boundaries because not only is there an upper and lower temperature limit in which a plant will grow but also a temperature at which plants develop most rapidly. This is the **cardinal temperature**, a level not easily defined in specific terms because such critical temperatures vary over time with a particular plant's development.

Temperatures in agriculture are gauged using **growing degree days (GDDs)**, which estimate the number of days of the year that are favorable for various crops. The index is similar to the degree days used by heating and cooling engineers. Their calculation is derived by using a base temperature. The temperature base that is selected varies on the crop grown. For example, the base temperature to calculate GDDs for wheat, barley, and rye is 40°F (4.4°C), for sunflowers and potatoes it is 45°F (7.2°C), and for corn and soybeans it is 50°F (10°C). Using degrees Fahrenheit, for example, if the daily mean temperature is 70°F, for corn one would subtract 50°. That single day would count as 20 GDDs. Interestingly, modified GDDs take into account that growth might be limited when higher temperatures prevail. For example, it is assumed that corn growth is limited once the temperature is above 86°F (30°C) so that no additional days are accumulated above this value.

From the foregoing account, the role of climate in agriculture is enormous. So important is the relationship between the two that a specialized discipline called agricultural climatology, or **agroclimatology**, has been created. This many-faceted field of study has recently been extended to include the potential impact of climate change on farming in different parts of the world. It is a topic about which much more will be researched and written in the years ahead.

INDUSTRY

Industrial activity can be considered in terms of primary, secondary, and tertiary industries. Primary industries are those concerned with the exploitation of raw materials and foodstuffs directly from the physical environment. Agriculture, fishing, forestry, and mining fall into this category. The secondary industries use resources gained from primary activities for further processing. Secondary activities range from iron and steel works to foodstuff processing. Tertiary industries are the service industries. These activities supply the services required by people who are employed in any level of industrial activities (e.g., professional services, trading, tourism). As indicated in the following pages, which provide examples of each of these three levels, climate plays a significant role in industrial activities.

The significance of climate in one of the major primary industries—agriculture—has already been discussed, and it is clear that climate is of outstanding importance in the nature and distribution of agricultural activity. The role of climate in another primary activity—the extractive industry—is not so clear because many of Earth's resources are not readily affected by differing climates. A mineral or fossil fuel deposit may be the result of paleoclimatic events, but the prevailing climatic conditions have no effect on their distribution. Although at the location-level climate is not a significant factor, its impact is certainly felt at the operational level. For example, the iron ore extraction in the subarctic climate of northern Scandinavia has higher costs because of the high wages necessary to attract workers and from the higher costs of heat and light in the long, dark hours of high-latitude winters.

Industrial Location

The role of climate becomes visible in the secondary and tertiary activities. The basic factors contributing to a given industrial location are the historical influence, provision of raw materials, availability of fuel and power resources, supply of labor and market considerations, and transport facilities. Although the word *climate* does not appear in any of these categories, a brief examination of each brings out the role that is played by climate.

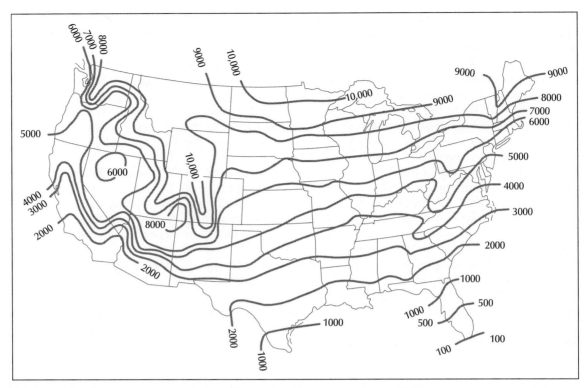

FIGURE 20.2 The average annual heating degree days (base 65°F) in the United States.

Although the role of climate in the development of some industries is difficult to assess, there are examples in which climate can be shown to play a major part. The rapid growth of industry in southern California illustrates the point for the development of the aircraft industry. This industry's development was initially directly related to climate. Early aircraft manufacturers were attracted to the location because of its mild winters and light winds (summed up by the term *flying weather*). A further factor of importance was that the large hangers required in aircraft manufacture did not require expensive heating. This latter point is well demonstrated when the number of heating degree days for California are compared with those of other areas of the United States. As described in Chapter 19, heating engineers use the degree day to estimate the amount of fuel likely to be required in given locations. As shown in Figure 20.2, the cumulative number of heating degree days is appreciably less in California than in many other parts of the country. Note, however, that with the advent of air conditioning, cooling costs also became significant.

An interesting example of the development of an industry because of climatic conditions is the motion picture industry in southern California. When camera equipment was not highly sophisticated, the long hours of daylight in southern California proved ideal for outdoor shooting of films. The location was further aided by the myriad of diverse climates and related vegetation associations that occur within a fairly small area around southern California. It was not necessary to travel far to film desert conditions. Nearby mountains provided a backdrop for stories associated with everything from Yukon miners to avid Alpinists. This historic impetus was negated by the development of sophisticated equipment. Lightweight cameras and improved transport meant that the situation being filmed no longer had to be simulated. Instead, the area about which the film was made could actually be visited. Technology and development of other film-making areas led to the decline of

Hollywood as the preeminent film capital of the world. Interestingly, and perhaps because of the concentration of expertise and facilities, television production films remain installed in Hollywood.

The most obvious influence of the role of climate on industrial location as conditioned by raw materials concerns the industries that are based on agricultural and forest products. In many cases, large industrial processing plants appear where raw materials are found. The meat-packing and grain-processing industries of the American Midwest and the location of pulp and paper mills in the forests of Georgia and Canada provide apt examples. With many primary processing activities, transportation costs are key to industrial location. Of course, there are variations from this pattern, many of them of historic origin.

There is good reason for the treatment of some products away from their areas of production. For example, although West Africa is the major producer of cocoa, western European countries are among the main producers of chocolate and chocolate products. It does not take an expert economist to work out why this is so. The hot, humid conditions of the growing areas would necessitate high-cost facilities to stop chocolate products from melting, and air-conditioned storage and ships would be required to stop its spoiling.

Energy Production

It has already been noted that present-day climate has little bearing on the location of fossil fuels. However, climate today does influence the location of hydroelectric power generation, solar energy production, and wind energy.

Hydroelectric power production relies on a sustainable flow of water over a change in elevation. This feat is most often achieved by construction of dams to create large reservoirs. To fill and maintain appropriate reservoir levels, there must be a supply of water, and hence a precipitation source. The source can be quite remote from the power-generation site, as illustrated by the location of the Hoover Dam, which impounds Lake Mead on the Colorado River in a dry region. Much of the water flowing into the reservoir is from precipitation falling in the Rockies.

The potential for generation of usable energy through solar and wind energy rests largely on the prevailing climate at a location. These sources currently contribute but a tiny fraction of global energy needs. Yet with a growing world population, environmental concerns, and diminishing fossil fuels, this will change in the years ahead.

Solar energy is used by passive solar collectors in most homes. A south-facing window will permit the sunlight to pass through, be absorbed, and eventually warm the room. In contrast, an active solar collector is often a roof-mounted dark box covered in glass. The heat energy collected in these is transferred to the home by circulating air or fluids through a pipe system and is often used for water heating. In dry, sunny areas, these are quite efficient. For example, it is estimated that in Israel, some 90% of all homes have a solar water heater.

The old idea of focusing the sun's rays to produce high temperatures is now used on a large scale. Pioneered by the French at Odeillo, the object is to place rotating mirrors that follow the sun during the day and continually reflect sunlight onto a receiving tower. The facility near Barstow in the Mojave Desert uses the concentrated sunlight to heat pressurized water, which is transferred to turbines.

Of major importance for the future is the use of **photovoltaic** (solar) cells that convert the sun's energy directly to electricity. When sunlight strikes a photovoltaic cell, electrons are dislodged, creating an electrical current. There is a strong market today in developing countries to provide rural electrification with solar panels where there is abundant sunlight and a large rural population without the proper infrastructure to develop an electrical grid. Photovoltaics are also making inroads as supplementary power for utility customers already served by the grid. Although still costly compared to most conventional sources of power, such as coal and oil, more individuals, companies, and communities are choosing to use photovoltaics for reasons other than cost, including the desire to develop a clean, sustainable energy source. Germany has become the leading photovoltaic market, with installed capacity rising from 100 mW in 2000 to more than 4000 mW at the end of 2007. Demand for photovoltaic cells is growing at an annual rate of 40%, and this trend is expected to continue.

Wind energy has been used for centuries to power sailing vessels and windmills. A wind-driven water pump remains a common site in the American West. Interest in wind power has been revived in the United States, and the Department of Energy has set up experimental wind farms. The wind farm in Altamont Pass, California, has 7000 wind turbines and supplies almost 1% of California's energy. As of 2008, wind turbines generated about 26 billion kW of electricity each year in the United States.

Although the use of the wind has much promise, there are problems related to its widespread development. Clearly, the location selected should have steady winds a good portion of time. The intermittent nature of the resource would mean that a back-up would be required to overcome production shortages. To harness the wind economically also requires numerous wind generators, and this requires large amounts of space. In densely populated areas, this could create a problem. However, the recent location of a huge facility near Lake Benton in Minnesota along with new development in other states suggests that wind power will continue to become more important in the future.

Besides its importance for providing a basic source of power, climate also directly influences the amount of energy required to maintain industrial plants at their optimum working temperatures. Furthermore, climate contributes to the problems in the transmission of power from generating stations to factories and homes. For any industry to function, it must have reliable and trained employees. In a modern, mobile society such as North America, the climate of a location can prove of some significance in attracting workers. The rapid growth of the sun-belt states is indicative of the attractiveness of a warm, almost winter-free environment.

Although the external climatic environment is important, it should be remembered that indoor workers also require a comfortable environment. Table 20.2 depicts the suggested optimum indoor operating conditions of selected industries. It shows quite clearly that in many of the highly industrialized areas of the world, it is necessary to either heat or cool the factory area and artificially modify the humidity conditions.

TRANSPORTATION

Air Transport

All phases of air travel are influenced by atmospheric conditions. The effects are felt at all levels of operation—from the construction of a landing strip to the trip *en route* and the landing and take-off conditions. The significance of atmospheric studies in relation to air travel is illustrated by the simultaneous development of the two. As air transport improved, it became necessary to understand flying weather. As planes flew higher, altitudinal variations of the atmosphere became an area of study. Development in one led to research in the other. Unfortunately, the importance of atmospheric conditions in air travel is also illustrated by the number of air disasters associated with weather conditions.

Many of the air transportation problems are concerned with conditions that will be encountered at a given time. As such, aviation weather is of more immediate significance than aviation climatology. *En route*, aircraft face many problems. Most are

TABLE 20.2 Optimum Indoor Temperatures for Industrial Activity

Industry	Temperature (°F)	Relative Humidity (%)
Textile industry		
Cotton	68–77	60
Wool	68–77	70
Silk	71–77	75
Nylon	85	60
Orlon	70	55–60
Food industry		
Milling	65–68	60–80
Flour storage	60	50–60
Bakery	77–81	60–75
Candy	65–68	40–50
Process cheese production	60	90
Miscellaneous industries		
Paper manufacturing	68–75	65
Paper storage	60–70	40–50
Printing	68	50
Drug manufacturing	68–75	60–75
Rubber production	71–76	50–70
Cosmetics manufacturing	68	55–60
Photographic film manufacturing	68	60
Cosmetics storage	50–60	50
Electric equipment manufacturing	70	60–65

Source: From Maunder (1970), after Landsberg and Grundke.

concerned with meteorological conditions and include such factors as icing, turbulence, and selection of optimum cruise altitude in relation to winds aloft.

Icing refers to the formation of ice on lift-producing surfaces (e.g., wings, control surfaces, propellers) so that the smooth flow of air over such airfoils is interrupted. This increases drag and reduces lift to modify the aircraft's flight capabilities. Icing occurs in clouds where air temperature ranges from slightly above freezing to about –20°C (–4°F) when supercooled water droplets exist in the cloud. These droplets will freeze on contact with a surface. In cumuliform clouds, the large size of the droplets allows the formation of clear ice. Stratiform clouds result in the formation of rime ice, where supercooled droplets are much smaller. Clear ice is heavier, and thus it can be more hazardous than rime, although most structural ice is often a combination of the two forms.

Aviation personnel are well aware of the dangers of icing, and many methods are available to combat its effects through anti-icing and de-icing processes. The mechanical boots on the leading edges of wings are composed of an inflatable rubber sheath that swells and causes ice to break up and fall from the aircraft. Some thermal devices are activated from the cockpit to melt ice in flight. Before take-off, anti-icing fluids are commonly used to remove ice from the aircraft while it is still on the ground. For smaller, general aviation aircraft, the avoidance of potential icing conditions is the best defense.

Turbulence is the irregular motion of air over short distances in the atmosphere. Its effect can pass from a slight bumpiness to structural damage of an aircraft. There is no direct measure available for turbulence, so flight forecasters and pilots use their knowledge of meteorology to predict turbulence. Clearly, uneven air flow will occur as a result of surface influences ranging from uneven heating of the ground to air flow over mountains. Wind-shear turbulence results from air moving at different speeds or direction in short distances and can occur anywhere in the atmosphere.

The deadliest among these in-flight wind events is the microburst, which is a localized downburst from a cumulonimbus, or thunder cloud, with a surface damage diameter of 4 km (2.5 mi) or less. Microbursts are especially deadly during take-offs and landings because at low altitudes there is very little space or time to recover. Between 1964 and 1985, there were 26 commercial aircraft accidents and 620 deaths attributed to microbursts. However, since 1995, all major airlines in the United States have been required to adopt wind shear detection equipment, as have numerous airports, which has greatly reduced the accident rate.

Clear-air turbulence (CAT), which normally is not visible to pilots, poses a severe safety hazard because of the violent motions it induces. Furthermore, aircraft that experience CAT are usually flying at high speeds that will increase the potential for damage or injury. The probability of encountering CAT increases with increasing altitude, being most typical at 9 to 13 km (6 to 8 mi). It is most often associated with the presence of strong vertical wind shear—that is, locations where large vertical gradients of wind speed or direction occur. These abrupt changes often occur near the boundaries of jet streams, so CAT is experienced there. Significant temperature differences exist between the two air masses on either side of the shear. To allow detection of CAT, microwave radar, optical (laser) radar, and infrared sensors have been used, but careful forecasting and pilot knowledge remain the best safety methods.

Air transport also is plagued by the problem of fog. Many flight delays, particularly during the local cool season, are related to fogged-in airports in a different part of the country. The ripple effect from flight delays, diversions, and cancellations across the system can be dramatic. A single flight cancellation can cause 15 to 20 delays while the decision to divert a flight can result in as many as 50 flight delays. This is particularly problematic when widespread advection fog forms. Such a problem is shared by other forms of transportation, such as highway, rail, and marine transport.

Shipping

The struggle between people and the sea is an age-old one described in literature by many famous writers. Given that most of the problems faced by mariners are related to atmospheric conditions, a study of the marine climate is basic to shipping.

The cause and frequency of fogs have already been discussed in an earlier chapter, but are mentioned here because they are so important to shipping. Many of the fogs are due to advection. The expansion of the infamous fogs off the Newfoundland Banks, for instance, is related to the movement of warm air off the warm waters of the North Atlantic Drift and its sudden chilling as it passes over the colder waters of the Labrador Current. Despite the adoption of modern navigation techniques, accidents still occur in fogs. The distribution of fogs also clearly shows that the coastal deserts of the world are among the foggiest that occur anywhere. It is little wonder that they have been designated by a special symbol (n) in the original Köppen climatic classification.

Violent storms at sea are a hazard facing shipping. Probably the most dangerous storms are hurricanes because the effects are felt over thousands of square miles of the oceans. Hurricanes originate over the warm seas in low latitudes. Thereafter, in the northern hemisphere, they typically take a westward and then a northeastward path. Deriving their energy from the oceans over which they originate, they soon dissipate once they pass over land. However, as well shown in the book and film *The Perfect Storm*, which described a storm off the northeast coast of United States, violent storms can prove fatal wherever there are large bodies of water and people traversing them. Similarly, in the song "The Wreck of the Edmund Fitzgerald," the sinking of an iron ore carrier and her crew in Lake Superior occurred when the gales of November came early.

Successful ocean trade requires that ships can dock in a location that provides and protects their cargo, and an ice-free port is of major importance. The role of climate in relation to this is well shown by considering North Atlantic ports. The 32°F (0°C) isotherm is located almost at 40° N over the northeast United States. In Scandinavia, it is located north of 60° N. Most of the ports located north of the isotherm thus experience subfreezing temperatures for much of the winter and many will be ice bound. The location of the freezing isotherm is, of course, a partial result of the influence of heat transport of ocean currents. Norway is affected by the warm North Atlantic Drift while waters off Newfoundland are associated with the cold Labrador Current.

The economic benefits of ice-free ports are illustrated by the use of Norwegian ports, such as Narvik, for export from Sweden. Swedish ports on the Baltic Sea, an inland sea, are frozen in winter. Iron ore from the northern ore fields in Sweden passes through the ice-free ports of Norway in winter.

Land Transportation

For most persons living in a midlatitude climate, the significance of climate in relation to land transportation is usually quite clear. Roads covered by snow or ice, or obscured by fog or blowing dust, are events most people face at one time or another. The frequency of these events, however, is climate based, and for each hazard a frequency probability may be derived. As an example, one might consider whether a location may have snow during a holiday period. Figure 20.3 shows the probability that there will be 1 in. (25 mm) of snow on the ground in the United States on Christmas Day. The effects of latitude and altitude on climate are clearly evident.

The quality of road transportation is also influenced by climate. The effect of freezing and thawing on roads is widespread, and most temperate regions experience some degree of frost. Frost action on pavements can cause frost heave and settlement that results in loss of compaction, rough surfaces, and deterioration of the surfacing material. The resulting thaw provides both drainage problems and restriction of subsurface drainage.

In frost-susceptible soils, ice may grow in the form of lenses or veins. As the ice crystallizes, water from below is attracted to the freezing mass and increases the volume of ice. The resulting surface raising is termed *frost heave*. The heave is most problematic when it is not uniform, leading to extremely uneven surfaces. This occurs when subgrade material varies, from sand to silt for example, where drains, culverts, or utility lines break the uniformity of the subgrade material. During the thawing period, water released by the topmost melting portion cannot drain through the still

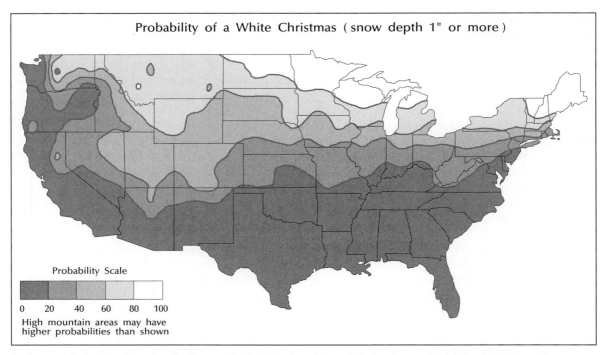

FIGURE 20.3 The probability of a "white Christmas" when 2.5 cm (1 in.) or more of snow is on the ground.

frozen lower soils. It therefore tends to move upward to emerge through cracks at the surface and to further weaken the roadbed.

TOURISM AND COMMERCE

The tertiary industries cover many sectors of the economic environment. To illustrate how climate plays a role in these sectors, tourism and commerce serve as two good examples.

Tourism

Tourism is the world's fastest growing industry, and some economists have suggested that fairly soon it will become the largest industry in the world. In some locations, it is already the main source of national income. In The Bahamas and Cayman Islands, for example, money from tourism constitutes approximately 80% of the income. In recent decades, tourists became aware of their impact on pristine wilderness sites while admonished to take only pictures and leave only footprints. Today, governments and individuals are taking notice of the carbon footprint they leave behind on Earth. In light of the increasing amount of incontrovertible evidence of climate change, people are becoming ever more concerned about contributing to this change and gradually more inquisitive about actions that will curtail carbon output.

Climate is part of the environment that makes a location a successful tourist center. To be sure, the cultural attraction of places like Paris and London, or the religious centers such as Rome or Mecca, do not rely on the prevailing climate, but the selection of a site to spend a holiday is most often weather related. This is particularly true in climate-dependent tourism, where travel is generated by the reliability of climate as part of the perceived attractiveness of the destination. For the United States, the Caribbean and Hawaiian Islands provide such outlets, whereas in Europe, the Canary Islands or the Riviera coast are popular with those from more northern realms. In each case, sun and sand are the basis for selection of these and many similar sites.

But there are other activities whose success rests on the climatic regime. This is especially true of specialized sports that require a given set of conditions. Perhaps the most significant of these are those grouped collectively as winter sports. Some locations, such as mountain resorts of the Alps and Rockies, exist in their present form because of the way they cater to visitors. Note, however, that the vagaries of winter weather can lead to low snowfall or warm temperatures in some locations. Despite the ability of making snow when conditions are appropriate, resort operators may lose entire winter seasons.

Other sports also require weather conditions that make them favorite locations. **Thermals** are needed for gliding, while onshore winds often provide suitable conditions for surfing. Locations that have longer playing seasons for golf have a weather advantage, as do courses that are open earlier following the winter layoff. Sailing, hot-air ballooning, and hiking also have a climate component in activity location.

Spectator sports in the United States have undergone great changes in recent years, with indoor stadiums providing a comfortable climate-controlled environment. Such is not always the case, as seen in the storied frigid conditions found in Green Bay and Chicago. It is noteworthy that sites selected for the Super Bowl and College Bowl games are invariably either indoor stadiums or sun-belt cities.

Commerce

Perhaps the most easily observed local impact of weather and climate on commerce is at the local mall or shopping center. The overall regional climate determines the demand for the type of goods sold while weather conditions influence the day-to-day variations in spending.

Retail managers are well aware of the needs of people living in the climate divisions that they serve. The change of seasons influences the types of clothing needed, and stores make the choices available long before the actual seasonal change occurs. For those living in snow areas, the needs of winter—ranging from snow shovels to snowmobiles—are considered. It is not uncommon to find season-sensitive stores, such as those selling ice cream or fishing supplies, to close during the off season. In those areas where winter is generally little more than a slight decline in temperature, the changes are less radical.

Both good and bad weather can influence retail sales. The weather might actually prevent people from reaching stores, with snowfall being a good example. Similarly, the weather may be such that people simply do not want to go out to a store on a miserable day. Those times when heat waves cause extensive discomfort can lead to extra time spent in shopping malls. In contrast, good weather, such as especially sunny, warm days, can cause people to enjoy the weather by going to the beach or working in their gardens. As a result, less money is spent shopping.

Of the many other commercial activities closely related to climate and weather is insurance, particularly that concerning agriculture. Crop and livestock insurance requires an appreciable amount of climatic information to assess risk. Further thought would identify many others.

Summary

Climate plays a significant part in the economic activities of people and societies. Both modern and traditional agriculture provides many examples because climate places limitations on what crops may be successfully grown at a given location. In many cases, a predominant crop of a region reflects the original natural vegetation cover as indicated by the biome. Similarities of climate in world biomes have allowed crops to be produced in areas where they did not exist.

To overcome climatic problems, methods for protection against frost have been introduced. Some of the methods use latent heat, others rely on air mixing, while still others use protective covers over susceptible plants. Irrigation is used to meet the problem of drought. Essentially, the goal is to add as much water as needed to meet the evapotranspiration requirement. The methods to do this are varied. Both frost protection and irrigation are part of maintaining conditions for the highest crop yield. Sunlight, necessary for photosynthesis, is ultimately the key to success while photoperiodism and temperature limits are a significant part of an agricultural venture's success.

Many examples can be cited to show that climate plays a role in industrial development and location. The movie industry is one such case. Agricultural and processing serves as another illustration. Similarly, some energy resources are more suitable for some climatic regions than others. Wind and solar energy are certainly applicable examples.

Transportation—whether by air, sea, or land—is susceptible to many problems relating to weather and climate. Turbulence and icing are troublemakers for air travelers, while fog has an adverse effect on all three methods of transport. Storm frequency, a function of climate, is obviously of major importance. Examination of other tertiary industries, particularly tourism and commerce, shows that climate influences the successful functioning of numerous economic activities.

Key Terms

Agroclimatology, *349*
Cardinal temperature, *349*
Clear-air turbulence (CAT), *354*

Frost, *345*
Drought, *347*
Growing degree days (GDDs), *349*

Icing, *353*
Photoperiod, *348*
Photosynthesis, *348*

Photovoltaic, *351*
Thermal, *357*

Review Questions

1. How is the distribution of major wheat-growing areas related to climate and natural vegetation?
2. Why do some crops introduced into new global areas, to which they are not native, produce good yields?
3. Outline the main methods used to protect crops against frost.
4. Describe the three types of drought identified in this chapter. What differentiates them?
5. Describe the role of climate in rates of photosynthesis and photoperiod.
6. How has the climate of California influenced the early location of some industries?
7. What are the major problems associated with the development of solar energy? Wind energy?
8. Describe an example of the economic impact of wind-driven ocean currents.
9. Provide examples of how climate influences sporting activities.
10. How do retail sales reflect the prevailing climate?

APPENDIX/METRIC UNITS

TABLE A.1 Temperature Conversion Table (To find either the Celsius or the Fahrenheit equivalent, locate the known temperature in the center column. Then read the desired equivalent value from the appropriate column.)

°C		°F	°C		°F	°C		°F	°C		°F
−40.0	−40	−40	−17.2	+1	33.8	5.0	41	105.8	27.2	81	177.8
−39.4	−39	−38.2	−16.7	2	35.6	5.6	42	107.6	27.8	82	179.6
−38.9	−38	−36.4	−16.1	3	37.4	6.1	43	109.4	28.3	83	181.4
−38.3	−37	−34.6	−15.4	4	39.2	6.7	44	111.2	28.9	84	183.2
−37.8	−36	−32.8	−15.0	5	41.0	7.2	45	113.0	29.4	85	185.0
−37.2	−35	−31.0	−14.4	6	42.8	7.8	46	114.8	30.0	86	186.8
−36.7	−34	−29.2	−13.9	7	44.6	8.3	47	116.6	30.6	87	188.6
−36.1	−33	−27.4	−13.3	8	46.4	8.9	48	118.4	31.1	88	190.4
−35.6	−32	−25.6	−12.8	9	48.2	9.4	49	120.2	31.7	89	192.2
−35.0	−31	−23.8	−12.2	10	50.0	10.0	50	122.0	32.2	90	194.0
−34.4	−30	−22.0	−11.7	11	51.8	10.6	51	123.8	32.8	91	195.8
−33.9	−29	−20.2	−11.1	12	53.6	11.1	52	125.6	33.3	92	197.6
−33.3	−28	−18.4	−10.6	13	55.4	11.7	53	127.4	33.9	93	199.4
−32.8	−27	−16.6	−10.0	14	57.2	12.2	54	129.2	34.4	94	201.2
−32.2	−26	−14.8	−9.4	15	59.0	12.8	55	131.0	35.0	95	203.0
−31.7	−25	−13.0	−8.9	16	60.8	13.3	56	132.8	35.6	96	204.8
−31.1	−24	−11.2	−8.3	17	62.6	13.9	57	134.6	36.1	97	206.6
−30.6	−23	−9.4	−7.8	18	64.4	14.4	58	136.4	36.7	98	208.4
−30.0	−22	−7.6	−7.2	19	66.2	15.0	59	138.2	37.2	99	210.2
−29.4	−21	−5.8	−6.7	20	68.0	15.6	60	140.0	37.8	100	212.0
−28.9	−20	−4.0	−6.1	21	69.8	16.1	61	141.8	38.3	101	213.8
−28.3	−19	−2.2	−5.6	22	71.6	16.7	62	143.6	38.9	102	215.6
−27.8	−18	−0.4	−5.0	23	73.4	17.2	63	145.4	39.4	103	217.4
−27.2	−17	+1.4	−4.4	24	75.2	17.8	64	147.2	40.0	104	219.2
−26.7	−16	3.2	−3.9	25	77.0	18.3	65	149.0	40.6	105	221.0
−26.1	−15	5.0	−3.3	26	78.8	18.9	66	150.8	41.1	106	222.8
−25.6	−14	6.8	−2.8	27	80.6	19.4	67	152.6	41.7	107	224.6
−25.0	−13	8.6	−2.2	28	82.4	20.0	68	154.4	42.2	108	226.4
−24.4	−12	10.4	−1.7	29	84.2	20.6	69	156.2	42.8	109	228.2
−23.9	−11	12.2	−1.1	30	86.0	21.1	70	158.0	43.3	110	230.0
−23.3	−10	14.0	−0.6	31	87.8	21.7	71	159.8	43.9	111	231.8
−22.8	−9	15.8	0.0	32	89.6	22.2	72	161.6	44.4	112	233.6
−22.2	−8	17.6	+0.6	33	91.4	22.8	73	163.4	45.0	113	235.4
−21.7	−7	19.4	1.1	34	93.2	23.3	74	165.2	45.6	114	237.2
−21.1	−6	21.2	1.7	35	95.0	23.9	75	167.0	46.1	115	239.0
−20.6	−5	23.0	2.2	36	96.8	24.4	76	168.8	46.7	116	240.8
−20.0	−4	24.8	2.8	37	98.6	25.0	77	170.6	47.2	117	242.6
−19.4	−3	26.6	3.3	38	100.4	25.6	78	172.4	47.8	118	244.4
−18.9	−2	28.4	3.9	39	102.2	26.1	79	174.2	48.3	119	246.2
−18.3	−1	30.2	4.4	40	104.0	26.7	80	176.0	48.9	120	248.0
−17.8	0	32.0									

TABLE A.2 Wind-Conversion Table (Wind speed units: 1 mph = 0.868391 knot = 1.609344 kph = 0.44704 m/s)

Miles per Hour	Knots	Meters per Second	Kilometers per Hour	Miles per Hour	Knots	Meters per Second	Kilometers per Hour
1	0.9	0.4	1.6	51	44.3	22.8	82.1
2	1.7	0.9	3.2	52	45.2	23.2	83.7
3	2.6	1.3	4.8	53	46.0	23.7	85.3
4	3.5	1.8	6.4	54	46.9	24.1	86.9
5	4.3	2.2	8.0	55	47.8	24.6	88.5
6	5.2	2.7	9.7	56	48.6	25.0	90.1
7	6.1	3.1	11.3	57	49.5	25.5	91.7
8	6.9	3.6	12.9	58	50.4	25.9	93.3
9	7.8	4.0	14.5	59	51.2	26.4	95.0
10	8.7	4.5	16.1	60	52.1	26.8	96.6
11	9.6	4.9	17.7	61	53.0	27.3	98.2
12	10.4	5.4	19.3	62	53.8	27.7	99.8
13	11.3	5.8	20.9	63	54.7	28.2	101.4
14	12.2	6.3	22.5	64	55.6	28.6	103.0
15	13.0	6.7	24.1	65	56.4	29.1	104.6
16	13.9	7.2	25.7	66	57.3	29.5	106.2
17	14.8	7.6	27.4	67	58.2	30.0	107.8
18	15.6	8.0	29.0	68	59.1	30.4	109.4
19	16.5	8.5	30.6	69	59.9	30.8	111.0
20	17.4	8.9	32.2	70	60.8	31.3	112.7
21	18.2	9.4	33.8	71	61.7	31.7	114.3
22	19.1	9.8	35.4	72	62.5	32.2	115.9
23	20.0	10.3	37.0	73	63.4	32.6	117.5
24	20.8	10.7	38.6	74	64.3	33.1	119.1
25	21.7	11.2	40.2	75	65.1	33.5	120.7
26	22.6	11.6	41.8	76	66.0	34.0	122.3
27	23.4	12.1	43.5	77	66.9	34.4	123.9
28	24.3	12.5	45.1	78	67.7	34.9	125.5
29	25.2	13.0	46.7	79	68.6	35.3	127.1
30	26.1	13.4	48.3	80	69.5	35.8	128.7
31	26.9	13.9	49.9	81	70.3	36.2	130.4
32	27.8	14.3	51.5	82	71.2	36.7	132.0
33	28.7	14.8	53.1	83	72.1	37.1	133.6
34	29.5	15.2	54.7	84	72.9	37.6	135.2
35	30.4	15.6	56.3	85	73.8	38.0	136.8
36	31.3	16.1	57.9	86	74.7	38.4	138.4
37	32.1	16.5	59.5	87	75.5	38.9	140.0
38	33.0	17.0	61.2	88	76.4	39.3	141.6
39	33.9	17.4	62.8	89	77.3	39.8	143.2
40	34.7	17.9	64.4	90	78.2	40.2	144.8
41	35.6	18.3	66.0	91	79.0	40.7	146.5
42	36.5	18.8	67.6	92	79.9	41.1	148.1
43	37.3	19.2	69.2	93	80.8	41.6	149.7
44	38.2	19.7	70.8	94	81.6	42.0	151.3
45	39.1	20.1	72.4	95	82.5	42.5	152.9
46	39.9	20.6	74.0	96	83.4	42.9	154.5
47	40.8	21.0	75.6	97	84.2	43.4	156.1
48	41.7	21.5	77.2	98	85.1	43.8	157.7
49	42.6	21.9	78.9	99	86.0	44.3	159.3
50	43.4	22.4	80.5	100	86.8	44.7	160.9

GLOSSARY

Absolute humidity   The ratio of the mass of water vapor per unit volume of air; for example, grams per cubic meter.

Acid precipitation   Precipitation that is more acidic than natural precipitation, which has an average pH of about 5.5.

Adiabatic   Heating or cooling of gases due strictly to their expansion and contraction.

Advection   Mass motion in the atmosphere; in general, horizontal movement of the air.

Aerology   The study of the free atmosphere through its vertical extent.

Aerosols   Solid or liquid particles dispersed in a gas; dust and fog are examples.

Air mass   A large body of air that is characterized by homogeneous physical properties at any given altitude.

Agroclimatology   The scientific study of the effects of climate on crops.

Albedo   The reflectivity of an object, generally measured as a percentage of incoming radiation.

Analemma   A graphical representation of the ephemeris of the sun.

Angstrom   A unit of length equal to 10^8 cm, used in measuring electromagnetic waves.

Anticyclone   An area of above-average atmospheric pressure characterized by a generally outward flow of air at the surface.

Aphelion   The point in the earth's orbit that is farthest from the sun.

Apparent temperature   What the air temperature feels like for the average person.

Arctic hurricane   An intense low-pressure system that develops along the edge of the pack ice in the Arctic. Arctic hurricanes produce high winds and heavy precipitation and represent a major hazard to shipping along their path.

Astronomical twilight   The period after sunset or before sunrise, ending or beginning when the sun is 18° below the horizon.

Atmosphere   The mixture of gases that surrounds the earth.

Atmospheric brown clouds   Brown-colored clouds of atmospheric pollution found over many areas of the world. The best known and largest are those found over eastern Asia.

Atmospheric circulation   The motion within the atmosphere that results from inequalities in pressure over the earth's surface. When the average of the entire globe is considered, the motion is referred to as the general circulation of the atmosphere.

Aurora   A display of colored light seen in the polar skies; called aurora borealis in the northern hemisphere and aurora australis in the southern hemisphere.

Beaufort wind scale   A system for estimating wind velocity (named for its inventor).

Biome   A well-demarcated environment that contains a complex of organisms defining the ecology of the region; tundra and rain forest are examples.

Black body   A substance or body that is a perfect absorber and a perfect radiator.

Blocking   The retardation of eastward-moving pressure systems by stagnation of a high-pressure system.

Blizzard   High winds accompanied by blowing snow; usually associated with winter cold fronts in midlatitudes.

Bora   A cold, dry wind blowing down off the highlands of Yugoslavia and affecting the Adriatic coast.

Boundary layer   The layer of air above the earth's surface wherein the motion of the air is influenced by friction with the earth's surface features. Also called the atmospheric boundary layer or the friction layer.

Buoyant   Less dense than the surrounding medium and thus able to float.

Calorie   A measurement equal to the amount of heat needed to raise 1 g water 1°C; equal to 4.19 joules.

Carbon monoxide   A poisonous gas with no color and little odor, formed when there is incomplete combustion of fossil fuels.

Carcinogen   A substance or an agent that produces or incites cancerous growth.

Castellanus   Towering clouds resembling castle turrets.

Centrifugal force   The apparent force exerted outward on a rotating body or on an object traveling on a curved path.

Chikungunya   A mosquito-borne viral disease whose symptoms are high fever, skin rashes, headache, and joint pain. Also known as chicken guinea.

Chinook   A warm, dry wind blowing down off the Rocky Mountains of western North America.

Chlorofluorocarbons   A class of compounds whose major ingredients are chlorine, fluorine, and carbon. These compounds are instrumental in the destruction of the stratospheric ozone layer.

Cholera   Any of several diseases that result in severe gastrointestinal symptoms.

Chromosphere   A relatively thin layer of burning gases located just above the photosphere.

Cirrostratus   A thin, whitish veil of cloud that forms at high altitudes.

Climate   All the types of weather that occur at a given place over time.

Climatic Optimum   A term originally applied to Scandinavia when temperatures were warm enough to favor more varied flora and fauna.

Climatic regime   The set of annual cycles associated with various climatic elements; for example, the thermal regime is the seasonal patterns of temperature, and the moisture regime is the seasonal patterns of precipitation.

Climatotherapy   The therapeutic utilization of prevailing climate.

Cloud streets   Long, thin lines of clouds forming in the trade winds when winds are steady and of low velocity over helical currents.

Cold front   The leading edge of a cold air mass that displaces a warmer air mass.

Condensation   The change of state from a gas to a liquid.

Conduction   Energy transfer directly from molecule to molecule. Conduction takes place most readily in solids in which molecules are tightly packed.

Continentality   A measure of the remoteness of a land area from the influence of the ocean.

Convection   Mass movement in a fluid or vertical movements in the atmosphere.

Convergent   Moving toward a central point of an area; coming together.

Cordillera   A group or system of mountain ranges, including the intervening valleys.

Coriolis effect   An apparent force caused by the earth's rotation. The Coriolis effect is responsible for deflecting winds clockwise in the northern hemisphere and counterclockwise in the southern hemisphere.

Crepuscular rays   Streaks of light that appear to emanate from the sun shortly before or after sunset. The rays result from the sun shining through a break in the clouds and illuminating particulate matter in the atmosphere.

Cyclone   A rotating low-pressure system. Hurricanes and typhoons are types of cyclones.

Deflation   The lifting and removal of Earth particles by the action of the wind.

Dendrochronology   The study of tree rings.

Dengue fever   An acute infectious disease characterized by the sudden onset of headache, severe aching of the joints, and rash. Dengue fever is caused by a virus transmitted by a mosquito and is found primarily in the tropics.

Desertification   The spread of desert-like conditions in a given location.

Determinism   The philosophical doctrine that every event has a cause and thus no events are intrinsically uncertain.

Dew point   The temperature at which saturation would be reached if an air mass were cooled at constant pressure without altering the amount of water vapor present.

Die back   In the case of vegetation, to die from the top down. When a population dies back, its numbers decrease.

Diurnal   Occurring in the daytime or having a daily cycle.

Divergence   The condition that exists when the distribution of winds within a given region results in a net horizontal outflow of air from the region. At lower levels, the resulting deficit is compensated for by a downward movement of air from aloft; hence, areas of divergent winds are unfavorable to cloud formation and precipitation.

Doldrums   An area of very ill-defined surface winds associated with the equatorial convergence zone.

Doppler radar   A type of radar that detects the relative direction of motion; used in locating rotation in clouds and tornadoes.

Drainage basin   A part of the surface of the earth that is occupied by a drainage system of rivers and streams.

Drought   A time of unusually low water supply, due to reduced rainfall, reduced snowmelt, or low levels of streams.

Dust devil   A small cyclonic circulation, or dust swirl, produced by intense surface heating. Dust devils, which are most common in arid regions, resemble miniature tornadoes.

Easterly wave   A weak, large-scale convergence system that is part of the secondary circulation of the tropics.

Eccentricity   The deviation from an established pattern. In astronomy, the difference in the orbit of a body from that of a circle.

Effluents   The outflow of water or waste liquid from an underground source or conduit.

Electromagnetic spectrum   The range of energy that is transferred as wave motions and that does not require any intervening matter to make the transfer. Electromagnetic waves travel at the speed of light (186,000 mi/s).

Electromagnetic waves   Waves characterized by variations of electric and magnetic fields.

El Niño   An abnormal warming of surface ocean water, most pronounced in the eastern tropical Pacific Ocean.

ENSO (El Niño–Southern Oscillation)   A reversal of pressure and wind patterns accompanying El Niño.

Evapotranspiration   The combined processes of evaporation and transpiration.

Exosphere   The outermost region of the atmosphere, from which particles may escape to space. The first interaction of solar radiation with the atmosphere occurs in the exosphere.

Famine   The scarcity of food over a large region, resulting in malnutrition or starvation for large numbers of people.

Fluid   A substance capable of flowing easily.

Föhn wind   A central European wind that is the same wind as the Chinook wind of North America; also called a leste wind or Foehn wind.

Friction layer   The atmospheric layer above the earth's surface in which wind speed and direction are influenced by friction.

Front   The boundary between two different air masses.

Fujita Scale   A scale that rates the relative severity of tornadoes based on the extent of damage incurred.

Gas law   A law which states that the pressure exerted by a gas is proportional to its density and absolute temperature.

Geostrophic wind   An idealized wind aloft flowing parallel to the isohypses, with the pressure gradient and the Coriolis effect in balance.

Geosynchronous   As applied to a satellite, having an orbit that remains fixed over the same point at the earth's equator.

Gradient wind   A wind that blows parallel to curved isohypses such that centrifugal, Coriolis, and pressure-gradient forces are in balance.

Greenhouse effect   The process by which the heating of the atmosphere is compared to the heating of a common greenhouse. Sunlight (short-wave radiation) passes through the atmosphere to reach the earth. The energy reradiated by the earth is at a longer wavelength, and its return to space is inhibited by atmospheric greenhouse gases. This process acts to increase the temperature of the lower atmosphere.

Green-up   The time in the spring when trees and shrubs leaf out.

Hadley cell   A convectional cell operating as part of the general circulation, located approximately between the equator and the tropics in each hemisphere.

Hantavirus   A pulmonary syndrome characterized by flulike symptoms such as fever, muscle aches, shortness of breath, and coughing. Hantavirus is carried by rodents.

Harmattan   A dry, dust-laden wind blowing south from the Sahara Desert.

Heat index   A value that describes the combined effect of high temperature and high humidity on the body. The apparent temperature.

Heat island   An area of higher air temperature, compared with the surrounding area, in an urban region.

Heat wave   Any unseasonably warm spell, which can occur any time of the year.

Hectare   A metric unit of area equal to 2.47 acres.

Heterosphere   The upper of a two-part division of the atmosphere, based on its composition, in which gases are no longer uniformly mixed.

High pressure   An area of the atmosphere where barometric pressure is higher than the pressure of the surrounding area.

Histogram   A graphical representation that uses rectangles to show the frequency distribution of selected variables.

Homosphere   The lower of a two-part division of the atmosphere, based upon its composition, in which there is a uniform mixture of gases.

Horse latitudes   The former name for belts of high pressure located at about 35° N and 35° S.

Hurricane   A tropical cyclone that develops in the Atlantic Ocean.

Hydrologic cycle   The processes by which water is cycled through the environment. The hydrologic cycle involves changes of state (solid, liquid, and gas) of water as well as the transport of water from place to place.

Hygroscopic   The attribute of being able to attract water from the atmosphere.

Hypothermia   The state resulting from a dangerous decrease in the body temperature of humans or other animals.

Ice cores   Samples derived by driving hollow tubes into carefully selected ice cap sites where the original ice deposit has been minimally disrupted.

Ice sheets   Large ice masses consisting of many joined glaciers and located on land. Ice sheets are found on Antarctica and Greenland.

Ice shelves   Large masses of floating ice attached to ice sheets. They are found most extensively around the Antarctic continent.

Infrared radiation   Radiation in the range just longer than red. Most sensible heat radiated by the earth and other terrestrial objects is in the form of infrared waves.

Intertropical convergence zone   The seasonally migrating low-pressure zone located approximately at the equator, wherein the northeast and southeast trade winds converge. Composed largely of moist and unstable air, the intertropical convergence zone (ITCZ) provides copious precipitation.

Inversion   A reversal of the normal atmosphere temperature regime; for example, an increase in temperature with height.

Ionosphere   A zone of the upper atmosphere characterized by gases that have been ionized by solar radiation.

Isobar   A line on a map or chart connecting points of equal barometric pressure.

Isotopes   Elements that have the same number of electrons and protons but have a different number of neutrons.

Jet stream   A high-speed flow of air that occurs in narrow bands of the upper air westerlies.

Katabatic wind   Any air blowing downslope as a result of the force of gravity.

Laminar flow   A flow in which a liquid or gas moves smoothly in parallel layers.

Langley   A measure of radiation intensity equal to 1 cal/cm^2.

La Niña   A strengthening of the normal circulation over the southern Pacific Ocean.

Latent energy   Energy involved in the change of state of matter that is temporarily stored or concealed, such as the heat contained in water vapor.

Laws of motion   Laws which state that a body will change its velocity of motion only if acted on by an unbalanced force.

Leveche   A dry, dust-laden wind blowing from the Sahara Desert into Spain.

Little Ice Age   The period from about 1650 to 1850, when glaciers in the northern hemisphere enlarged to the extent that some Alpine villages were overwhelmed by ice.

Low pressure   An area of the atmosphere where barometric pressure is lower than the surrounding area.

Malaria   A disease produced by a parasitic protozoan that multiplies in and destroys red blood cells. Transmitted by mosquitoes, malaria is endemic in the tropics.

Maunder minimum   A term applied to a time of minimum sunspot activity that occurred from about 1645 to 1700.

Mean   A measure of central tendency that is the sum of individual values divided by the number of items in the data set.

Median   A measure of central tendency that is the value of the middle item when items are arranged according to size.

Melanoma   A tumor of high malignancy that starts in a dark mole and metastasizes rapidly and widely. Also referred to as malignant melanoma.

Meridional circulation   Air flowing in a circulation pattern that is essentially aligned parallel to meridians of longitude.

Mesocyclone   A very large thunderstorm in which rotation has developed. The rotation often results in the formation of a tornado.

Mesopause   The upper limit of the mesosphere.

Mesosphere   The layer of the atmosphere above the stratosphere in which temperatures drop fairly rapidly with increasing height.

Mesothermal   A climate in which the mean temperature of the coolest month is above –3°C (26.6°F).

Microthermal   A climate in which the mean temperature of the coolest month is below –3°C (26.6°F).

Metabolism The set of chemical processes that sustains organisms.

Meteorology The study of phenomena of the atmosphere, with a view toward understanding and predicting weather and climate.

Methane A colorless, odorless, flammable atmospheric gas that absorbs Earth radiation quite efficiently.

Microclimate The climate of a small area, such as a forest floor or small valley.

Midlatitude cyclone A low-pressure system with frontal boundaries that occurs in the midlatitudes.

Milankovich cycles A series of long-term changes in Earth's climate resulting from changes in planetary orbital parameters.

Millibar A unit of pressure equal to 1000 dynes/cm^2.

Mistral A cold, dry gravity wind blowing down off the Alps and affecting the French and Italian Rivieras.

Mode A measure of central tendency that is the most frequent value that occurs in a set of data.

Monsoon Atmospheric circulation typified by a change in prevailing wind direction from one season to another.

Nimbostratus A low, dark gray cloud layer that is precipitating.

Nor'easter A large low-pressure system in the atmosphere that travels north along the east coast of the United States. It may result in heavy precipitation and flooding along the coast. In winter, nor'easters produce large amounts of snowfall.

Obliquity The angle between the plane of the earth's equator and the orbit of the earth around the sun.

Occluded front When a cold front overtakes a warm front and lifts from the surface the warm sector.

Opaqueness The degree to which light will not pass through a substance.

Orographic precipitation Precipitation that results from the lifting of air over some topographic barrier such as a coastline, hills, or mountains.

Ozone Oxygen in the triatomic form (O_3); highly corrosive and poisonous.

Ozone layer The layer of ozone, 40.25 km (25 mi) above the earth's surface, that absorbs ultraviolet radiation from the sun.

Paleoclimatology A branch of climatology that uses surrogate data to study climate prior to the period of direct measurement.

Palynology The study of pollen grains or spores.

Particulates Solid particles found in the atmosphere.

PDO (Pacific Decadal Oscillation) A climatic pattern similar to that of the ENSO that can persist for decades and tends to take place primarily in the northern Pacific.

Perihelion The point in the earth's orbit around the sun when the earth is closest to the sun.

Periodic Occurring or appearing at regular intervals, such as the sun's rising and setting.

Permafrost Ground that is permanently frozen at some depth. The surface may melt in the summer, but the soil remains frozen beneath the surface.

Persistence The continuation of a set of existing conditions.

Photochemical Having to do with a chemical change that either releases or absorbs radiation.

Photoperiod The period of each day when direct solar radiation reaches the earth's surface; approximately sunrise to sunset.

Photosphere The bright outer layer of the sun which emits most of the radiation, particularly visible light.

Photosynthesis The process by which sugars are manufactured in plant cells, using water and carbon dioxide in the presence of sunlight.

Plague A disease that reached pandemic proportions in the past. The plague is carried by fleas and has a high fatality rate. In the Middle Ages, it was known as the Black Death.

Pluvial Pertaining to precipitation.

Polar front The frontal zone of a storm that separates air masses of polar origin from air masses of tropical origin.

Polar orbiting satellites Satellites that orbit the earth in a north-to-south direction. They periodically pass over all parts of the planet.

Polar stratospheric clouds Very high, thin clouds that form in the stratosphere, particularly over the Antarctic continent.

Potential evaporation The maximum evaporation rate from a water surface; depends on the vapor pressure of the water.

Pressure gradient The amount of pressure change occurring over a given distance.

Proxy data Data that serve as a substitute, or proxy, for direct observation. Examples include ice cores and tree rings.

Radiation The process by which electromagnetic radiation is propagated through space.

Radiosonde A package of weather instruments sent aloft by balloon or rocket. The collected data are transmitted to a ground receiving station by radio.

Rain forest A forest ecosystem that develops in regions with high rates of precipitation year-round. Rain forests consist mainly of evergreen vegetation, with trees dominant.

Rayleigh scattering The scattering of solar radiation by particles in the earth's atmosphere.

Relative humidity The ratio of the amount of water present in the air to the amount of water vapor the air can hold, multiplied by 100.

Remote sensing The process of acquiring data or information about an object or a phenomenon when the instrumentation performing the measurement is not in physical contact with the object.

Respiration The physical and chemical processes by which an organism supplies itself with oxygen.

Ridge An elongated area in which pressure is higher than in the surrounding region.

Rossby waves   Waves in the middle and upper troposphere of the midlatitudes with wavelengths of 4000 to 6000 km; named for C. G. Rossby, the meteorologist who developed the equations for parameters governing the waves.

Saffir-Simpson Scale   A relative scale that utilizes wind velocity, storm surge, and atmospheric pressure to categorize the strength of tropical cyclones.

Sand drift   The movement of sand over an area of dunes in a desert. Sand drift is due to the heating of the sand by the sun and is primarily a daytime phenomenon. The sand moves along the surface and is responsible for a large part of the movement of the dunes.

Santa Ana   A Chinook wind occurring in southern California and northern Mexico.

Sastrugi   Ripples produced in snow by persistent gravity winds in Antarctica.

Saturation vapor pressure   The pressure exerted by water vapor when the atmosphere is saturated.

Savanna   Tropical grassland interspersed with trees and shrubs.

Scattering   The process by which small particles and molecules of gases diffuse part of the radiation striking them in different directions.

Sensible temperature   The sensation of temperature the human body feels, in contrast to the actual heat content of the air recorded by a thermometer.

Sirocco   A hot, dry wind blowing north across the Mediterranean Sea.

Snow pack   The depth of accumulated snow, most commonly measured in mountain regions as an indicator of projected spring runoff.

Solar constant   The mean rate at which solar radiation reaches the earth's outer atmosphere.

Southern oscillation   The shifting of atmospheric circulation over the southern Pacific region that usually accompanies El Niño.

Specific gravity   The ratio of a unit mass of a substance to a unit mass of water.

Specific heat   The amount of heat needed to raise 1 g of a substance 1°C.

Specific humidity   The ratio of the mass of water vapor in the air to a unit of air that includes the water vapor, such as grams of water vapor per kilogram of wet air.

Standard atmosphere   A conventional vertical profile of temperature, pressure, and density within the atmosphere.

Stationary front   A cold or warm front that has ceased to move; the boundary between two stagnant air masses.

Stefan-Boltzmann Law   A law which states that the amount of radiant energy emitted by a black body is proportional to the fourth power of the absolute temperature of the body.

Stratopause   The upper boundary of the stratosphere.

Stratosphere   A thermal division of the earth's atmosphere located between the troposphere and the mesosphere. The primary zone of ozone formation.

Sublimation   The transition of water directly from the solid state to the gaseous state, without passing through the liquid state.

Subsidence   Descending or settling, as in the air associated with high pressure.

Sunspots   Dark areas on the sun that are cooler that the surrounding chromosphere.

Supercell   An extremely large thunderstorm, often reaching as high as the tropopause and sometimes extending into the stratosphere.

Surface ozone   Ozone that forms near the ground as a result of sunlight acting on incompletely combusted hydrocarbons. Surface ozone occurs most often and most intensely in urban areas of the industrial world.

Synoptic climatology   A study of climatology that relates local and regional climates to atmospheric circulation patterns.

Taiga   The coniferous forests of Siberia; also, other coniferous forests of the northern hemisphere.

Terminal fall velocity   The velocity at which a particle will fall through a fluid when the acceleration due to gravity is balanced by friction.

Terrestrial   Pertaining to the land, as distinguished from the sea or air.

Thermal   A small-scale rising current of warm air.

Thermodynamics   The science of the relationship between heat and mechanical work.

Thermohaline circulation   The global ocean circulation that is caused by differences in the temperature (*thermo*) and salinity (*haline*) of sea water.

Thermosphere   The atmospheric zone that includes the ionosphere and exosphere.

Thunderstorm   A convective cell characterized by vertical cumuliform clouds.

Tornado   An intense vortex in the atmosphere with abnormally low pressure in the center and a converging spiral of high-velocity winds.

Trade winds   Two belts of winds that blow almost constantly from easterly directions and are located on the equatorward sides of the subtropical highs.

Transmissivity   The proportion of the solar radiation ultimately passing through the atmosphere.

Transpiration   The process by which water leaves a plant and changes to vapor in the air.

Tropical cyclone   A large, rotating, low-pressure storm that develops over tropical oceans, called a hurricane in the Atlantic Ocean, a typhoon in the Pacific Ocean, and a cyclone in the Indian Ocean.

Tropopause   The upper boundary zone of the troposphere, marked by a discontinuity of temperature and moisture.

Troposphere   The lower layer of the atmosphere, marked by decreasing temperature, pressure, and moisture with height; the layer in which most day-to-day weather changes occur.

Trough   An elongated area of low pressure relative to the pressure of the surrounding area.

Tundra   The treeless plains of the Arctic; a climatic region exhibiting tundra vegetation.

Turbulent flow   A flow in which irregular and seemingly random fluctuations occur.

Twilight   The period before sunrise and after sunset in which refracted sunlight reaches the earth.

Typhoon   A large, rotating, low-pressure storm that develops over the tropical Pacific Ocean.

Ultraviolet radiation   Radiation of a wavelength just shorter than violet. The invisible ultraviolet radiation is largely responsible for sunburn.

Upslope   Moving uphill, as does a breeze or fog.

Upwelling   A vertical current of water in the ocean. Upwellings are common along the eastern sides of the oceans in midlatitudes.

Van Allen belts   Two zones of charged particles existing around the earth at very high altitudes and associated with the earth's magnetic field.

Vapor pressure   The partial pressure of the total atmospheric gaseous mixture that is due to water vapor; also called vapor tension.

Virga   A thin veil of rain seen hanging from a thunderstorm but not reaching the ground because the droplets are evaporated.

Viscosity   The internal friction in fluids that offers resistance to flow.

Volcanic explosivity index   A measure of the role of volcanoes in climatic change.

Vortex   A whirling or rotating fluid with low pressure in the center.

Walker circulation   The atmospheric circulation over the southern Pacific Ocean. The Walker circulation follows the general model of the Hadley cell, with westward drift of the trade winds along the surface near the equator and counterflow aloft.

Warm front   A zone along which a warm air mass displaces a colder one.

Waterspout   A tornado occurring at sea that touches the surface and picks up water.

Water year   A 12-month period used in hydrology that corresponds to the annual precipitation or flow of streams. In the United States, the water year runs from October 1 to September 30. October 1 is the time of year when the water supply in the soil and streams is usually at its lowest.

Wavelength   The linear distance between the crests or the troughs of adjacent waves.

Weather   The state of the atmosphere at any one point in time and space.

West Nile virus   A mosquito-borne virus that can infect birds and humans. It has spread to many of the conterminous 48 states and can cause encephalitis.

Williwaw   A bora wind in Alaska, Greenland, or coastal Antarctica.

Windchill   The physiological effects resulting from the combined effect of low temperature and wind.

Wind rose   A type of diagram used to show prevailing wind speed and direction.

Wind shear   A change in wind speed and/or direction with height.

Yellow fever   An acute infectious disease characterized by the sudden onset of fever, jaundice, albuminuria, and, often, hemorrhage. Yellow fever is caused by a virus transmitted by a mosquito found primarily in the tropics.

Zebra mussel   A freshwater mollusk accidently introduced into the Great Lakes of North America. The zebra mussel competes with native fish for food and proliferates rapidly. It has become a problem in that it clogs pipes taking water from the lake.

Zenith A point in space directly above a person's head; a point on a line passing through the zenith.

Zonal circulation The approximate flow of air along a parallel of latitude.

Zonda A hot, dry wind blowing down the east side of the Andes Mountains of South America. A zonda is formed in the same manner as the Chinook in the United States and Canada.

BIBLIOGRAPHY

Addison D., Weatherwise, © 1994, *Building for Comfort and Safety*, June/July, pp. 14–18.

Aguado E. & Burt J. E., *Understanding Weather and Climate*, 4th ed.,© 2007, Pearson Prentice Hall, Upper Saddle River, N.J.

Ahrens C. D., *Essentials of Meteorology*, 4th ed., © 2004, Brooks/Cole, Pacific Grove, Calif.

Appenzeller T., Big thaw, © 2007, *National Geographic*, June.

Arya S. P., *Air Pollution Meteorology and Dispersion*, © 1998, Oxford University Press, New York.

Barry R. G. & Chorley R. J., *Atmosphere, Weather, and Climate*, 7th ed., © 1998, Routledge, New York.

Bird E. C. F., *The Effects of Rising Sea Level on Coastal Environments*, © 1993, John Wiley, Chichester, U.K.

Bligh W., *A Narrative of the Mutiny on Board the HMS Bounty*, © 1838, Hodder and Stoughton, Ltd., Kent, England.

Brown, L. R., *Plan B, Rescuing a Planet Under Stress and a Civilization in Trouble*, © 2003, Earth Policy Institute, W.W. Norton, New York.

Brasseur G. P., Orlando J. J., & Tyndall G. S., *Atmospheric Chemistry and Global Change*, © 1999, Oxford University Press, New York.

Bridgeman H. A. & Oliver J. E., *The Global Climate System*, © 2006, Cambridge University Press, New York.

Christopherson R. W., *Geosystems*, 7th ed., © 2009, Pearson Prentice Hall, Upper Saddle River, N.J.

Clawson D. L., *World Regional Geography: A Development Approach*, 7th ed., © 2001, Prentice Hall, Upper Saddle River, N.J.

Dana R. H., Jr., *Two Years Before the Mast*, © 1840, Harper & Row, New York.

Edgerton L. T., *The Rising Tide: Global Warming and World Sea Levels*, © 1991, Island Press, San Francisco.

Finkel M., Bedlam in the blood, © 2007, *National Geographic*, July.

Environmental Protection Agency (EPA), *Nitrous Oxide: Sources and Emissions*, © 2006, online, www. epa. gov.

Gelbspan R., *Boiling Point*, © 2004, Basic Books, New York.

Goody R., *Principles of Atmospheric Physics and Chemistry*, © 1995, Oxford University Press, New York.

Gore A., *An Inconvenient Truth*, © 2007, Viking, New York.

Henson R., *The Rough Guide to Climate Change: Symptoms, Science, Solutions*, © 2007, The Penguin Group, London.

Hidore J. J., *Weather and Climate*, © 1985, Park Press, Champagne, Ill.

Hidore J. J., *Global Environmental Change*, © 1996, Prentice Hall, Upper Saddle River, N.J.

Hidore J. J., Why climates change, In W. A. Dando, ed., *Climate Change and Variation: A Primer for Teachers*, Vol. 1, © 2007, National Council for Geographic Education, Washington, D.C.

Intergovernmental Panel on Climate Change (IPCC), *Summary for Policymakers*, © 2007, online, www.ipcc.ch.

Intergovernmental Panel on Climate Change (IPCC), Working Group 1, *The Physical Basis of Climate Change*, © 2007, online, www.ipcc.ch.

Kump L. R., Kasting J. F., & Crane R. G., *The Earth System*, © 1999, Prentice Hall, Upper Saddle River, N.J.

Johansen B. E., *The Global Warming*, © 2002, Greenwood Press, Westport, Conn.

Lutgens F. K. & Tarbuck E. J., *The Atmosphere*, 10th ed., © 2007, Pearson Prentice Hall, Upper Saddle River, N.J.

Mackenzie F. T., *Our Changing Planet*, 2nd ed., © 1998, Prentice Hall, Upper Saddle River, N.J.

Mann M. E. & Kump L. R., *Dire Predictions: Understanding Global Warming*, © 2008, Pearson Prentice Hall, Upper Saddle River, N.J.

McKibben B., Carbon's new math, © 2007, *National Geographic*, October.

McKnight T. L. & Hess D., *Physical Geography: A Landscape Appreciation*, 8th ed., © 2005, Pearson Prentice Hall, Upper Saddle River, N.J.

Moran J. M. & Morgan M. D., *Meteorology: The Atmosphere and the Science of Weather*, 5th ed., © 1997, Prentice Hall, Upper Saddle River, N.J.

Morgan M. D. & Moran J. M., *Weather and People*, © 1997, Prentice Hall, Upper Saddle River, N.J.

NASA, Laboratory for Atmospheres, *Ozone Hole Watch*, © 2008, online, http://ozonewatch.gsfc.nasa.gov (retrieved November 16, 2008).

NASA, *NASA Eyes Effects of a "Giant Brown Cloud" Worldwide*, © 2004, online, www.nasa.gov/vision/earth/environment/brown_cloud.html.

National Aeronautics and Space Administration, NASA, www.nasa.gov.

National Climate Data Center, www.ncdc.noaa.gov.

National Oceanographic and Atmospheric Administration, www.noaa.gov.

National Research Council, *Acid Deposition: Long-Term Effects*, © 1986, National Academies Press, Washington, D.C.

Nicklen P., Vanishing sea ice, © 2007, *National Geographic*, June.

NOAA, *Arctic Report Card 2008, Sea Ice Cover*, NOAA, Washington, D.C.

Oliver J. E., Climate and art: An organizational survey, © 2004, *Occasional Publications in Geography*, Kansas State University, Manhattan, Kansas.

Oliver J. E. & Hidore J. J., *Climatology*, © 1984, Merrill Publishing Co., Columbus, Ohio.

Oliver J. E, ed., *Encyclopedia of World Climatology*, © 2005, Springer, New York.

Reynolds R. W., A real-time global sea surface temperature analysis, © 1988, *Journal of Climate*, 1, pp. 75–86.

Poole R. M., The Ethiopia Campaign, © 2007, *Smithsonian*, June, pp. 87–96.

Shell E. R., Malaria: Resurgence of a deadly disease. © 1997, *The Atlantic Monthly*, August.

Snow R. & Snow M., Transportation, In W. A. Dando, ed., *Climate Change and Variation: A Primer for Teachers*, Vol. 1, © 2008, National Council for Geographic Education, Washington, D.C.

Snow R., Snow M., & Oliver J., *Exercises in Climatology*, © 2003, Prentice Hall, Upper Saddle River, N.J.

Thornthwaite C. W., An approach toward a rational classification of climate, © 1948, *Geographical Review*, 38, pp. 55–94.

Turco R. P., *Earth Under Siege: From Air Pollution to Global Change*, © 1996, Oxford University Press, New York.

Turekian K. K., *Global Environmental Change*, © 1996, Prentice Hall, Upper Saddle River, N.J.

Van Loon G. W. & Duffy S. J., *Environmental Chemistry*, © 2000, Oxford University Press, New York.

Warrick R. A., Barrow E. M., & Wigley T. M., *Climate and Sea Level Change*, © 1993, Cambridge University Press, New York.

Wayne R. P., *Chemistry of Atmospheres*, 3rd ed., © 1999, Oxford University Press, New York.

INDEX

Note: page numbers followed by b, f, or t indicate boxes, figures and tables, respectively

Dew-point Temperature (°C)

DRY-BULB TEMP (°C)	WET-BULB DEPRESSION (°C)														
	0.5	1.0	1.5	2.0	2.5	3.0	3.5	4.0	4.5	5.0	7.5	10.0	12.5	15.0	17.5
−10.0	−12.1	−14.5	−17.5	−21.3	−26.6	−36.3	—	—	—	—	—	—	—	—	—
−7.5	−9.3	−11.4	−13.8	−16.7	−20.4	−25.5	−34.4	—	—	—	—	—	—	—	—
−5.0	−6.6	−8.4	−10.4	−12.8	−15.6	−19.0	−23.7	−31.3	−78.6	—	—	—	—	—	—
−2.5	−3.9	−5.5	−7.3	−9.2	−11.4	−14.1	−17.3	−21.5	−27.7	−41.3					
0.0	−1.3	−2.7	−4.2	−5.9	−7.7	−9.8	−12.3	−15.2	−18.9	−23.9	—	—	—	—	—
2.5	1.3	0.1	−1.3	−2.7	−4.3	−6.1	−8.0	−10.3	−12.9	−16.1	—	—	—	—	—
5.0	3.9	2.8	1.6	0.3	−1.1	−2.6	−4.2	−6.1	−8.1	−10.4	−47.7	—	—	—	—
7.5	6.5	5.5	4.4	3.2	2.0	0.7	−0.8	−2.3	−4.0	−5.8	−21.6	—	—	—	—
10.0	9.1	8.1	7.1	6.0	4.9	3.8	2.5	1.2	−0.2	−1.8	−12.8	—	—	—	—
12.5	11.6	10.7	9.8	8.8	7.8	6.7	5.6	4.5	3.2	1.9	−6.8	−28.2	—	—	—
15.0	14.2	13.3	12.5	11.6	10.6	9.6	8.6	7.6	6.5	5.3	−1.9	−14.5	—	—	—
17.5	16.7	15.9	15.1	14.3	13.4	12.5	11.5	10.6	9.6	8.5	2.3	−7.0	−35.1	—	—
20.0	19.3	18.5	17.7	16.9	16.1	15.3	14.4	13.5	12.6	11.6	6.1	−1.4	−14.9	—	—
22.5	21.8	21.1	20.3	19.6	18.8	18.0	17.2	16.3	15.5	14.6	9.6	3.2	−6.3	−37.5	—
25.0	24.3	23.6	22.9	22.2	21.4	20.7	19.9	19.1	18.3	17.5	12.9	7.3	−0.2	−13.7	—
27.5	26.8	26.2	25.5	24.8	24.1	23.3	22.6	21.9	21.1	20.3	16.1	11.1	4.7	−4.7	−31.7
30.0	29.4	28.7	28.0	27.4	26.7	26.0	25.3	24.6	23.8	23.1	19.1	14.5	9.0	1.6	−11.1
32.5	31.9	31.2	30.6	29.9	29.3	28.6	27.9	27.2	26.5	25.8	22.1	17.8	12.8	6.6	−2.4
35.0	34.4	33.8	33.1	32.5	31.9	31.2	30.6	29.9	29.2	28.5	24.9	21.0	16.4	11.0	3.9
37.5	36.9	36.3	35.7	35.1	34.4	33.8	33.2	32.5	31.9	31.2	27.7	24.0	19.8	14.9	8.9
40.0	39.4	38.8	38.2	37.6	37.0	36.4	35.8	35.1	34.5	33.9	30.5	26.9	23.0	18.5	13.3

Note: To determine the dew-point temperature (the saturation temperature), you need the current temperature and the wet-bulb temperature. The difference between the two temperatures is the wet-bulb depression. Find the number in the row containing the current temperature and the column under the wet-bulb depression. For example, if the current temperature is 20°C, and the wet-bulb depression is 3°C, the dew-point temperature is 15.3°C.